DATE DUE

MAR 0 7 1999

RECEIVED

MAR 1 5 1999

MAR 2 6 2002

RECEIVED

APR 2 6 2002

UPI 201-9509   PRINTED IN USA

ADVANCES IN REGULATION OF CELL GROWTH

Volume 2

Advances in Regulation of Cell Growth Series

VOLUMES IN THE SERIES

ADVANCES IN REGULATION OF CELL GROWTH

Volume 2

Cell Activation: Genetic Approaches

Editors

James J. Mond, M.D., Ph.D.

*Department of Medicine
Uniformed Services University
of the Health Sciences
Bethesda, Maryland*

John C. Cambier, Ph.D.

*Department of Pediatrics
National Jewish Center for
Immunology and Respiratory Medicine
Denver, Colorado*

Arthur Weiss, M.D., Ph.D.

*Department of Medicine
University of California at San Francisco
San Francisco, California*

Raven Press New York

Raven Press Ltd., 1185 Avenue of the Americas, New York, New York 10036

© 1991 by Raven Press, Ltd. All rights reserved. This book is protected by copyright. No part of it may be reproduced, stored in a retrieval system, or transmitted, in any form or by any means, electronical, mechanical, photocopying, or recording, or otherwise, without the prior written permission of the publisher.

Made in the United States of America

Library of Congress Catalog Card Number

Library of Congress Cataloging-in-Publication Data

Cell activation : genetic approaches / editors, James J. Mond, John C.
  Cambier, Arthur Weiss.
       p.   cm. — (Advances in regulation of cell growth : v. 2)
    Includes bibliographical references and index.
    ISBN 0-88167-819-8
    1. Cellular signal transduction.   2. Growth factors.   3. Second
  messengers (Biochemistry)   I. Mond, James J.   II. Cambier, John C.,
  1948-   .   III. Weiss, Arthur, 1952-   .   IV. Series.
    [DNLM: 1. Cell Differentiation—genetics.   2. Lymphocyte
  Transformation—genetics.   3. Signal Transduction—genetics.   W1
  AD83N v. 2 / QH 607 C393]
  QP517.C45C45   1991
  574.87 '612—dc20
  DNLM/DLC
  for Library of Congress   91-20473
                                                                 CIP

Contents

Contributors

Kristin M. Abraham, Ph.D.
Howard Hughes Medical Institute
Department of Biochemistry,
Immunology and Medicine
(Medical Genetics),
University of Washington School of
Medicine,
Seattle, Washington 98195

William E. Biddison, Ph.D.
Molecular Immunology Section
Neuroimmunology Branch
National Institute of Neurological
Disorders and Stroke
National Institutes of Health
Bethesda, Maryland 20892

Flavia Borellini, Ph.D.
Department of Pharmacology
Georgetown University School of
Medicine
3900 Reservoir Road, N.W.
Washington, D.C. 20007

Michael P. Cooke
Howard Hughes Medical Institute
Departments of Biochemistry,
Immunology and Medicine
(Medical Genetics),
University of Washington School of
Medicine,
Seattle, Washington 98195

Richard A. Firtel, Ph.D.
Department of Biology
Center for Molecular Genetics
University of California at San Diego
9500 Gilman Drive
LaJolla, California 92093-0634

Robert I. Glazer, Ph.D.
Department of Pharmacology
Georgetown University School of
Medicine
3900 Reservoir Road, N.W.
Washington, D.C. 20007

Mark A. Goldsmith, M.D., Ph.D.
University of California at
San Francisco
Third and Parnassus Avenues
San Francisco, California 94143-0724

Martha Graber, M.B.
University of California at
San Francisco
Third and Parnassus Avenues
San Francisco, California 94143-0724

Sunil K. Gupta, Ph.D.
Department of Basic Sciences
National Jewish Center for
Immunology and Respiratory
Medicine
1400 Jackson Street
Denver, Colorado 80206

Joel F. Habener, M.D.
Laboratory of Molecular
Endocrinology
Massachusetts General Hospital
Howard Hughes Medical Institute
Boston, Massachusetts 02114

Harvey R. Herschman, Ph.D.
Department of Biological Chemistry
Laboratory of Biomedical and
Environmental Sciences
University of California at Los Angeles
School of Medicine
Los Angeles, California 90024

Gary L. Johnson, Ph.D.
Department of Basic Sciences
National Jewish Center for
  Immunology and Respiratory
  Medicine
1400 Jackson Street
Denver, Colorado 80206

Carl H. June, M.D.
Department of Immunobiology and
  Transplantation
Naval Medical Research Institute
8901 Wisconsin Avenue
Bethesda, Maryland 20814-5055

John Klingensmith
Department of Genetics
Harvard Medical School
25 Shattuck Street
Boston, Massachusetts 02115

Brian K. Kobilka, M.D.
Department of Molecular and Cellular
  Physiology
Stanford University Medical Center
157 Berkman Center
Stanford, California 94305

Gary A. Koretzky, M.D., Ph.D.
University of California at
  San Francisco
Third and Parnassus Avenues
San Francisco, California 94143-0724

Kelvin Lee, M.D.
Howard Hughes Medical Institute,
  Ann Arbor, Michigan
Department of Internal Medicine
Division of Hematology/Oncology
University of Michigan
1150 West Medical Center Drive
Ann Arbor, Michigan 48109

Steven D. Levin
Howard Hughes Medical Institute
Departments of Biochemistry,
  Immunology and Medicine
  (Medical Genetics),
University of Washington School of
  Medicine
Seattle, Washington 98195

Dan R. Littman, M.D., Ph.D.
Howard Hughes Medical Institute
Departments of Microbiology and
  Immunology, and Biochemistry and
  Biophysics
University of California at
  San Francisco
Third and Parnassus Avenues
San Francisco, California 94143

Joseph M. Lowndes, Ph.D.
Department of Basic Sciences
National Jewish Center for
  Immunology and Respiratory
  Medicine
1400 Jackson Street
Denver, Colorado 80206

Sandra K. O. Mann
Department of Biology
Center for Molecular Genetics
University of California at San Diego
9500 Gilman Drive
LaJolla, California 92093

Terry E. Meyer, Ph.D.
Laboratory of Molecular
  Endocrinology
Massachusetts General Hospital
Howard Hughes Medical Institute
Boston, Massachusetts 02114

John G. Monroe, Ph.D.
Department of Pathology and
  Laboratory Medicine
University of Pennsylvania School of
  Medicine
240 John Morgan Building
Philadelphia, Pennsylvania 19104-6064

Shoji Osawa, Ph.D.
Department of Basic Sciences
National Jewish Center for
  Immunology and Respiratory
  Medicine
1400 Jackson Street
Denver, Colorado 80206

Roger M. Perlmutter, M.D., Ph.D.
Howard Hughes Medical Institute
Departments of Biochemistry,
Immunology and Medicine
(Medical Genetics)
University of Washington School of
Medicine
Seattle, Washington 98195

Norbert Perrimon, Ph.D.
Department of Genetics
Harvard Medical School
25 Shattuck Street
Boston, Massachusetts 02115

Joel Picus, M.D.
University of California at
San Francisco
Third and Parnassus Avenues
San Francisco, California 94143-0724

Jacalyn H. Pierce, Ph.D.
Laboratory of Cellular and Molecular
Biology
National Cancer Institute
Bethesda, Maryland 20892

Vicki L. Seyfert, Ph.D.
Department of Pathology and
Laboratory Medicine
University of Pennsylvania School of
Medicine
240 John Morgan Building
Philadelphia, Pennsylvania 19104-6064

Thomas E. Smithgall, Ph.D.
Department of Pharmacology
Georgetown University School of
Medicine
3900 Reservoir Road, N.W.
Washington, D.C. 20007

Craig B. Thompson, M.D.
Howard Hughes Medical Institute,
Ann Arbor, Michigan
Department of Internal Medicine
Division of Hematology/Oncology
University of Michigan
1150 West Medical Center Drive
Ann Arbor, Michigan 48109

Julia M. Turner, Ph.D.
Department of Pathology
University of Cambridge
Tennis Court Road
Cambridge CB2 1QP, United Kingdom

Arthur Weiss, M.D., Ph.D.
Department of Rheumatology
University of California at
San Francisco
Third and Parnassus Avenues
San Francisco, California 94143-0724

Christine C. Winter
Molecular Immunology Section
Neuroimmunology Branch
National Institute of Neurological
Disorders and Stroke
National Institutes of Health
Bethesda, Maryland 20892

Gang Yu, M.D.
Laboratory of Biological Chemistry
National Cancer Institute
National Institutes of Health
Bethesda, Maryland 20892

Preface

In studying signal transduction in various cell types, many overlapping pathways of activation have been described. Whether all or some of these newly stimulated biochemical events are required for the initiation and maintenance of cell growth and differentiation, is presently unknown. In Volume 2 of this series, the authors utilize various molecular biologic and genetic approaches to further their understanding of the role of cell surface receptors, and the intracellular molecules linked to these receptors, in regulating cell growth and activation.

James J. Mond
John C. Cambier
Arthur Weiss

Acknowledgment

We would like to thank Mary Chase for her excellent editorial assistance.

ADVANCES IN REGULATION OF CELL GROWTH

Volume 2

Advances in Regulation of Cell Growth, Volume 2;
Cell Activation: Genetic Approaches, edited by
James J. Mond, John C. Cambier, and Arthur Weiss.
Raven Press, Ltd., New York.

1

Use of HLA-A2 Mutant Molecules for the Analysis of the Structural Requirements for Presentation of Viral Peptides to Cytotoxic T Lymphocytes

William E. Biddison and Christine C. Winter

Molecular Immunology Section, Neuroimmunology Branch, National Institute of Neurological Disorders and Stroke, National Institutes of Health, Bethesda, MD 20892

The activation of T lymphocytes through engagement of the antigen-specific T cell receptor (TCR) in general requires the recognition of a peptide derived from a foreign protein that is presented in the context of a class I or class II major histocompatibility complex (MHC) molecule. The purpose of this review is to summarize work from our laboratory that has focused on defining the structural requirements for presentation of an influenza viral antigen by the HLA class I molecule HLA-A2.1 to antigen-specific cytotoxic T lymphocytes (CTL).

HLA class I molecules consist of two chains, an α (heavy) chain encoded by the HLA-A2.1 gene that is noncovalently associated with a light chain (β-2 microglobulin)that is encoded by a non-HLA gene (1,2). The α chain consists of three outer membrane domains (designated α1, α2, and α3), a transmembrane domain, and a small intracytoplasmic tail. These molecules are highly polymorphic, with almost all of the polymorphic residues located in the α1 and α2 domains (3,4). The HLA-A2.1 molecule is the first MHC molecule for which a X-ray crystallographic structure was solved (5,6). The α1 and α2 domains fold together to form a groove of approximately 25 by 10 A that is bounded by two α-helices with a β-pleated sheet floor (Fig. 1) and functions in peptide binding. The membrane proximal domains are composed of the α3 domain and β-2 microglobulin. A more refined structure of HLA-A2.1 (7) has revealed the existence of pockets that extend from the peptide binding groove. One prominent pocket is located below the α1 α-helix and is composed of amino acids 24, 26, 34, 45, and 67, and has been termed the 45 pocket (7).

The current view of peptide binding by class I molecules is that these molecules bind intracellularly generated peptides during assembly in the endoplasmic reticulum (8,9,10). The capacity and specificity with which class I molecules bind is thought to be due to the

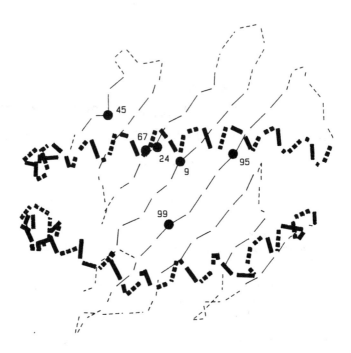

FIG. 1. View of the peptide binding groove of HLA-A2.1. Adapted from Bjorkman et al. (5).

function of the amino acid side chains that point into the peptide binding groove. Although some of these amino acid side chains may also be available for binding by the TCR, those which lie in the floor of the groove or in pockets that extend from the groove are not available for binding by the TCR, especially when the peptide binding groove is occupied by a peptide. We have attempted to define the role of some of the amino acid side chains on the floor of the groove and in the 45 pocket in the presentation of an influenza viral matrix peptide to peptide-specific CTL by analyzing the effects of single amino acid substitutions in the HLA-A2 molecule.

MATERIALS AND METHODS

Characterization of variant and construction of mutant HLA-A2 molecules. The single amino acid substituted variant and mutant HLA-A2 molecules that have been used in these studies are listed in Table I and their locations in the peptide binding groove of the HLA-A2 molecule are illustrated in Fig. 1. The naturally occurring variants of HLA-A2 which have substitutions at positions 9 (9FY) and 99 (99YC) were characterized by DNA sequencing using the polymerase chain reaction (11,12). While multiple substitutions are found in variants HLA-A2.2F and -A2.2Y, both have a Val to Leu substitution at position 95 (13,14). A single amino acid HLA-A2 mutant with a Leu substituted for a Val at position 95 (95VL, Table I) was produced by oligonucleotide site-directed mutagenesis and

Table 1. Sequences of HLA-A2 mutant genes.

			Codon			
	9	24	45	67	95	99
A2.1	TTC(F)	GCA(A)	ATG(M)	GTG(V)	GTC(V)	TAT(Y)
9FY	TAC(Y)					
9FH	CAC(H)					
95VL					CTC(L)	
95VI					ATC(I)	
99YC						TGT(C)
24AS		TCA(S)				
45MT			ACG(T)			
67VS				TCC(S)		

shuffling of genomic fragments of the HLA-A2.2F gene (11). Five additional single amino acid mutants in which the residues found in HLA-A2.1 were replaced with the corresponding amino acids found in HLA - B37 were produced using the Muta - Gene Phagemid mutagenesis system (Bio - Rad Laboratories, Richmond, CA) as previously described (15). The rationale for introducing these particular substitutions was based on the observation that HLA-A2 and HLA-B37 have largely non-overlapping peptide binding specificities as assessed by peptide competition assays (16,17). Thus, at position 9, His replaced Phe and at position 95, Ile replaced Val. The three remaining substitutions were introduced at the three polymorphic residues within the 45 pocket of HLA-A2 (15): Ala to Ser at position 24 (24AS); Met to Thr at position 45 (45MT); and Val to Ser at position 67 (67VS). All site-directed mutant HLA-A2 molecules were transfected into the human HLA-A and-B negative mutant plasma cell line HMy.C1R (18, kindly provided by Dr. Peter Cresswell, Duke Univ., Durham, NC) by the lipofectin technique as described by the manufacturer (BRL, Bethesda, MD) (the 95VL genomic mutant was co-transfected with a plasmid containing the neomycin-resistance gene) and selection performed in 600 ug/ml Geneticin (GIBCO Laboratories, Grand Island, NY). Cell surface expression of transfected genes was monitored by HLA-A2-specific monoclonal antibodies and FACS analysis (12,15). Levels of cell surface expression of all transfectants were within a decade of each other on a log scale of fluorescence.

Generation and assay of CTL. Human CTL lines specific for the influenza virus matrix peptide M1 57-68 were generated as previously described (11,12,16). The M1 57-68 synthetic peptide has the sequence KGILGFVFTLTV. M1 peptides with individual substitutions were also synthesized that have I replaced with either F (F59) or K (K59) at position 59. Cytolytic activity was measured on ^{51}Cr-labelled target cells that were pulsed with synthetic peptides as described (11,12).

RESULTS

Effects of amino acid substitutions in the floor of the peptide binding groove. Amino acids 9, 95, and 99 are located on beta strands on the floor of the peptide binding groove (Fig. 1) and their side chains point up into the groove where they could be available to contribute to peptide binding (5,6). Substitution of Tyr for Phe at position 9 produced an HLA-A2 molecule (9FY) that could only present the influenza viral peptide M1 57-68 to 2 out of 36 M1 peptide-specific CTL lines tested (11,12,Table II). Substitution of His for Phe at position 9 (9FH) had a similar effect: only four of fourteen M1 peptide-specific CTL lines tested could recognize the M1 peptide presented by the 9FH molecule (Table II). Thus, the amino acid at position 9 significantly contributes to the conformation of the M1 peptide when it is bound in the peptide binding groove, but both substitutions at position 9 produced HLA-A2 molecules that retained the ability to bind the M1 peptide.

Table 2. Summary of M1 peptide-specific CTL recognition of peptide-pulsed mutant HLA-A2 molecules.

Mutant HLA-A2 Molecule: # Positive/# of CTL Lines Tested

Mutant HLA-A2 Molecule	# Positive/# of CTL Lines Tested
9FY	2/36
9FH	10/14
95VL	36!/36
95VI	12/12
99YC	1/36
24AS	20/27
45MT	0/27
67VS	9/27

All 36 M1 peptide-specific CTL lines tested could recognize the 95VL transfectant with a Val to Leu substitution at position 95 (12). However, the 95VL molecule demonstrated enhanced kinetics of M1 peptide presentation and the ability to be sensitized by at least 10-fold lower amounts of peptide (12). Thus, the replacement of Val with Leu at position 95 enhanced the avidity of the HLA-A2 molecule for the M1 peptide, and demonstrates that the side chain of amino acid residue 95 can have an important role in peptide binding. The specificity of the enhancing effect of the Val to Leu substitution at position 95 was investigated by comparing the amount of M1 peptide required to sensitize the 95VL versus the 95VI transfectants. The results (Fig.2) demonstrate that at a higher peptide concentra-

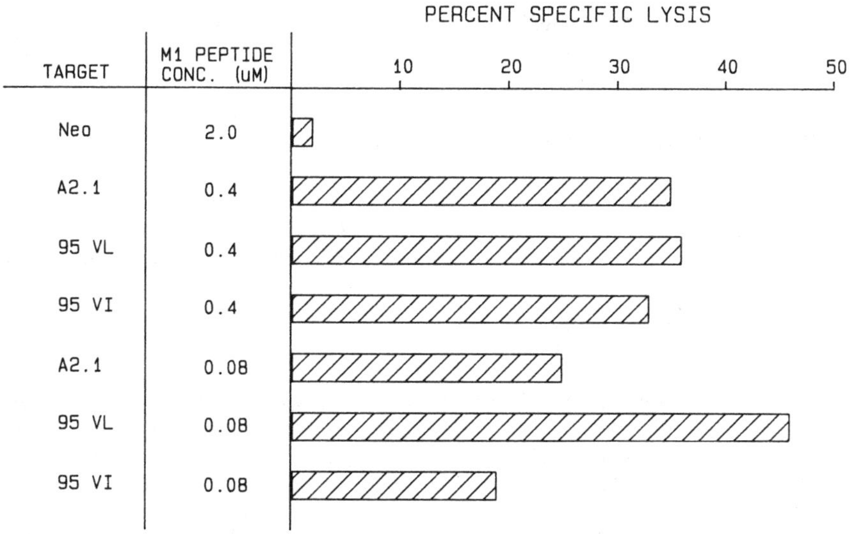

PERCENT SPECIFIC LYSIS

TARGET	M1 PEPTIDE CONC. (uM)					
		10	20	30	40	50
Neo	2.0					
A2.1	0.4					
95 VL	0.4					
95 VI	0.4					
A2.1	0.08					
95 VL	0.08					
95 VI	0.08					

FIG. 2. Enhancement of M1 peptide presentation by the 95VL but not 95VI molecule.

tion (0.4 uM) no differences are seen between the levels of lysis of the A2.1, 95VL, and 95VI transfectants, whereas at the lower peptide concentration (0.08 uM) the 95VL but not the 95VI transfectant shows an enhanced level of lysis relative to the A2.1 tranfectant. These results indicate that the enhancement of presentation observed with the 95VL substitution is highly specific.

Replacement of Tyr with Cys at position 99 (99YC) produced a very dramatic effect on presentation of the M1 peptide: only one out of 36 M1 peptide-specific CTL lines could recognize the peptide pulsed cells that express the naturally occurring 99YC mutant molecule (12, Table II). Thus, the side chain of amino acid 99 in the floor of the peptide binding groove can also contribute to the conformation of the bound M1 peptide.

Another approach that can demonstrate the effects of particular amino acids in the determination of the conformation of a peptide bound in the peptide binding groove is to examine the effects of amino acid substitutions on the presentation of peptides with different sequences. An example of how this approach can demonstrate that the amino acid at position 9 can affect peptide conformation is given in Table III (12). The M1 peptide and two M1 analogues (F59 and K59) are presented by wild-type HLA-A2.1 and the mutant 9FY to two different M1 peptide-specific CTL lines, TIIB5 and 17B5. CTL line TIIB5 can recognize the M1 and K59 peptides presented by A2.1 and 9FY, but it can only recognize the F59 peptide presented by A2.1 and not 9FY. In contrast, the 17B5 CTL line can recognize the M1 and F59 peptides presented by both A2.1 and 9FY, but it can only recognize the K59 peptide presented by A2.1 and not by 9FY. Thus, the substitution of Tyr for Phe at position 9 can affect the conformation of the bound peptide, and the detection of that conformation depends on the specificity of the antigen-specific TCR on individual CTL.

Table 3. Summary of recognition of M1, F59, and K59 peptides presented by HLA-A2.1 and 9FY molecules.

HLA-A2 Molecule	CTL line and peptide					
	TIIB5			17B5		
	M1	F59	K59	M1	F59	K59
A2.1	+	+	+	+	+	+
9FY	+	-	+	+	+	-

Effects of amino acid substitutions in the 45 pocket. The 45 pocket of HLA-A2.1 extends from the peptide binding groove below the alpha1-domain alpha-helix and is comprised of five amino acids (7), three of which differ between HLA-A2.1 and HLA-B37. Site-directed mutagenesis was used to replace the hydrophobic residues at positions 24, 45, and 67 in the 45 pocket of HLA-A2.1 with the hydrophilic residues found in these positions in HLA-B37 (12,Table I). These mutants (24AS, 45MT, and 67VS) weretransfected into HMy cells and assayed for their ability to be recognized by a panel of 27 M1 peptide-specific HLA-A2.1-restricted CTL lines (15,Table II). Twenty out of 27 of these CTL lines recognized M1 peptide-pulsed 24AS transfectants, 9 out of 24 recognized the 67VS transfectants, and zero out of 27 recognized the 45MT transfectant. A similar pattern of recognition was seen when the mutant transfectants were infected with a recombinant vaccinia virus that expressed the influenza M1 gene, indicating that these mutations in the 45 pocket can also affect presentation of the naturally processed and endogenously presented M1 peptide (15). These results demonstrate that each of the single amino acid substitutions at positions 24 and 67 in the 45 pocket of HLA-A2.1 can affect the conformation of the M1 peptide bound by the HLA-A2 molecule. The 45MT mutant clearly has had the most drastic affect on M1 presentation, and it is currently unclear whether the 45MT molecule has retained the capacity to bind the M1 peptide.

DISCUSSION

The results summarized in this review demonstrate that amino acid side chains in the floor of the peptide binding groove and the 45 pocket of the HLA-A2.1 molecule can play a critical role in viral peptide presentation by affecting the conformation of the bound peptide as recognized by the antigen-specific TCR. Due to their location deep within the peptide binding groove and 45 pocket, the amino acid residues at positions 9, 95, 99, 24, 45, and 67 are probably not accessible for direct contact with the T cell receptor (19).

Although it is formally possible that these single amino acid substitutions could induce changes in the conformation of other parts of the class I molecule that affect TCR recognition, such effects would not be predicted based on the results of comparing the alpha-carbon backbones of HLA-A2.1 and HLA-Aw68 in which amino acid differences produce only local changes (7). Thus, our findings suggest that the conformation of the M1 peptide that is bound in the binding groove of HLA-A2.1 is determined by the composite effect of at least five amino acids located on the beta-pleated sheet floor and within the 45 pocket. It is also possible that the changes in T cell recognition that are induced by amino acid substitutions are due to quantitative differences in peptide binding. Methods to quantitate assembly of HLA-A2.1-peptide complexes are currently under investigation.

Approximately 37 amino acid residues constitute the floor of the peptide binding groove (3,4) and five amino acids make up the 45 pocket (7). Eight of these 42 positions are sites of allelic polymorphism (9, 24, 45, 67, 95, 99, 114, and 116)(3). We have so far evaluated six of these eight sites for their effects on presentation of the M1 peptide and have found that each of them has a significant effect on M1 presentation. These findings strongly indicate that each of these polymorphic positions is involved in M1 binding and presentation, and constitute clear evidence that polymorphism in HLA class I molecules creates different peptide binding grooves that have distinct specificities for binding and presentation of peptides to T cells.

REFERENCES

1. Ploegh HL, Orr HT, Strominger JL. Major histocompatibility antigens: the human (HLA-A,B,C) and murine (H-2K,H-2D) class I molecules. Cell 1981;24:287-299.
2. Lopez de Castro JA, Barbosa JA, Krangel MS, Biro PA, Strominger JL. Structural analysis of the functional sites of class I HLA antigens. Immunol. Rev. 1985;6:149-168.
3. Parham P, Lomen CE, Lawlor DA, Ways JP, Holmes N, Coppin HL, Salter RD, Wan AM, Ennis PD. Nature of polymorphism in HLA-A, -B,and -C molecules. Proc. Natl. Acad. Sci. USA 1988;85:4005-4009.
4. Parham P, Lawlor DA, Lomen CE, Ennis PD. Diversity and diversification of HLA-A,B,C alleles. J. Immunol. 1989;142:3937-3950.
5. Bjorkman PJ, Saper MA, Samraoui B, Bennett WS, Strominger JL, Wiley DC. Structure of the human class I histocompatibility antigen, HLA-A2. Nature 1987;329:506-512.
6. Bjorkman PJ, Saper MA, Samraoui B, Bennett WS, Strominger JL, Wiley DC. The foreign antigen binding site and T cell recognition regions of class I histocompatibility antigens. Nature 1987:329;512-518.
7. Garrett TPJ, Saper MA, Bjorkman PJ, Strominger JL, Wiley DC. Specificity pockets for the side chains of peptide antigens in HLA-Aw68. Nature 1989;342:692-696.

8. Nuchtern JG, Bonifacino JS, Biddison WE, Klausner RD. Brefeldin A implicates egress from endoplasmic reticulum in class I restricted antigen presentation. Nature 1989;339:223-226.
9. Yewdell JW, Bennick JR. Brefeldin A specifically inhibits presentation of protein antigens to cytotoxic T lymphocytes. Science 1989;240:637-640.
10. Townsend A, Ohlen C, Basten J, Ljunggren H-G, Foster L, Karre K. Association of class I major histocompatibility heavy and light chains induced by viral peptides. Nature 1989;340:443-448.
11. Mattson DH, Shimojo N, Cowan, EP, Baskin JJ, Turner RV, Shvetsky BD, Coligan JE, Maloy WL, Biddison WE. Differential effects of amino acid substitutions in the beta-sheet floor and alpha-2 helix of HLA-A2 on recognition by alloreactive and viral peptide-specific cytotoxic T lymphocytes. J. Immunol. 1989;143:1101-1107.
12. Shimojo N, Anderson RW, Mattson DH, Turner RV, Coligan JE, Biddison WE. The kinetics of peptide binding to HLA-A2 and the conformation of the peptide-A2 complex can be determined by amino acid side chains on the floor of the peptide binding groove. Int. Immunol. 1990;2:193-200.
13. Mattson DH, Handy DE, Bradley DA, Coligan JE, Cowan EP, Biddison WE. DNA sequences of the genes that encode the CTL-defined HLA-A2 variants M7 and DK1. Immunogenetics 1987;26:190-192.
14. Holmes N, Ennis P, Wan AM, Denney DW, Parham P. Multiple genetic mechanisms have contributed to the generation of the HLA-A2/A28 family of class I MHC molecules. J. Immunol. 1987;139:936-941.
15. Winter CW, Carreno BM, Turner RV, Koenig S, Biddison WE. The 45 pocket of HLA-A2.1 plays a role in presentation of influenza virus matrix peptide and alloantigens. (Submitted).
16. Shimojo N, Maloy W, Anderson R, Biddison W, Coligan J. Specificity of peptide binding by the HLA-A2.1 molecule. J. Immunol. 1989;143:2939-2945.
17. Carreno B, Anderson R, Coligan J, Biddison W. HLA-B37 and HLA-A2.1 molecules bind largely nonoverlapping sets of peptides. Proc. Natl Acad. Sci USA 1990;87:3420-3424.
18. Storkus W, Howell D, Salter R, Dawson J, Cresswell P. NK susceptibility varies inversely with target cell class I HLA antigen expression. J. Immunol. 1987;138:1657-1663.
19. Davis M, Bjorkman P. T cell antigen receptor genes and T cell recognition. Nature 1988;334:395-402.

*Advances in Regulation of Cell Growth, Volume 2;
Cell Activation: Genetic Approaches,* edited by
James J. Mond, John C. Cambier, and Arthur Weiss.
Raven Press, Ltd., New York © 1991.

2

Regulation of Gene Expression and Cell-Type Differentiation via Signal Transduction Processes in *Dictyostelium*

Sandra K. O. Mann and Richard A. Firtel

*Department of Biology, Center for Molecular Genetics, University of California,
San Diego, 9500 Gilman Drive, La Jolla, California, 92093-0634*

During development and morphogenesis of a multicellular organism, various signals must be perceived and responded to by the individual cells, so as to allow proper organization and growth of the organism as a whole. The process by which a cell receives a signal, whether it be a hormone, a chemotactic agent, or an inducing or growth factor, and then effects the proper intracellular response is presently a subject of intense investigation in many developmental systems. Many such signals are known to bind to receptors on the surface of a cell and induce a specific intracellularsignal transduction pathway, ultimately causing cellular changes and alteration in the transcription of developmentally regulated genes.

The cellular slime mold *Dictyostelium discoideum* provides an elegant system in which to investigate signal transduction processes and the ensuing temporal and spatial regulation of gene expression (1,2). Individual amoebae grow vegetatively until the bacterial food source is depleted. When a population of cells begins to starve, they initiate a 24-hour developmental cycle. The life cycle of this simple organism can be divided into two general phases, the early stage of aggregation and the later multicellular stage of morphogenesis and culmination, with temporal patterns of differential gene expression occurring throughout the developmental cycle and coupled tightly to morphogeneis. Approximately 10^5 cells stream together to form an aggregate that then undergoes a series of defined morphogenic changes involving cell movement and differentiation (see Figure 1). A loose mound is formed by 9 hours, which produces a defined tip and elongates to form a finger-like structure. This falls onto the substratum and develops into a migrating slug or pseudoplasmodium. This forms a second finger structure that "sits down" upon itself. Culmination then begins, resulting in the formation of a fruiting body comprised of a sorocarp, or "head" of spores, borne atop a slender stalk (see ref. 3 for a detailed discussion of culmination). These spores can then await the return of favorable conditions before germinating and once again beginning vegetative growth. The precursors of the mature spore and stalk cells, the prespore and prestalk cells, can first be identified during late aggregation. Because of the relative simplicity of the morphological patterns followed during development and the availability of molecular and cytological markers, *Dictyostelium* also lends itself to the study of spatial and cell-type-specific gene regulation. This manuscript reviews the mechanisms by which morphological differentiation is regulated in this organism. As described below, this involves the integration of cell-cell signal transduction pathways that not only control

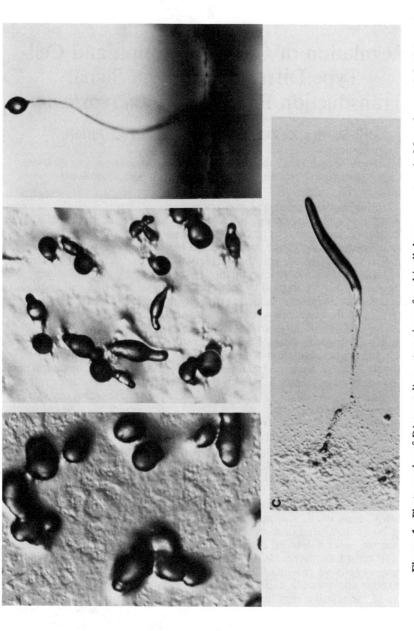

Figure 1. Photographs of *Dictyostelium* - top view of multicellular aggregates. A. Mound stage (~9-10 hours). B. Tipped aggregate (~11-12 hours). C. Migrating slug or pseudoplasmodium. D. Mature fruiting body.

morphogenesis and cell-type differentiation, but also regulate coordinate sets of genes whose expression is required for multicellular development (see Figure 2).

Aggregation and Early Development

Approximately 3 to 4 hours after removal of the food source, a small percentage of cells within the population initiate aggregation by emitting a pulse of cAMP. This cAMP binds to receptors on the surface of surrounding cells, causing them to respond in two ways (1,2,4,5). First, cells move towards the cAMP source, up the cAMP gradient. The cAMP emitted by a single cell, however, is only sufficient to chemotactically attract those cells in its immediate vicinity. Therefore, a second response is used to propagate or relay the signal by activating adenylate cyclase in the responding cell, resulting in the intracellular synthesis of cAMP and its subsequent release/secretion into the environment. Adjacent cells then respond to this signal by chemotaxing toward the now-forming aggregation center and by relaying the cAMP signal themselves (6,7). As in many signalling pathways, the response adapts; cells stop moving after ~1 minute and cease producing and releasing cAMP (6). Work of P. Devreotes has suggested that adaptation involves phosphorylation of the cAMP receptor (8,9; see below). Extracellular and cell surface phosphodiesterases clear the cAMP from the extracellular environment and the receptors become unoccupied and dephosphorylated, thereby regaining the ability to activate the downstream pathways in response to another pulse of cAMP (10,11). A new pulse of cAMP is then emitted from the center, resulting in another round of activation and adaptation. This process continues with the cAMP signal being propagated outward and the concurrent movement of cells towards the center (12). By 9 to 10 hours after the initiation of starvation, ~10^5 cells have moved together to form a mound.

The importance of the adaptation response can be understood by "viewing" what happens to a particular cell within a domain of aggregating cells. As the cAMP wave moves outward from the center, cells immediately proximal to our marker cell release cAMP. This cAMP-sensitive cell responds as described above, chemotaxing toward the center and relaying the cAMP signal, and then adapts. Cells more distal to this cell then respond and emit cAMP. If our marker cell were not adapted to the signal, it would respond by moving outward, toward this new cAMP source, rather than toward the center. Adaptation is thus essential for the directed movement of the cells during aggregation.

Studies from a number of labs have been responsible for biochemically dissecting the pathways controlling the aggregation process (see refs. 1,2,4 for reviews). Aggregation is mediated via the coordinate regulation of three intracellular signal transduction pathways that direct chemotaxis and the relay of the cAMP signal, as well as cause changes in the transcription of developmentally-regulated early genes. Binding of cAMP to the cell surface receptors transiently activates adenylate cyclase (13,14). This response is relatively slow, with intracellular cAMP levels begining to increase after ~20 seconds and peaking at ~1-2 minutes. At this time the response adapts and cAMP synthesis terminates (15). In addition, cAMP binding to the receptors transiently activates both guanylate cyclase (16) and phospholipase C (17-19, Okaichi and Firtel, manus. in prep.); these are designated the rapid responses. Cyclic GMP and inositol 1,4,5-trisphosphate (1,4,5-IP$_3$) levels peak at ~10 seconds, followed by adaptation of the pathways. However, the kinetics of adaptation are too rapid to be initially

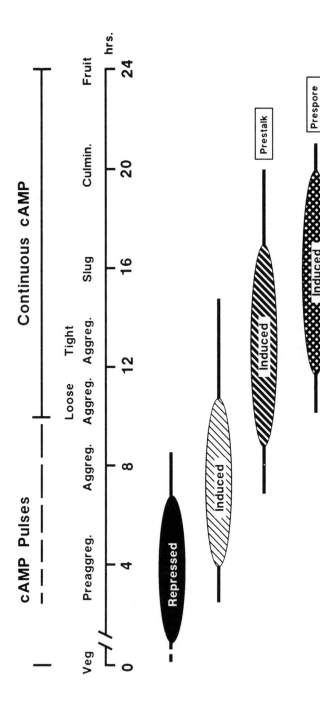

Figure 2. The regulation of gene expression by cAMP receptor-mediated signal transduction pathways. Presented are the temporal expression patterns and the relative levels of expression (indicated by the thickness of the bar) for four classes of cAMP-regulated genes. In chronological order of appearance, these are: pulse-repressed genes, pulse-induced genes, cAMP-induced prestalk genes, and cAMP-induced prespore genes. Also indicated are the morphological stages and the extracellular cAMP patterns present during the 24-hour developmental cycle. Cyclic AMP pulses initiate several hours after starvation. By the multicellular stages, cAMP levels become more continuous, although they continue to oscillate, with the wave initiating at the tip.

controlled by receptor phosphorylation. Production of cGMP has been linked to the initiation of chemotactic movement, and the increase in 1,4,5-IP3 has been implicated in gene regulation (20-24,26; see below). This is pictured in Figure 3.

Molecular studies have identified classes of coordinately regulated genes that are differentially expressed during development. One class is induced upon the

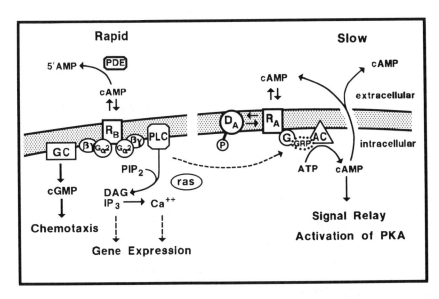

Figure 3. Model for signal transduction pathways active during early development. The responses are divided into a slow pathway on the right that involves the activation of adenylate cyclase (AC) and the rapid response pathways that include the activation of guanylate cyclase (GC) and phospholipase C (PLC). Activation of adenylate cyclase is mediated through the receptor (R) by a G protein (activation requires GTP *in vitro*) and a soluble factor GRP (see refs. 143 and 144). Cyclic AMP is secreted/released from the cells into the extracellular medium where it can feedback and activate unoccupied receptors on the same cell or activate receptors on adjacent cells, thus relaying the signal (relay pathway). The receptor is shown to be phosphorylated in response to ligand which is believed to result in the adaptation of the receptor. Activation of cAMP-dependent protein kinase A (PKA) is shown as a possible intracellular function. Coupling of the two pathways is indicated by dotted arrow.

The rapid response pathways induce chemotaxis and regulate gene expression. Cyclic AMP binds to receptors (R) that are coupled to a heterotrimeric G protein that includes the subunit Gα2. PLC and GC (both activated by GTPγS) are shown as possible direct effectors of Gα. Gene regulation is thought to be mediated by activation of phospholipase C and production of diacylglycerol (DAG) and 1,4,5-IP3, which may function by mobilizing Ca^{++} stores. *ras* has been suggested to be involved in the phosphotidyl inositol pathway as described in refs. 137 and 138. As in other systems, *ras* is associated with the plasma membrane in *Dictyostelium* (its location here is for cartoon puposes only). PLC is shown associated with the plasma membrane because of the location of the substrate (PIP2) and Gα2. Chemotaxis is thought to involve the activation of guanylate cyclase and the production of cGMP. Extracellular phosphodiesterases (PDE) clear the cAMP, thus allowing the receptors to de-adapt. See text and refs. 1, 2, and 4 for more complete details.

onset of development and is maximally expressed during the interphase period between starvation and aggregation (25-28). A second class is induced with the onset of cAMP pulses and is maximally expressed during aggregation, after which transcription is repressed. Many of this second class of genes encode components important for the signal transduction process or for cell-cell adhesion [e.g. cAMP receptor (29,30); Gα subunit (31-33); D2, a serine esterase (28); contact sites A, a cell adhesion molecule (see 25)]. Physiological studies in our lab have shown that these genes are induced by pulses of cAMP (pulse-induced genes), and, thus, their expression is coordinate with aggregation. The first class, conversely, is repressed by this same signal and are designated the pulse-repressed genes (26).

Molecular genetic analysis of aggregation

The molecular dissection of signal transduction pathways in *Dictyostelium* initiated with the cloning of genes that encode specific components of the pathways, including the receptors and coupled G proteins and, more recently, developmentally regulated kinases and phosphatases. The cell surface receptor that binds cAMP during aggregation (cAR1) has been cloned and characterized in the laboratories of P. Devreotes and A. Kimmel (30). It has the seven transmembrane domains typical of receptors coupled to G proteins. Support for the coupling of a G protein to the aggregation-stage receptor is provided by the observations that, in membranes isolated from aggregating cells, binding of ligand to cAMP receptors stimulates GTPγS binding and increases GTPase activity (4,13,16). In addition, guanine nucleotides decrease the affinity of cAMP receptors for the ligand. The cAR1 receptor has an intracellular carboxy-terminal tail that contains clusters of serine and threonine residues (30), some of which are phosphorylated in response to ligand binding. This is believed to cause the adaptation of the receptor, possibly by blocking the interaction of the receptor with its coupled G protein (9,29,30).

Because in most systems the specificity of a cellular response to an extracellular signal resides primarily in the receptor and the α subunit of the G protein, our laboratory undertook to isolate clones of Gα subunits. Using oligonucleotide probes complementary to conserved regions in Gα subunits characterized in other systems, we isolated cDNA clones for two Gα genes, *Gα1* and *Gα2*. cDNA clones for these two genes were independently isolated in P. Devreotes' laboratory using similar approaches, and subsequent characterizations occurred collaboratively. Gα1 and Gα2 show ~45% amino acid sequence identity to each other and to Gα proteins from other eukaryotes, with the homology found primarily within the regions highly conserved among all Gα subunits (31-33). Subsequently, three additional Gα proteins, Gα3 (Pupillo and Devreotes, unpublished observation), and Gα4 and Gα5 (J. Hadwiger, T. Wilkie, M. Strathmann, and R. Firtel, sub. for pub.) have been cloned by the two labs (see below).

Gα2 is not expressed in vegetative cells grown on bacteria. Its expression is induced early in development from a distal promoter and is maximal during aggregation, corresponding to the period of maximal cAR1 expression (31-33; Carrel and Firtel, unpub. obser.). mRNA levels for both Gα2 and cAR1 decrease rapidly as the aggregate forms (~10 hours). Interestingly, *Gα2* and *cAR1* are both induced for a second time during later development, from more proximal promoters (31-33; Carrel and Firtel, unpub. obser.). To examine the spatial expression of *Gα2* at this time, the *Gα2* proximal promoter was isolated and fused to the *E.coli lacZ* gene. Stable transformants were plated for development and stained *in situ* for β-gal activity at various developmental stages. Expression was strongly localized to the anterior, prestalk region of the migrating slug and to the very apical tip region of the culminant, suggesting that Gα2 has a specific function in these cells during culmination (Carrel and Firtel, unpub. obser.).

The recent development of methods for creating gene disruptions (null mutants) or gene replacements by the Spudich laboratory and subsequently our own, has allowed us to examine the function of specific genes (35-37). These approaches have required the isolation of genes that can be used to complement auxotrophic strains, and the subsequent use of those genes to disrupt a gene of interest (via homologous recombination) and to provide a selectable marker for use in the auxotrophic strain (38,39). The re-introduction of wild-type or mutant copies of the gene into the null mutant allows a more detailed examination of gene function or of structure/function relationships. These approaches have been particularly useful for examining the biochemical basis of cAMP-mediated responses during aggregation and for determining the roles of specific Gα subunits and cAMP receptors. Biochemical and physiological analyses of *gα2*- gene disruptants (37) and cAR1 antisense mutants (40) indicate that both these proteins are required for aggregation, and suggest that Gα2 is coupled to cAR1. *cAR1* and *Gα2* are both preferentially expressed during aggregation, and lack of expression of either one gene results in an inability to aggregate and the loss of all aggregation-stage, cAMP-activated responses. Membranes isolated from *gα2*- cells have only low-affinity cAMP binding sites, consistent with the absence of the cAR1-coupled G protein (37). These results confirm earlier analyses of *Frigid*A strains, a class of aggregation-deficient mutants isolated by M. Coukell (41). Results from his laboratory and from that of P. van Haastert demonstrated that these strains lack cAMP-activation of adenylate cyclase and guanylate cyclase and also guanine nucleotide-mediated loss of high-affinity cAMP binding sites (42). Work from our laboratory showed that these mutants lack cAMP-repression of the pulse-repressed genes and cAMP-induction of the pulse-induced genes (26; see below). Subsequent molecular genetic analyses carried out in our laboratory and that of P. Devreotes indicated that the *Frigid*A mutants are defective in the gene encoding Gα2 (30). In particular, the allele HC85 carries a deletion in the gene, and RNA blot analysis indicated that HC213 has little or no Gα2 expression.

Transformation of *gα2*- null cells with vectors expressing the wild-type *Gα2* results in the full restoration of all biochemical and physiological responses to cAMP (35). Expression of mutant Gα2 proteins in either wild-type cells or *gα2*- cells also indicates Gα2 function is directly required for induction of adenylate and guanylate cyclases (Okaichi and Firtel, unpub. obser.). It has not yet been formally established what the direct effector for Gα2 is; however, the most attractive candidates are phospholipase C and/or guanylate cyclase. Preliminary evidence indicates the presence of GTPγS-stimulated PLC activity in aggregating cells (18,19,43). It is possible that *Gα2* missense mutants currently being examined in our laboratory will allow us to test *in vitro* whether Gα2 regulates phospholipase C activity. Other results suggest that, although Gα2 function is essential for the cAMP receptor-activation of adenylate cyclase, it may not be the G protein directly activating this effector (42,44).

Developmental RNA blots indicate that *Gα1* is preferentially expressed during vegetative growth and early aggregation (31,33). It is not known to what receptor Gα1 is coupled, nor what its effector is. Overexpression of Gα1 during vegetative growth results in cells that are large and multinucleate and that appear to be defective in cytokinesis (32). However, mutants lacking Gα1 (antisense mutants or $g\alpha1^-$ null strains constructed by disruption of the $G\alpha1$ gene) have no discernible phenotype during growth or development, suggesting that Gα1 function is not required for proper growth and development, or that there is more than one gene encoding this function (37). In addition, biochemical analysis suggests that Gα1 does not play a role in either cAMP- or folic acid-mediated signalling pathways.

Developmental mutants that are altered in the aggregation response have also been instrumental in the dissection of the signalling pathways controlling the expression of genes during early *Dictyostelium* development, such as the pulse-induced genes studied in our lab. *Synag* mutants are aggregation-deficient, but they chemotax properly toward exogenous cAMP and can aggregate synergistically when mixed with wild-type cells to form normal fruiting bodies (13,14,45-47). Physiological and biochemical analyses of these mutants have shown that they are blocked in the ability to activate adenylate cyclase. However, when exogenous pulses of cAMP are supplied, these cells show a normal cAMP-mediated activation of guanylate cyclase. *Synag* cells do not turn off pulse-repressed genes nor do they express the pulse-induced genes to their normal high levels (26). However, pulsing *Synag* cells with cAMP restores normal repression of pulse-repressed genes and full expression of the pulse-induced class of genes. These analyses, therefore, suggest the activation of adenylate cyclase and the concomitant increase in intracellular cAMP is not required for either chemotactic movement or the induction of gene expression. These observations on *Synag* mutants and $g\alpha2^-$ null mutants indicate that the activation of adenylate cyclase acts primarily as a step in the cell-cell relay pathway [although there is some evidence to suggest that the activation of adenylate cyclase may also regulate other intracellular events, such as the repression of gene *M4-1* (46,47)], and supports the conclusion that the receptor-mediated rapid response pathways are essential in regulating expression of both the pulse-repressed and pulse-induced classes of genes.

An important conclusion from these studies is that genes encoding a number of proteins essential for the signal transduction pathways, such as cAR1 and Gα2, are induced by those very pathways. Therefore, it is not immediately apparent how the initial stages of this regulatory loop are controlled. The analysis of gene expression in *Synag7* (47) allowed us to address this apparent conundrum by uncoupling cAMP pulsing from the early stages of starvation and aggregation. Our results indicate that, shortly after starvation initiates, this class of genes is induced to a low level by a mechanism independent of cAMP pulsing, which allows the initial establishment of the signal transduction pathways at a low level. This in turn enables the cells to begin responding to and relaying the cAMP signal, which brings about a second phase, during which these genes are induced to a higher level in response to cAMP pulses (27). Such an autoregulatory mechanism allows the cells to fine-tune the level of transcription for each of these genes, since they are each differentially induced by the two phases.

The first, cAMP-independent, inductive phase requires the synthesis of a protein factor after the initiation of starvation. A likely candidate for this factor is CMF, or Conditioned Medium Factor, an 80-kD secreted glycoprotein (50-52). This factor is believed to be required for development and probably functions as a density-determining factor. Cells at a very low density do not efficiently enter early development and activate early gene expression. Cells starved at low density also cannot be stimulated to induce prestalk or prespore cell differentiation in response to cAMP. Providing CMF extracellularly overcomes the effects of low density. Our results suggest that CMF must be present in the extracellular environment above a threshold level for development to initiate. It may therefore serve as a mechanism by which *Dictyostelium* determines whether the cell density is sufficient for the developmental program to proceed successfully. The first phase of induction for the pulse-induced class of genes would likely be an early step in that program.

Another mode of autoregulation of early gene expression is most eloquently demonstrated by analysis of the genes for phosphodiesterase (PDE) and PDE inhibitor in the laboratory of R. Kessin. PDE, which regulates the level of extracellular cAMP, is induced by either pulses or continuous levels of cAMP (10,11,53-55). As cAMP levels rise, more PDE is synthesized in order to clear the extracellular cAMP. If PDE levels are too high (*i.e.* the cAMP level falls below a certain threshold), the PDE inhibitor is induced. This extracellular protein forms a tight complex with PDE, which is not dissociable under physiological conditions. This complex has a substantially higher Km for cAMP, resulting in an effective decrease in PDE activity. The differential regulation of PDE and its inhibitor by cAMP effectively controls the level of active extracellular PDE, so as to maintain the proper cell-cell cAMP relay necessary for aggregation.

Other signalling pathways

The further dissection of the signal transduction pathways that function early in the *Dictyostelium* developmental cycle is the subject of investigation in several laboratories. The intermediate steps and components in these pathways are at this time unknown, though several candidates have been proposed. Protein kinases, both those that phosphorylate tyrosine residues and those that phosphorylate serine/threonine residues, play roles in signal transduction processes in other systems. We and others have investigated the role of cAMP-dependent protein kinase A (PKA) during early development. The work of Veron and Leichtling has shown that in *Dictyostelium*, the holoenzyme is comprised of one regulatory (R) and one catalytic (C) subunit (RC), in contrast to R_2C_2 in metazoans; however, the *Dictyostelium* C subunit can form a complex with the mammalian R subunit (56-61). Binding of two molecules of cAMP to two sites (A and B) on the R subunit results in the dissociation and consequent activation of the C subunit (62). We have effectively obstructed function of this enzyme by overexpressing an altered mammalian regulatory subunit (63), which lacks the A and B sites and therefore does not bind cAMP (64,65). Expression of this cAMP-unresponsive regulatory subunit results in a dominant negative phenotype, in which the mutant R constitutively inactivates the catalytic subunit. Cells in which this overexpression occurs do not aggregate or develop, indicating that intracellular cAMP levels and proper function of this enzyme are involved in developmental regulation (63). In these cells, cAMP induction of pulse-induced gene expression is normal, as is the cAMP-activation of adenylate and guanylate cyclase, indicating PKA is not required for these processes. These cells do express PDE to an abnormally high level. Whether this is involved in the aggregation-less phenotype is not known. Simon *et al* (66) have achieved a similar agg- phenotype by overexpressing the *Dictyostelium* R subunit. It is possible that the excess R subunits titrate out the cAMP, thus maintaining inactive RC dimers.

Genes encoding several other protein kinases have been cloned in a number of laboratories (6,7; J. Williams, P. Devreotes, and C. Reymond, pers. commun.), as well as our own (see below). Some of these kinases may function as intermediates in signal transduction, though this has not yet been established for any of them. One serine/threonine protein kinase, DdPK3, the gene for which has been characterized in our laboratory, appears to be required for proper aggregation (S. Mann and R. Firtel, sub. for pub.). Cells in which *DdPK3* has been disrupted do not aggregate or develop, though they appear to express the pulse-induced class of genes normally. These *Ddpk3⁻* null cells can aggregate synergistically when mixed with wild-type cells, so they appear able to respond properly to cAMP pulses. To examine this further, we have marked these *Ddpk3⁻* cells by electroporating in a vector carrying a fusion of the *Actin6* (constitutive) promoter and the *E.coli lacZ* gene using a construct and protocols worked out by Dingermann *et al.* (68). When these "tagged" *Ddpk3⁻* cells are mixed with wild-type cells and the population is allowed to develop, *in situ* staining for β-gal activity enables us to distinguish the *Ddpk3⁻* cells, which stain blue. As development of the mixed aggregate proceeds, the *Ddpk3⁻* cells do not participate normally in fruiting body formation, but remain as a clump of cells near the base of the culminant (S. Mann and R. Firtel, sub. for pub.). The *DdPK3* gene has a fairly complex pattern of expression during normal development of wild-type cells. Transcripts are detectable by 2.5 hours after starvation initiates, with levels increasing until ~6 hours and then remaining fairly constant through culmination. There is a shift in transcript size from 2.8 kb to 2.6 kb between 2.5 and 5 hours that appears to be cAMP-pulse-dependent. Cells shaken in the presence of cAMP pulses experience the normal shift; cells shaken in the absence of exogenous cAMP pulses accumulate only the 2.8-kb transcript (S. Mann and R. Firtel, sub. for pub.). We have not yet determined whether this reflects the presence of two differentially-induced promoters. Exactly what role this and other kinases may play in signal transduction processes remains to be elucidated.

Other proteins that may be involved in signal transduction processes include protein phosphatases, which play an opposite and yet complementary role to that of protein kinases. It has been demonstrated that enzyme activation can be accomplished not only by phosphorylation, but in some cases by removal of phosphate. A gene encoding a phosphotyrosine phosphatase, *DdPTP1*, has been cloned and characterized in our laboratory (Howard and Firtel, sub. for pub.). Transcripts are developmentally regulated and preliminary results from gene disruption studies indicate that *DdPTP1* is required for normal development. Genes encoding two developmentally regulated protein tyrosine kinases, which could be considered as counterpart enzymes to protein tyrosine phosphatases, have been cloned in the laboratory of J. Spudich (69). The role of these proteins is currently not understood.

Differentiation and Morphogenesis in the Multicellular Organism

Once aggregation has been completed, the multicellular organism undergoes a defined series of morphogenetic changes that result in the formation of a mature fruiting body. First the aggregate forms a tip, then elongates upward. This vertical "finger" structure falls over to become a migrating pseudoplasmodium or slug by about 16 hours. The duration of this developmental stage depends on conditions of light, NH_4^+ concentration, and other environmental factors (70-72). Under appropriate conditions, the slug becomes stationary and forms a second vertical finger structure. As the organism culminates, the prestalk cells follow a type of "reverse fountain" pattern, moving to the tip of the structure, then down the center as they form a core of terminally differentiated cells. The prespore/spore cells form a sort of sphere that moves up this core, or stalk, until it is borne aloft by the slender stalk at the end of culmination (see 3). The mature fruiting body is comprised of ~80% spores and ~20% terminally differentiated stalk cells. These proportions are also reflected in the slug, in which the anterior ~15% is comprised of prestalk cells and the posterior ~85% is comprised almost entirely of prespore

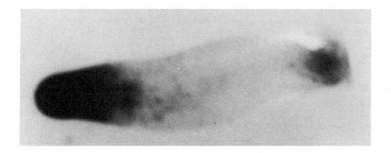

Figure 4. Cytological staining of *Ddras/lac*Z transformants developed to the slug stage. X-gal is the substrate. Dark regions are β-gal positive cells and are dark blue in the original color photographs. Note the lighter cone-shaped region in the center of the stained domain. These lighter cells represent the prestalk B subpopulation. See ref. 85 for details.

cells. The initial spatial patterns of these two functionally-distinct cell types can be distinguished at the time of aggregate formation (see below) and were first identified using cytological markers and an antibody against an epitope found on mature spores (73).

Early work in our laboratory identified cDNA clones complementary to mRNAs preferentially expressed late in development (74). Northern blot hybridization studies using RNA isolated from prestalk and prespore cells crudely fractionated on the basis of density demonstrated the presence of three classes of genes: those preferentially expressed in prestalk cells; those preferentially expressed in prespore cells; and cell-type-nonspecific genes (74,75). Further analysis indicated that many of the cell-type-specific RNAs were inducible in single-cell culture in response to cAMP (50,76; see discussion of regulation by cAMP below). These cell-type-specific genes have enabled us to investigate the regulatory pathways controlling late gene expression and cell-type differentiation. Over the last several years, promoters for several of these genes have been isolated and *cis*-acting sequences and *trans*-acting factors involved in their regulation have been identified (see below). These genes have also been used as markers to follow the spatial patterning of the two cell types during morphogenesis.

To examine the spatial patterning of prestalk and prespore cells, we and others have used antibodies against prestalk or prespore fusion proteins made in *E. coli,* or monoclonal antibodies against cell-type-specific proteins (77-85). Promoter/reporter gene fusions have also been used to specifically express a reporter protein in a subset of cells. In earlier studies, we used *E. coli* β-glucuronidase as the reporter and followed cells expressing this reporter gene using an anti-β-glucuronidase antibody (77). Recently we and others have transformed cells with constructs carrying either a prestalk- or prespore-specific promoter fused to the *E. coli lac*Z gene, which encodes b-galactosidase (82,84,85). Cells are fixed at appropriate stages in development and stained for β-galactosidase activity, thus allowing us to determine the spatial pattern of expression for that specific promoter (see Figures 4 and 5). As described in more detail below, the promoters for a number of these genes are preferentially induced in either prestalk or prespore cells. These constructs therefore allow us to follow the ontogeny and morphogenesis of the specific cell types during multicellular development.

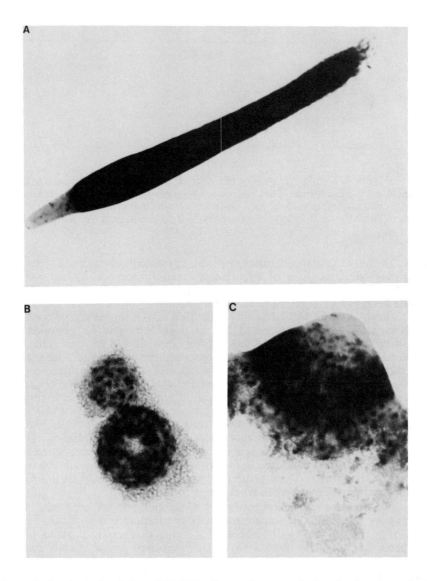

Figure 5. Cytological staining of *SP60/lacZ* transformants during development. A. A migrating slug. B. An aggregation center at the mid-aggregation/early mound stage (9-10 hours), showing a ring of stained cells. Cells entering the aggregate, which were not stained, were lost in the dissection; however, a skirt of unstained cells is seen at theriphery. The very center of the aggregate shows no staining. The smaller aggregate is at the early aggregate stage, with the mound just starting to form. C. A side view of a mound/early tipped aggregate (11 hours). Note that some cells stain more intensely than the rest. See ref.84 for details.

In our laboratory, we have used the gene encoding the prespore protein SP60 as the prespore marker and *Ddras* as the prestalk marker. *Dictyostelium* has two *ras* genes: *DdrasG*, which is preferentially expressed during growth (86,87), and *Ddras* (85,88,90), which is expressed at a low level during growth from a distal promoter, turns off, and then is expressed in a cell-type-specific manner starting at the aggregation stage. This second induction occurs from the distal promoter, and slightly later from a middle and proximal promoter as well. The two genes, *SP60* and *Ddras*, are expressed in the slug in a complementary fashion (84,85). *SP60/lacZ* is expressed in the previously defined prespore domain, which comprises 80-85% of the slug and includes most of the posterior region (73,91). In the mature fruiting body, only the spore cells in the sorocarp are stained. *Ddras/lacZ* is expressed in the anterior 10-15% of the slug, which has been previously defined as the prestalk region, and also in the posterior 2-5% of the slug. In the fruiting body, the stalk, basal cells, and "cap-like" regions on the top and bottom of the sorocarp are stained (see cartoon in Figure 6A, also Figure 4). In addition, we have used *pst-cath/CP2/lacZ* to complement earlier *pst-cath/CP2/β-glucuronidase* studies (Gaskins and Firtel, unpub. obser.). [*Pst-cath/CP2* encodes a prestalk-specific cysteine protease (77,92).] The laboratory of G. Williams has used *D19*, which encodes a prespore-specific extracellular matrix protein, and also *Dd56* and *Dd63*, which encode two prestalk-specific extracellular matrix components, for similar studies (82). *SP60*, *D19*, and *Ddras* are all induced by cAMP (74,75,82,88,93,94), whereas *Dd56* and *Dd63* are induced in response to the morphogen DIF (93,95), identified by R. Kay's group (97-100).

These markers have been used to examine the ontogeny of the two regions or tissues within the developing aggregate. As the aggregate forms, we have shown that *Ddras/lacZ* is expressed in ~10-15% of the cells, which are randomly scattered throughout the loose mound. A similar pattern is observed with *Dd63*, except that its expression appears to be induced slightly later than *Ddras* (82,95). As the tip forms, the *Ddras/lacZ*-expressing cells appear to sort to the anterior/apical region and form the tip (85), as has also been proposed by Takeuchi (101) and Williams (82). Some photographs suggest a spiral pattern of sorting by the *Ddras*-expressing cells. We have proposed that this differential cell sorting occurs in response to spiral patterns of cAMP waves emitted from the forming tip (see also 102). Other results from the laboratory of M. Coukell have indicated that prestalk cells in fact have a higher rate of chemotaxis to cAMP than prespore cells (103). This could be the basis for tip formation by prestalk cells. In contrast, *SP60/lacZ*-expressing cells initially appear in the main body of the aggregate within the developing mound (84) (see Figure 5 and 6B). During later aggregation, as the mound continues to form, they appear as a thick ring or, in some aggregates, as a spiral that coalesces into a ring. No β-gal-staining cells are seen in the center of this ring, which becomes the emerging tip, nor are they seen in the "skirt" of cells moving into the developing mound, although ~80% of these cells will become prespore cells. These results, combined with those of Williams *et al.* (80) and Krefft *et al.* (79), as well as our earlier work using anti-*SP60* antibodies (78), suggest that in the developing aggregate, the initial induction of prespore cell differentiation is spatially restricted. These studies, however, are limited by the fact that all analyses are "static", with large populations of aggregates being fixed and stained at specific morphological stages.

Prespore and prestalk patterning is more complicated than previously thought. The J. Williams laboratory has analyzed *Dd56* expression patterns and found that it is preferentially expressed in a subregion of the anterior prestalk zone,

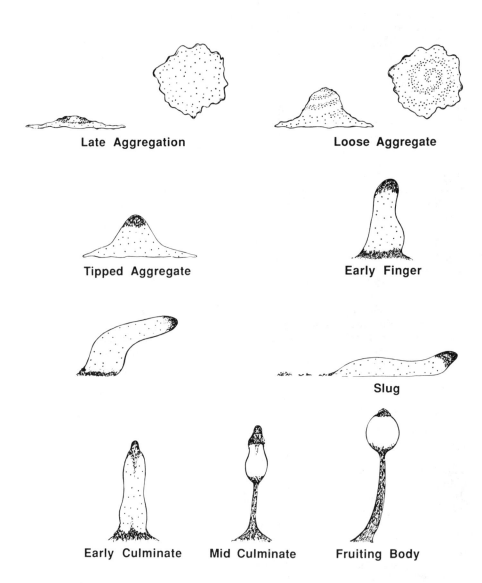

Figure 6. Cartoon of the spatial patterning of *Ddras* and *SP60* expression. A. *Ddras* expression pattern. For the late aggregation and loose aggregate stages, side(left) and top(right) views are shown. The dots representing scattered β-gal-expressing cells within the prespore zone of the tipped aggregate through early culminant stages presumably indicate the presence of anterior-like cells.

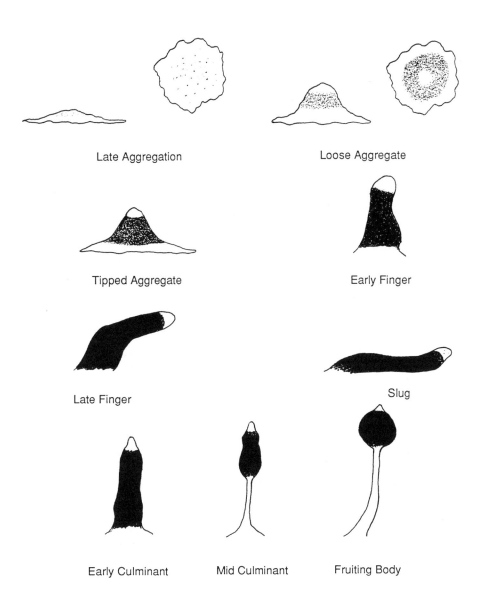

Late Aggregation

Loose Aggregate

Tipped Aggregate

Early Finger

Late Finger

Slug

Early Culminant

Mid Culminant

Fruiting Body

B. *SP60* expression pattern. Little or no *SP60* expression is detectable until the mound begins to form. By the tipped aggregate stage, 70-80% of cells in the multicellular organism express *SP60*, in a pattern complementary to that observed for *Ddras* expression. See ref.84 and 85 for details.

as a cone of cells within the anterior of the slug (83). This defines the prestalk B spatial expression pattern. In general, *Ddras* follows the prestalk A spatial pattern of expression; however, it is expressed differentially from the three promoters active during the multicellular stages (see above). The distal promoter is highly responsive to cAMP, while the middle and proximal promoters are less responsive (85). Deletion of the distal promoter results in a loss of expression in the very posterior region of the slug and a reduction of expression in the anterior region, but causes an increase in relative expression in scattered cells within the prespore region previously designated "anterior-like" cells (104,105). These results suggest that the spatial expression patterns of some genes do not necessarily fall into a simple anterior/prestalk and posterior/prespore pattern. Differences also appear to exist with regard to the pattern of prespore gene expression. Analysis of another prespore gene, *14-E6* (74), reveals some reproducible differences in the patterning within the prespore zone, using *14-E6/lacZ* fusions (Powell and Firtel, unpub. observ.). This suggests that the *14-E6* and *SP60* promoters differentially respond to the signals regulating prespore gene expression.

The observation that *Ddras* and *Dd63* expression initiates in a seemingly randomly population of cells within the forming aggregate suggests that some earlier mechanism functions in the initial determination of prestalk cell ontogeny. Studies from a number of laboratories have indicated that cells that are at a stage early in the cell cycle [S and early G_2; there is no detectable G_1 in *Dictyostelium* (106)] at the time of starvation become localized to the prestalk region, while those cells starved during mid to late G_2 become localized to the prespore region (107-110). These results were obtained by mixing synchronized cells from particular stages in the cell cycle with unsynchronized cells. In other studies using time-lapse video microscopy, we followed single cells during growth and recorded when a specific cell divided (111). Growth medium was removed and replaced with non-nutrient buffer containing CMF (see above), then cAMP was added after 6 hours. At 20 hours, cells were examined for the expression of prestalk or prespore cell markers, using antibodies against pst-cathepsin/CP2 (prestalk) or SP70, another spore coat protein (74,78, see also 76). Our results indicated that cells respond in a cell-autonomous manner, with those starved early in the cell cycle showing a propensity to differentiate into prestalk cells, and cells starved late in the cell cycle inducing prespore markers. No cells were observed to express both markers. Other results have indicated that it is cells starved early in the cell cycle that preferentially initiate cAMP signalling at the onset of development, forming the aggregation centers (112). Cells starved late in the cell cycle can also initiate signalling, but not until a longer time after starvation. Thus, while all cells do have an equal developmental capacity, cells starved early in the cell cycle seem to initiate the responses early and to preferentially induce prestalk markers. Other studies by Schaap and co-workers have shown that these cells have a higher relative level of signalling components.

Signalling pathways controlling late gene expression and patterning

The morphogens involved in the spatial patterning and movements of the two cell types include cAMP and differentiation-inducing factor (DIF) (97-100,113), and probably adenosine (91,114) and NH_4^+ (115) as well. The mechanisms by which the majority of these morphogens act and the signal transduction systems involved are not well understood. It has been demonstrated that cAMP is required for induction of prespore-specific genes and many prestalk-specific genes (49-51,74,94,115). Analyses using cAMP analogs and pharmacological agents, as well as signal transduction mutants, have indicated that cAMP induction of these late genes is mediated via cell surface receptors and G protein-mediated signal transduction pathways (50,117-121). In contrast to pulse-induced gene expression, induction of these genes requires high continuous cAMP levels, conditions that would adapt and down-regulate the early receptors. These observations suggest that different cAMP receptors regulate the responses at different stages in development. Although prespore genes and one class of prestalk genes are induced by cAMP, results from Blumberg's and Schaap's labs using inhibitors of signalling pathways in mammalian systems have, in fact, suggested that the intracellular signalling pathways regulating cAMP-induced expression in these two cells types is different (114,122,123). Other work by A. Kimmel's lab has shown that 1,4,5-IP$_3$ and diacylglycerol can induce expression of these two classes of genes, suggesting that activation of phospholipase C mediates the expression of both classes (124). It is also now clear that the DIF and cAMP signalling pathways play reciprocal roles in regulating some aspects of prestalk and prespore gene expression and thus the differentiation of the respective cell types. DIF inhibits the transcription of prespore genes (125, 126), while cAMP inhibits one of the DIF-induced prestalk genes, *Dd56* (see 96, 127). In contrast, both DIF and cAMP are required for the maximal expression of *Dd63* (126). Other evidence suggests that DIF may be preferentially expressed in a subset of prestalk cells and that DIF responsiveness may require previous exposure of the cells to cAMP (128-131).

In addition to the cAMP receptor that is active during aggregation (cAR1), clones for two other cAMP receptors have been isolated and characterized by the Devreotes and Kimmel laboratories (34,132). Transcripts of cAR2 are induced fairly late in development, between 10 and 15 hours, and are present at a level more than 10-fold higher in prestalk cells than in prespore cells. cAR3 is expressed beginning at ~5 hours, with message levels peaking at ~10 hours, partially overlapping the expression of cAR1. Transcripts of cAR3 are 3- to 4-fold more abundant in prestalk cells than prespore cells. Both of these receptors have the seven-transmembrane-domain motif typical of receptors coupled to G proteins, and both are candidates for mediating cAMP-induced signal transduction events during the later stages of development. cAR2 and cAR3 share ~60% sequence identity with each other and with cAR1 over the transmembrane domains and the three extracellular and three intracellular loops. However, they share no identity over the terminal ~100 amino acids, and neither cAR2 nor cAR3 have the serine/threonine-rich regions found in cAR1. This suggests that they are not phosphorylated/desensitized and would therefore not adapt in response to the continuous levels of extracellular cAMP that are found in the multicellular organism and are used to regulate late, cell-type-specific gene expression. DIF is another inducing factor active in later development, and a search for its receptor has led to the identification of a DIF binding protein by the laboratory of R. Kay (133). Whether or not this is the DIF receptor responsible for mediating DIF activity has not been determined.

As part of the effort to dissect signal transduction pathways active in later development, our laboratory has cloned genes for two Gα subunits, *Gα4* and *Gα5*, that are expressed during the multicellular stages (J. Hadwiger, T. Wilkie, M. Strathmann, and R. Firtel, sub. for pub.). Either or both of these may be coupled to a cAMP receptor and may mediate signal transduction processes during differentiation. *Gα4* is induced in multicellular aggregates, between 9 and 12 hours after the initiation of starvation, and remains at a fairly high, constant level through culmination. $G\alpha5$ is induced between 12 and 15 hours and also remains at a constant level thereafter. Cells in which $G\alpha4$ has been disrupted do not develop normally, indicating that $G\alpha4$ is required for multicellular development.

The *ras* protein has been implicated in signal transduction pathways in vertebrates and has been shown to regulate adenylate cyclase in *S. cerevisiae* (134). The presence of a cell-type-specific *ras* gene in *Dictyostelium* suggested a specific role for *ras* in prestalk cell differentiation. We examined this by expressing a missense mutation (gly_{12} --> thr_{12}) known to be oncogenic in mammalian cells due to a reduced intrinsic GTPase activity (135). Cells overexpressing *Ddras*(G12T) produce aggregates with multiple tips that do not proceed further in development (see Figure 7). Physiological analysis indicates that adenylate cyclase activity is normal in these cells, but that cAMP activation of guanylate cyclase adapts early, resulting in a reduced level of cGMP (135,136). Since cGMP levels have been linked to chemotaxis, the results suggest that these cells may have altered morphogenic movements. Both Newell's and van Haastert's groups have also shown that *Ddras*(G12T) cells have altered phosphotidyl inositol and inositol phosphate pathways, suggesting a role for *ras* in these essential signal transduction pathways (137,138). Other studies in our lab using antisense mutagenesis have suggested that *ras* function is essential for growth in *Dictyostelium* (90).

Figure 7. Comparative phenotypes of *ras* transformants. Transformed cells carrying a *ras*-Gly12 (G) construct (wild-type *ras*) or a *ras*-Thr12 (T) (ras carrying a gly-->thr change at position 12) construct were plated for development and photographed at the times indicated (16, 19, and 25 hours). Cells carrying the *ras*-Thr12 construct produced aggregates that formed multiple tips but did not differentiate further. Taken from ref. 135. See ref. 135 for details.

Molecular characterization of cell-type-specific regulatory elements and effects on spatial patterning.

In order to further determine the mechanisms by which signal transduction processes regulate gene expression, our laboratory has isolated genomic clones for a number of cell-type-specific genes. These genes have been characterized, including their *cis*-acting regulatory regions and potential *trans*-acting factors that bind to alter transcription. Most extensively studied is the prespore-specific gene *SP60*, which encodes one of the three major spore coat proteins (84). In order to dissect the 5' regulatory region of this gene, ~1.3 kb of upstream sequence was fused in-frame to the firefly *luciferase* reporter gene and carried on a vector conferring G418 resistance. Cells transformed with this construct exhibit proper temporal, spatial, and cAMP-induced expression relative to the endogenous *SP60* gene. The 5' flanking sequence includes three homologous CA-rich sequences ($CACACAC_TC_TC_AACACAC$), termed CAEs (CA-rich Elements) found at positions -599 (box1), -516 (box2), and -402 (box3) relative to the start of transcription. Deletion analysis indicates that the level of expression, as quantitated by luciferase activity, decreases dramatically as each CAE is removed. Regulation appears to be at the level of transcription, as the luciferase results correlate with changes in the mRNA levels for both the endogenous and fusion genes and with nuclear run-on results for the endogenous gene. The region containing all three CAEs is not sufficient to confer proper regulation, as constructs missing the sequences between -383 and -154 show little luciferase activity in multicellular aggregates in response to cAMP. Therefore, in addition to the CAEs, more proximal sequences are also required for expression. Analysis of other internal deletions suggests that both of the two most 5' CAEs are involved in controlling the level of developmentally- or cAMP-induced expression (Haberstroh and Firtel, sub. for pub.). Quantitative data for luciferase activity indicate that the effects of the CAEs are multiplicative, suggesting that the CAEs may be recognized by transcription factors that work cooperatively to induce transcription of *SP60*.

In order to determine the spatial pattern of *SP60* expression, we fused the 5' regulatory sequences in-frame to the *lacZ* gene from *E. coli* in the same vector background used for the luciferase experiments (84). Stable transformants were plated for development and, at various stages, were stained for β-gal activity. As described above, the posterior 85% of a migrating slug stains uniformly and strongly, whereas the anterior remains unstained except for a few random cells. Fruiting bodies stain in the sorocarp, which contains the spores. To determine the effect of 5' deletions on the spatial patterning of *SP60* expression, many of the same deleted sequences used in the quantitative luciferase studies were fused to the *lacZ* gene. As before, stable transformants were plated for development and stained at the slug and early culminant stages. Cells carrying constructs lacking all three CAEs did not stain. Those missing the first and/or second CAE stain less intensely, as expected from the luciferase activity. Unexpectedly, with the sequential 5' deletions, a decreasing gradient of expression from the anterior to the posterior of the prespore zone is observed. Constructs lacking the first CAE show staining at the prespore/prestalk boundary of the slug, with a gradient of staining extending posteriorly to include approximately 1/2 to 3/4 of the prespore region. For constructs missing both the first and second CAEs, staining is restricted to a collar of cells at the anterior of the prespore zone along the prespore/prestalk boundary. Deletion of only the most proximal CAE (CAE-3) by an internal deletion also results in a reduced level of expression and a gradient of b-gal staining; however, in this case, the gradient is in the opposite direction as with the 5' deletions of the first two CAEs (L. Haberstroh and R. Firtel, sub. for pub.). In these slugs, staining is strongest in the posterior and very light in the anterior of the prespore domain, with little if any boundary visible between the prespore and prestalk regions. These studies have allowed us to visualize the presence of a previously undetected spatial gradient within the prespore zone. This could be a gradient of an essential transcription factor, possibly one that binds the CAE. Such a gradient may in turn reflect an anterior-to-posterior gradient of a positive morphogen such as cAMP, or a posterior-to-anterior gradient of an inhibiting morphogen such as DIF, which is known to inhibit prespore gene expression. This model is outlined in Figure 8.

Using gel mobility shift assays, binding activities for the CAE have been identified (Haberstroh and Firtel, sub. for pub.). CAE-1 and CAE-2 preferentially bind an activity that is developmentally regulated and that first appears in extracts from 6-hour cells (aggregation stage), the time at which *SP60* expression is first induced. This activity is induced in shaking culture by cAMP, suggesting it may be a factor responsible for developmental and cAMP-induced expression. This binding activity is specific and is competed by specific oligonucleotides, but not by those containing point mutations that inactivate the CAEs *in vivo*. Interestingly, CAE-3 binds the developmental and cAMP-induced factor fairly poorly, but does form some other specific complex. This suggests that the biological differences in the staining patterns observed with the CAE-1/CAE-2 deletions verses the CAE-3 deletion may result from the involvement of different *trans*-acting factors.

Promoter analyses have also been undertaken for the cAMP-induced genes *pst-cath/CP2*, *CP1*, and for the gene encoding UDPG pyrophosphorylase (77,92,139-146). *Pst-cath/CP2* and *CP1* are expressed during multicellular development with kinetics similar to those for the prestalk gene *Ddras*. The gene encoding UDPG pyrophosphorylase is expressed during growth and then reinduced with similar kinetics to *pst-cath/CP2* and *CP1*. All three genes are induced by cAMP. In all three cases, a GT/CA-rich element has been shown to be important for cAMP regulation. 5' or internal deletions that specifically remove the GT-rich element result in a loss of activity. Addition of the element into a deletion construct restores cAMP and developmental inducibility of the promoter. A detailed internal deletion analysis was performed on *pst-cath/CP2* (141-144). These data indicate that the GT-rich element (designated G-box regulatory element or GBRE) is not sufficient to confer developmental regulation; other sequences are also required. Some of the internal mutations differentially affect developmental and cAMP-induced expression, indicating that both cAMP-dependent and cAMP-independent factors are involved in controlling expression of this gene (141). To examine whether different GT/CA-rich elements from *CP1*, *pst-cath/CP2*, the gene encoding UDPG pyrophosphorylase, and from *DG17*, another cAMP-inducible gene (147), can complement the function of the GBRE, our group, in association with J. Williams' lab, examined the effect of wild-type and mutant GT/CA-rich elements on a *pst-cath/CP2* deletion lacking the GBRE (142,144). The data suggest that all of the elements tested can function to activate gene expression in this context.

Using gel mobility shift assays, we have identified a factor (GBF) that binds the *pst-cath/CP2* GBRE (143,144). GBF activity is developmentally regulated and cAMP induced, and is believed to be the essential factor responsive to cAMP-mediated signal transduction pathways. Experiments using cycloheximide to inhibit protein synthesis have implied that GBF or a protein required for GBF activity (*e.g.* a kinase) is expressed in response to cAMP. This indicates that the

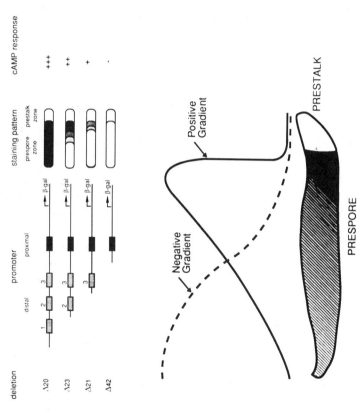

Figure 8. Model for the graded expression of *SP60* within the prespore zone. (Top) The four promoter constructs, with CAEs 1, 2, and 3 indicated, are shown, along with a cartoon of the *SP60/lacZ* staining response (intensity and localization of expression) for each. Also presented are the cAMP and developmental response for each construct. (Bottom) Model showing two possible gradients within the prespore zone to explain *SP60/lacZ* responses. Positive gradient could be transcription factors or parts of the cAMP receptor/intracellular signal transduction pathway. Negative gradient could be DIF, adenosine, or some other morphogen or metabolite that negatively affects expression. Taken from ref. 84. See ref. 84 for details.

mechanism of cAMP regulation of *pst-cath/CP2* is, in part, at the level of the expression of a putative *trans*-acting factor and/or at the level of post-translational modification of the factor, possibly by phosphorylation. Other results indicate that GBF interacts with the various GT/CA-rich binding sites (for the cAMP-inducible genes described above) with relative affinities corresponding to the ability of an oligonucleotide containing that sequence to complement the *pst-cath/CP2* internal GBRE deletion. Since this binding activity interacts with a variety of moderately divergent sequences, it is possible that GBF activity is comprised of a number of related activities. Preliminary experiments indicate that the GBF activity is different than the one described for the *SP60* CAE-1/CAE-2 elements (Schnitzler, Haberstroh, and Firtel, unpub. obser.); however, the same oligonucleotide sequences form complexes with both partially purified activities.

Conclusions

One of the intriguing aspects of development in *Dictyostelium discoideum* is that cAMP plays an essential role throughout the developmental cycle. This is in contrast to the more primitive species *D. minutum* (148), in which cAMP does not regulate chemotaxis but is involved in regulating later morphogenesis. Our studies indicate that the four classes of cAMP-regulated genes examined in our lab are controlled by the changing pattern of cAMP levels during *D. discoideum* development (see Figure 2). As starvation initiates the developmental cycle, the pulse-repressed genes are induced. With the establishment of the cAMP signal relay system, these genes are then repressed and expression of the pulse-induced genes initiates. As the aggregate forms, the levels of cAMP rise and become more continuous, causing repression of the pulse-induced genes, possibly due to cAR1 adaptation and down-regulation. This rise in cAMP is, at least in part, responsible for the induction of cAMP-induced prestalk and prespore genes in different subpopulations of cells.

Cyclic AMP-induced signal transduction systems control chemotaxis and aggregation early in the developmental cycle and control morphogenic movements during patterning in the multicellular organism. They also regulate the transcriptional expression of genes required for these processes. Components of the signal transduction pathways, such as cell surface cAMP receptors, G proteins coupled to these receptors, and other putative components, are themselves developmentally regulated. Thus it appears that different signal transduction systems are established at different times in development and also in the two general cell types during multicellular development. Presumably the changing pattern of cAMP levels and concomitant gene expression is paralleled by changes in the cAR and Gα subunit complement through development.

It remains to be determined what the intermediate components and steps in the various signal transduction pathways are, and work progresses in many laboratories toward this end. The metabolism of inositol phosphates and activation of a presumed phospholipase C are being investigated, as are the activation of guanylate cyclase and increases in cGMP levels. In addition, several developmentally-regulated protein kinases have been identified, including serine/threonine and tyrosine kinases, and are currently being characterized. Other potential components include a developmentally-regulated phosphotyrosine phosphatase and cAMP-dependent protein kinase A.

In order to define the mechanism by which transcription of a developmentally-regulated gene is altered, this being the end result of the signal transduction process, several cell-type-specific genes have been isolated and characterized. The *cis*-acting regulatory sequences have been identified for some of these genes, and *trans*-acting factors that presumably bind these sequences are being investigated. Purification of these DNA-binding factors should allow the isolation of the genes encoding them. Regulation of these genes could then in turn be investigated, with the identification of *trans*-acting factors involved in that regulation. Thus it should be possible to work backward along a signal transduction pathway to identify components of that pathway and to understand the mechanism by which the cAMP signal is transduced to regulate gene expression.

ACKNOWLEDGEMENTS

We would like to thanl Jennifer Roth for assistance in preparing the manuscript. This work was supported by USPHS grants to R.A.F.

REFERENCES

1. Devreotes, P. 1989. *Dictyostelium discoideum:* A model system for cell-cell interactions in development. *Science* 245:1054-1058.
2. Firtel, R.A., P.J.M. van Haastert, A.R. Kimmel, and P. Devreotes. 1989. G-protein linked signal transduction pathways in development: *Dictyostelium* as an experimental system. *Cell* 58:235-239.
3. Schaap, P. 1986. Regulation of size and pattern in the cellular slime molds. *Differentiation* 33:1-16.
4. Janssens, P.M.W. and P.J.M. van Haastert. 1987. Molecular basis of transmembrane signal transduction in *Dictyostelium discoideum. Microbiol. Rev.* 51:396-418.
5. Gerisch, G. 1987. Cyclic AMP and other signals controlling cell development and differentiation in *Dictyostelium. Annu. Rev. Biochem.* 56:853-879.
6. Dinauer, M.C., T.L. Steck, and P.N. Devreotes. 1980. Cyclic 3',5'-AMP relay in *Dictyostelium discoideum.* Adaptation of the cAMP signaling response during cAMP stimulation. *J. Biol. Chem.* 86:554-561.
7. Roos, W. and Gerisch, G. 1976. Receptor mediated adenylate cyclase activation in *Dictyostelium discoideum. FEBS Lett.* 68:170-172.
8. Klein, P., R. Vaughan, J. Borleis, and P.N. Devreotes. 1987. The surface cyclic AMP receptor in *Dictyostelium.* Levels of ligand-induced phosphorylation, solubilization, identification of primary transcript, and developmental regulation of expression. *J. Biol. Chem.* 262:358-364.
9. Vaughan, R. and Devreotes, P.N. 1988. Ligand-induced phosphorylation of the cAMP receptor from *Dictyostelium discoideum. J. Biol. Chem.* 263:14538-14543.
10. Franke, J., M. Faure, L. Wu, A.L. Hall, G.J. Podgorski, and R.H.Kessin. 1991. Cyclic nucleotide phosphodiesterase of *Dictyostelium discoideum* and its glycoprotein inhibitor: structure and expression of their genes. *Dev. Genet.* 12:104-112.
11. Kessin, R.H. 1988. Genetics of early *Dictyostelium discoideum* development. *Microbiol. Rev.* 52:29-49.
12. Tomchik, K. and P.N. Devreotes. 1981. Cyclic AMP waves in *Dictyostelium discoideum*: A demonstration by isotope detection fluorography. *Science* 212:443-446.
13. Theibert, A. and Devreotes, P.N. 1986. Surface receptor-mediated activation of adenylate cyclase in *Dictyostelium*: regulation by guanine nucleotides in wild-type cells and aggregation deficient mutants. *J. Biol. Chem.* 261:15121-15125.

14.   van Haastert, P.J.M., B.E. Snaar-Jagalska, and P.M.W. Janssens. 1987. The regulation of adenylate cyclase by guanine nucleotides in *Dictyostelium discoideum* membranes. *Eur. J. Biochem.* **162**:251-258.

15.   van Haastert, P.J.M. 1987. Down-regulation of cell-surface cyclic AMP receptors and desensitization of cyclic AMP-stimulated adenylate cyclase by cyclic AMP in *Dictyostelium discoideum*. *J. Biol. Chem.* **262**:7700-7704.

16.   van Haastert, P.J.M., R.J.W. de Wit, P.M.W. Janssens, F. Kesbeke, and J. DeGoede. 1986. G-protein mediated interconversions of cell-surface cAMP receptors and their involvement in excitation and desensitization of guanylate cyclase in *Dictyostelium discoideum*. *J. Biol. Chem.* **261**:6904-6911.

17.   Europe-Finner, G.N. and Newell, P.C. 1987. Cyclic AMP stimulates accumulation of inositol triphosphate in *Dictyostelium*. *J. Cell Sci.* **87**:221-229.

18.   Europe-Finner, G.N. and P.C. Newell. 1987. GTP analogues stimulate inositol triphosphate formation transiently in *Dictyostelium*. *J. Cell Sci.* **87**:513-518.

19.   van Haastert, P.J.M., M.J. de Vries, L.C. Penning, E. Roovers, J. van der Kaay, C. Erneux, and M.M. van Lookeren Campagne. 1989. Chemoattractant and guanosine 5'-[τ-thio]triphosphate induce the accumulation of inositol 1,4,5-triphosphate in *Dictyostelium* cells that are labelled with [^{3}H]inositol by electroporation. *Biochem. J.* **258**:577-586.

20.   Ross, F.M. and P.C. Newell. 1981. Streamers: chemotactic mutants of *Dictyostelium discoideum* with altered cyclic GMP metabolism. *J. Gen. Microbiol.* **127**:339-350.

21.   Europe-Finner, G.N. and P.C. Newell. 1986. Inositol 1,4,5-triphosphate and calcium stimulate actin polymerization in *Dictyostelium discoideum*. *J. Cell Sci.* **82**:41-51.

22.   Liu, G. and P.C. Newell. 1988. Evidence that cyclic GMP regulates myosin interaction with the cytoskelteton during chemotaxis of *Dictyostelium*. *J. Cell Sci.* **90**:123-129.

23.   Dharmawardhane, S., P. Warren, A. Hall, and J. Condeelis. 1989. Changes in the association of actin-binding proteins with the actin cytoskeleton during chemotactic stimulation of *Dictyostelium discoideum* amoebae. *Cell Motil. Cytoskel.* **13**:57-63.

24.   Hall, A., P. Warren, S. Dharmawarhane, and J. Condeelis. 1989. Transduction of the chemotactic signal to the actin cytoskeleton of *Dictyostelium discoideum*. *Dev. Biol.* **136**:517-525.

25.   Gerisch, G., H. Fromm, A. Huesgen, and U. Wick,. 1975. Control of cell-contact sites by cyclic AMP pulses in differentiating *Dictyostelium* cells. *Nature* **255**:547-549.

26.   Mann, S.K., C. Pinko, and R.A. Firtel. 1988. cAMP regulation of early gene expression in signal transduction mutants of *Dictyostelium*. *Dev. Biol.* **130**:294-303.

27.   Mann, S.K.O. and R.A. Firtel. 1989. Two-phase regulatory pathway controls cAMP receptor-mediated expression of early genes in *Dictyostelium*. *Proc. Natl. Acad. Sci. USA* **86**:1924-1928.

28.   Mann, S.K.O. and R.A. Firtel. 1987. Cyclic AMP regulation of early gene expression in *Dictyostelium discoideum*: Mediation via the cell surface cyclic AMP receptor. *Mol. Cell. Biol.* **7**:458-469.

29.   Klein, P., C. Fontana, B. Knox, A. Theibert, and P.N. Devreotes. 1985. cAMP receptors controlling cell-cell interactions in the development of *Dictyostelium*. *Cold Spring Harbor Symp. Quant. Biol.* **50**:787-799.

30.   Klein, P., T.J. Sun, C.L. Saxe, A.R. Kimmel, and P.N. Devreotes. 1988. A chemoattractant receptor controls development in *Dictyostelium discoideum*. *Science* **241**:1467-1472.

31.   Pupillo, M., A. Kumagai, G.S. Pitt, R.A. Firtel, and P.N. Devreotes. 1989. Multiple α subunits of guanine nucleotide binding proteins in *Dictyostelium*. *Proc. Natl. Acad. Sci. USA* **86**:4892-4896.

32. Kumagai, A., M. Pupillo, R. Gundersen, R. Miake-Lye, P.N. Devreotes, and R.A. Firtel. 1989. Regulation and function of Gα protein subunits in *Dictyostelium*. *Cell* **57**:265-275.

33. Kumagai, A., S.K.O. Mann, M. Pupillo, G. Pitt, P.N. Devreotes, and R. A. Firtel. 1989. A molecular analysis of G proteins and control of early gene expression by the cell surface cAMP receptor in *Dictyostelium*. *Cold Spring Harbor Symp. Quant. Biol.* **53**:675-685.

34. Saxe, C.L. III, R.L. Johnson, P.N. Devreotes, and A.R. Kimmel. 1991. Expression of a cAMP receptor gene of *Dictyostelium* and evidence for a multigene family. *Genes Dev.* **5**:1-8.

35. de Lozanne, A. and J.A. Spudich. 1987. Disruption of the *Dictyostelium* myosin heavy chain gene by homologous recombination. *Science* **236**:1086-1091.

36. Manstein, D.J., M.A. Titus, A. de Lozanne, and J.A. Spudich. 1989. Gene replacement in *Dictyostelium:* generation of myosin null mutants. *EMBO J.* **8**:923-932.

37. Kumagai, A., J. Hadwiger, M. Pupillo, and R.A. Firtel. 1991. Molecular analysis of two Gα protein subunits in *Dictyostelium*. *J. Biol. Chem.* **266**:1220-1228.

38. Dynes, J.L. and R.A. Firtel. 1989. Molecular complementation of a genetic marker in *Dictyostelium* using a genomic DNA library. *Proc. Natl. Acad. Sci. USA* **86**:7966-7970.

39. Kalpaxis, D., H. Werner, E.B. Marcotte, M. Jacquet, and T. Dingermann. 1990. Positive selection for *Dictyostelium* mutants lacking uridine monophosphate synthase activity based on resistance to 5-fluoro-orotic acid. *Dev. Genet.* **11**:396-402.

40. Sun, T.J., P.J.M. van Haastert, and P.N. Devreotes. 1990. Surface cAMP receptors mediate multiple responses during development in *Dictyostelium*: evidenced by antisense mutagenesis. *J. Cell Biol.* **110**:1549-1554.

41. Coukell, M.B., Lappano, S., and Cameron, A.M. 1983. Isolation and characterization of cAMP unresponsive (frigid) aggregation-deficient mutants of *Dictyostelium discoideum*. *Dev. Genet.* **3**:283-297.

42. Kesbeke, F., B.E. Snaar-Jagalska, and P.J.M. van Haastert. 1988. Signal transduction in *Dictyostelium fgd* A mutants with a defective interaction between surface cAMP receptor and a GTP-binding regulatory protein. *J. Cell Biol.* **197**:521-528.

43. Newell, P.C., G.N. Europe-Finner, N.V. Small, and G. Lui. 1988. Inositol phosphates, G-proteins and *ras* genes involved in chemotactic signal transduction of *Dictyostelium*. *J. Cell Sci.* **89**:123-127.

44. Snaar-Jagalska, B.E. and P.J.M. van Haastert. 1988. G-proteins in the signal-transduction pathway of *Dictyostelium discoideum*. *Dev. Genet.* **9**:215-226.

45. Darmon, M., P. Brachet, and L.H. Pereira da Silva. 1975. Chemotactic signals induce cell differentiation in *Dictyostelium discoideum*. *Proc. Natl. Acad. Sci. USA* **72**:3163-3166.

46. Juliani, M.H. and C. Klein. 1978. A biochemical study of the effects of cAMP on aggregationless mutants of *Dictyostelium discoideum*. *Dev. Biol.* **62**:162-172.

47. Frantz, C.E. 1980. Phenotype analysis of aggregation mutants of *Dictyostelium discoideum*. Ph.D. thesis, University of Chicago, Chicago.

48. Kimmel, A.R. and B. Carlislile. 1986. A gene expressed in undifferentiated vegetative *Dictyostelium* is repressed by developmental pulses of cAMP and reinduced during dedifferentiation. *Proc. Natl. Acad. Sci. USA* **83**:2506-2510.

49. Kimmel, A.R. 1987. Different molecular mechanisms for cAMP regulation of gene expression during *Dictyostelium* development. *Dev. Biol.* **122**:163-171.

50. Mehdy, M.C. and R.A. Firtel. 1985. A secreted factor and cyclic AMP jointly regulate cell-type-specific gene expression in *Dictyostelium discoideum*. *Mol. Cell. Biol.* **5**:705-713.

51. Gomer, R.H., S. Datta, M. Mehdy, T. Crowley, A. Sivertsen, W. Nellen, C. Reymond, S. Mann, and R.A. Firtel. 1985. Regulation of cell-type specific gene expression in *Dictyostelium*. *Cold Spring Harbor Symp. Quant. Biol.* **50**:801-812.

52. Gomer, R.H., I.S. Yuen, and R.A. Firtel. 1991. A secreted 80x10 Mr protein mediates sensing of cell density and the onset of development in *Dictyostelium*. *Development* **112**, in press.

53. Podgorski, G., J. Franke, M. Faure, and R.H. Kessin, R.H. 1989. The cyclic nucleotide phosphodiesterase gene of *Dictyostelium discoideum* utilizes alternate promotors and splicing for the synthesis of multiple mRNAs. *Mol. Cell. Biol.* **9**:3938-3950.

54. Faure, M., J. Franke, A.L. Hall, G.J. Podgorski, and R.H. Kessin. 1990. The cyclic nucleotide phosphodiesterase gene of *Dictyostelium discoideum* contains three promoters specific for growth, aggregation and late development. *Mol. Cell. Biol.* **10**:1921-1930.

55. Faure, M., G.J. Podgorski, J. Franke, and R.H. Kessin. 1988. Disruption of *Dictyostelium discoideum* morphogenesis by overproduction of cAMP phosphodiesterase. *Proc. Natl. Acad. Sci. USA* **85**:8076-8080.

56. Majerfeld, I.H., B.H. Leichtling, J.A. Meligeni, E. Spitz, and H.V. Rickenberg. 1984. A cytosolic cyclic AMP-dependent protein kinase in *Dictyostelium discoideum*. I. Properties. *J. Biol. Chem.* **259**:654-661.

57. de Gunzberg, J., D. Part, N. Guiso, and M. Veron. 1984. An unusual adenosine 3'/5' phosphate dependent protein kinase from *Dictyostelium discoideum*. *Biochemistry* **23**:3805-3812.

58. de Gunzberg, J. and M. Veron. 1982. A cAMP dependent protein kinase is present in differentiating *Dictyostelium discoideum* cells. *EMBO J.* **1**:1063-1068.

59. de Gunzberg, J., J. Franke, R.H. Kessin, and M. Veron. 1986. Detection and developmental regulation of the mRNA for the regulatory subunit of the cAMP-dependent protein kinase of *D. discoideum* by cell-free translation. *EMBO J.* **5**:363-367.

60. Leichtling, B.H., I.H. Majerfeld, E. Spitz, K.L. Scheller, C. Woffendin, S. Kakinuma, and H.V. Rickenberg. 1984. A cytosolic cyclic AMP-dependent protein kinase in *Dictyostelium discoideum*. II. Developmental regulation. *J. Biol. Chem.* **259**:661-668.

61. Chevalier, M., J. de Gunzberg, and M. Veron. 1986. Comparison of the regulatory and catalytic subunits of cAMP dependent protein kinase from *Dictyostelium discoideum* and bovine heart using polyclonal antibodies. *Biochem. Biophys. Res. Commun.* **136**:651-656.

62. Taylor, S.S., J.A. Buechler, L.W. Slice, D.K. Knighton, S. Durgerian, G.E. Ringheim, J.J. Neitzel, W.M. Yonemoto, J.M. Sowadski, and W. Dospmann. 1988. cAMP-dependent protein kinase: A framework for a diverse family of enzymes. *Cold Spring Harbor Symp. Quant. Biol.* **53**:121-130.

63. Firtel, R.A. and A.L. Chapman. 1990. A role for cAMP-dependent protein kinase in early *Dictyostelium* development. *Genes Dev.* **4**:18-28.

64. Clegg, C.H., L.A. Correll, G.G. Cadd, and G.S. McKnight 1987. Inhibition of intracellular cAMP dependent protein kinase using mutant genes of the regulatory type I subunit. *J. Biol. Chem.* **262**:13111-13119.

65. Woodfort, T.A., L.A. Correll, G.S. McKnight, and J.D. Corbin. 1989. Expression and characterization of mutant forms of type I regulatory subunit of cAMP-dependent protein kinase: The effect of defective cAMP binding on holoenzyme activation. *J. Biol. Chem.* **264**:13321-13328.

66. Simon, M.N., D. Driscoll, R. Mutzel, D. Part, J. Williams, and M. Veron. 1989. Overproduction of the regulatory subunit of the cAMP dependent protein kinase blocks the differentiation of *Dictyostelium discoideum*. *EMBO J.* **8**:2039-2043.

67. Haribabu, B. and R.P. Dottin. 1991. Homology cloning of protein kinase and phosphoprotein phosphatase sequences of *Dictyostelium discoideum. Dev. Genet.* **12**:45-49.
68. Dingermann, T., N. Reindl, I. Werner, M. Hildebrandt, W. Nellen, A. Harwood, J.G. Williams, and K. Nerke. 1989. Optimization and *in situ* detection of detection of *Escherichia coli* β-galactosidase gene expression in *Dictyostelium discoideum. Gene* **85**:353-362.
69. Tan, J.L. and J.A. Spudich. 1990. Developmentally regulated protein-tyrosine kinase genes in *Dictyostelium discoideum. Mol. Cell. Biol.* **10**:3578-3583.
70. Loomis, W.F., ed. 1982. *The Development of Dictyostelium discoideum.* New York:Academic Press.
71. Bonner, J.T., I.N. Feit, A.K. Selassie, and H.B. Suthers. 1990. Timing of the formation of the prestalk and prespore zones in *Dictyostelium discoideum. Dev. Genet.* **11**:439-441.
72. Feit, I.N., J.T. Bonner, and H.B. Suthers. 1990. Regulation of the anterior-like cell state by ammonia in *Dictyostelium discoideum. Dev. Genet.* **11**:442-446.
73. Takeuchi, I. 1963. Immunochemical and immunohistochemical studies on the development of the cellular slime mold *Dictyostelium discoideum. Dev. Biol.* **8**:1-26.
74. Mehdy, M., D. Ratner, and R.A. Firtel. 1983. Induction and modulation of cell-type-specific gene expression in *Dictyostelium. Cell* **32**:761-771.
75. Barklis, E. and H.F. Lodish. 1983. Regulation of *Dictyostelium discoideum* mRNAs specific for prespore and prestalk cells. *Cell* **32**:1139-1148.
76. Fosnaugh, K.L. and W.F. Loomis. 1989. The spore coat genes, SP60 and SP70, of *Dictyostelium discoideum. Mol. Cell. Biol.* **9**:5215-5218.
77. Datta, S., R.H. Gomer, and R.A. Firtel. 1986. Spatial and temporal regulation of a foreign gene by a prestalk-specific promoter in transformed *Dictyostelium discoideum.* cells *Mol. Cell. Biol.* **6**:811-820.
78. Gomer, R.H., S. Datta, and R.A. Firtel. 1986. Cellular and subcellular distribution of a cAMP-regulated prestalk protein and prespore protein in *Dictyostelium discoideum. J. Cell. Biol.* **103**:1999-2015.
79. Krefft, M., L. Voet, J.H. Gregg, H. Mairhofer, and K.L. Williams. 1984. Evidence that positional information is used to establish the prestalk-prespore pattern in *Dictyostelium discoideum* aggregates. *EMBO J.* **3**:201-206.
80. Wallace, J.S., J.H. Morrissey, and P.C. Newell. 1984. Monoclonal antibodies specific for stalk differentiation in *Dictyostelium discoideum. Cell Differentiation* **14**:205-211.
81. Tasaka, M., T. Noce, and I. Takeuchi. 1983. Prestalk and prespore differentiation in *Dictyostelium* as detected by cell-type-specific monoclonal antibodies. *Proc. Natl. Acad. Sci. USA* **80**:5340-5344.
82. Williams, J.G., K.T. Duffy, D.P. Lane, S.J. McRobbie, A.J. Harwood, D. Traynor, R.R. Kay, and K.A. Jermyn. 1989. Origins of the prestalk-prespore pattern in *Dictyostelium* development. *Cell* **59**:1157-1163.
83. Jermyn, K.A, K.T.I. Diffy, and J.G. Williams. 1989. A new anatomy of the prestalk zone in *Dictyostelium. Nature* **340**:144-146.
84. Haberstroh, L. and R.A. Firtel. 1990. A spatial gradient of expression of a cAMP-regulated prespore cell-type-specific gene in *Dictyostelium. Genes Dev.* **4**:596-612.
85. Esch, R.K. and R. A. Firtel. 1991. cAMP and cell sorting control the spatial expression of a developmentally essential cell-type-specific *ras* gene in *Dictyostelium. Genes Dev.* **5**:9-21.
86. Robbins, S.M., J.G. Weeks, K.A. Jermyn, G.B. Spiegelman, and G. Weeks. 1989. Growing and developing *Dictyostelium* cells express different *ras* genes. *Proc. Natl. Acad. Sci. USA* **86**:938-942.

87. Khosla, M., S.M. Robbins, G.B. Spegelman, and G. Weeks. 1990. Regulation of DdrasG gene expression during *Dictyostelium* development. *Mol. Cell. Biol.* **10**:918-922.

88. Reymond, C.D., R.H. Gomer, M.C. Mehdy, and R. A. Firtel. 1984. Developmental regulation of a *Dictyostelium* gene encoding a protein homologous to mammalian *ras* protein. *Cell* **39**:141-148.

89. Reymond, C.D. and N.A. Thompson. 1991. Analysis of the multiple transcripts of the Dd *ras* gene during *Dictyostelium discoideum* development. *Dev. Genet.* **12**:139-146.

90. Reymond, C., W. Nellen, and R.A. Firtel. 1985. Regulated expression of *ras* constructs in *Dictyostelium* transformants. *Proc. Natl. Acad. Sci. USA* **82**:7005-7009.

91. Schaap, P. and M. Wang. 1986. Interactions between adenosine and oscillatory cAMP signalling. *Cell* **45**:137-144.

92. Pears, C.J., H.M. Mahbubani, and J.G. Williams. 1985. Characterization of two highly diverged but developmentally co-regulated cysteine proteinase genes in *Dictyostelium discoideum*. *Nucleic Acids Res* **13**:8853-8866.

93. Williams, J.G. 1988. The role of diffusable molecules in regulating the cellular differentiation of *Dictyostelium discoideum*. *Development* **103**:1-16.

94. Chisholm, R.L., E. Barklis, and H.F. Lodish. 1984. Mechanism of sequential induction of cell-type specific mRNAs in *Dictyostelium* differentiation. *Nature* **310**:67-69.

95. Williams, J.G., A. Ceccarelli, S. McRobbie, H. Mahbubani, R.R. Kay, A. Early, M. Berks, and K.A. Jermyn. 1987. Direct induction of *Dictyostelium* prestalk gene expression by DIF provides evidence that DIF is a morphogen. *Cell* **49**:185-192.

96. Jermyn, K.A., M. Berks, R.R. Kay, and J.G. Williams. 1987. Two distinct classes of prestalk-enriched mRNA sequences in *Dictyostelium discoideum*. *Development* **100**:745-755.

97. Morris, H.R., G.W. Taylor, M.S. Masento, K.A. Jermyn, and R.R. Kay. 1987. Chemical structure of the morphogen differentiation inducing factor from *Dictyostelium discoideum*. *Nature* **328**:811-814.

98. Kopachik, W., A. Oohata, B. Dhokia, J.J. Brookman, and R.R. Kay. 1983. *Dictyostelium* mutants lacking DIF, a putative morphogen. *Cell* **33**:397-403.

99. Brookman, J.J., K.A. Jermyn, and R.R. Kay. 1987. Nature and distribution of the morphogen DIF in the *Dictyostelium* slug. *Devel.* **100**:119-124.

100. Kopachik, W.J., B. Dhokia, and R.R. Kay. 1985. Selective induction of stalk-cell-specific proteins in *Dictyostelium*. *Differentiation* **28**:209-216.

101. Tasaka, M. and I. Takeuchi. 1981. Role of cell sorting in pattern formation in *Dictyostelium discoideum*. *Differentiation* **18**:191-196.

102. Clark, R.L. and T.L. Steck. 1979. Morphogenesis in *Dictyostelium*: An orbital hypothesis. *Science* **204**:1163-1167.

103. Mee, J.D., C. Tortolo, and M.B. Coukell. 1986. Chemotaxis associated properties of separated prestalk and prespore cells. *Biochem. Cell. Biol.* **64**:722-732.

104. Sternfeld, J. and C.N. David. 1982. Fate and regulation of anterior-like cells in *Dictyostelium* slugs. *Dev. Biol.* **93**:111-118.

105. Devine, K., and W. Loomis. 1985. Molecular characterization of anterior-like cells in *Dictyostelium discoideum*. *Dev. Biol.* **107**:364-372.

106. Weijer, C.J., G. Duschl, and C.N. David. 1984. A revision of the *Dictyostelium discoideum* cell cycle. *J. Cell Sci.* **70**:111-131.

107. McDonald, S.A. and A.J. Durnston. 1984. The cell cycle and sorting behavior in *Dictyostelium discoideum*. *J. Cell Sci.* **66**:195-204.

108. Weijer, C.J., G. Duschl, and C.N. David. 1984. Dependence of cell-type proportioning and sorting on cell cycle phase in *Dictyostelium discoideum*. *J. Cell Sci.* **70**:133-145.

109. Weijer, C.J., G. Duschl, and C.N. David. 1984. Dependence of cell-type proportioning and sorting on cell cycle phase in *Dictyostelium discoideum* slugs: Evidence that cyclic AMP is the morphogenetic signal for prespore differentiation. *Development* **103**:611-618.
110. van Lookeren Campagne, M.M., G. Duschl, and N.C. David. 1984. Dependence of cell type proportioning and sorting on cell cycle phase in *Dictyostelium discoideum.. J. Cell. Sci.* **70**:133-145.
111. Gomer, R.H. and R.A. Firtel. 1987. Cell-autonomous determination of cell-type choice in *Dictyostelium* development by cell-cycle phase. *Science* **237**:758-762.
112. McDonald, S.A. 1986. Cell-cycle regulation of center initiation in *Dictyostelium discoideum. Dev. Biol.* **117**:546-549.
113. Kay, R.R. and K.A. Jermyn. 1983. A possible morphogen controlling differentiation in *Dictyostelium. Nature* **303**:242-244.
114. Spek, W., K. van Drunen, R. van Eijk, and P. Schaap. 1988. Opposite efffects of adenosine on two types of cAMP-induced gene expression in *Dictyostelium* indicate the involvement of at least two different intracellular pathways for the transduction of cAMP signals. *FEBS Lett.* **228**:231-234.
115. Williams, G.B., E.M. Elder, and M. Sussman. 1984. Modulation of the cAMP relay in *Dictyostelium discoideum* by ammonia and other metabolites: Possible morphogenic consequences. *Dev. Biol.* **105**:377-388.
116. Williams, J.G., A.S. Tsang, and H. Mahbubani. 1980. A change in the rate of transcription of a eukaryotic gene in response to cyclic AMP. *Proc. Natl. Acad. Sci. USA* **77**:7171-7175.
117. Schaap, P. and R. van Driel. 1985. The induction of post-aggregative differentiation in *Dictyostelium discoideum* by cAMP. Evidence for the involvement of the cell surface cAMP receptor. *Exp. Cell. Res.* **159**:388-398.
118. Oyama, M. and D.D. Blumberg. 1986. Interaction of cAMP with the cell-surface receptor induces cell-type-specific mRNA accumulation in *Dictyostelium discoideum. Proc. Natl. Acad. Sci. USA* **83**:4819-4823.
119. Gomer, R.H., D. Armstrong, B.H. Leichtling, and R.A. Firtel. 1986. cAMP induction of prespore and prestalk gene expression in *Dictyostelium* is mediated by the cell-surface cAMP receptor. *Proc. Natl. Acad. Sci. USA* **83**:8624-8628
120. Haribabu, B. and R.P. Dottin. 1986. Pharmacological characterizaiton of cyclic AMP receptors mediating gene regulation in *Dictyostelium discoideum. Mol. Cell. Biol.* **6**:2402-2408.
121. Schaap, P., M. van Lookeren Campagne, R. van Driel, W. Spek, P. J.M. van Haastert, and J. Pinas 1986. Postaggregative differentiation induction by cyclic AMP in *Dictyostelium*: Intracellular transduction pathway and requirement for additional stimuli. *Dev. Biol.* **118**:52-63.
122. Peters, D.J.M., M.M. van Lookeren Campagne, P.J.M. van Haastert, W. Spek, and P. Schaap. 1989. Lithium ions induce prestalk-associated gene expression and inhibit prespore gene expression in *Dictyostelium discoideum. J. Cell. Sci.* **93**:205-210.
123. Blumberg, D.D., J.F. Comer, and K.G. Higinbothem. 1988. A Ca^{2+}-dependent signal transduction system participates in coupling expression of some cAMP-dependent prespore genes to the cell surface receptor. *Dev. Genet.* **9**:359-369.
124. Ginsburg, G. and A.R. Kimmel. 1989 Inositol triphosphate and diaglycerol can differentially modulate gene expression in *Dictyostelium. Proc. Natl. Acad. Sci. USA* **86**:9332-9336.
125. Early, A.E. and J.G. Williams. 1988. A *Dictyostelium* prespore specific gene is transcriptionally repressed by DIF *in vitro. Development* **103**:519-524.
126. Wang, M., P.J.M. van Haastert, and P. Schaap. 1986. Multiple effects of differentiation-inducing factor on prespore differentiation and cyclic-AMP signal transduction in *Dictyostelium. Differentiation* **33**:24-28.

127. Berks, M. and R.R. Kay. 1990. Combinatorial control of cell differentiation by cAMP and DIF-1 during development of *Dictyostelium discoideum*. *Devel.* **110**:977-984.
128. Sobolewski, A., N. Neave, and G. Weeks. 1983. The induction of stalk differentiation in submerged monolayers of *Dictyostelium discoideum*: characterization of the temporal sequence for the molecular requirements. *Differentiation* **25**:93-100.
129. Kwong, L., A. Sobolewski, L. Atkinson, and G. Weeks. 1988. Stalk cell formation in monolayers from isolated prestalk and prespore cells of *Dictyostelium discoideum*: evidence for two populations of prestalk cells. *Development* **104**:121-127.
130. Kwong, L., A. Sobolewski, and G. Weeks. 1988. The effect of cyclic AMP on the DIF mediated formation of stalk cells in low density monolayers of *Dictyostelium discoideum*. *Differentiation* **37**:1-6.
131. Berks, M. and R.R. Kay. 1988. Cyclic AMP is an inhibitor of stalk cell differentiation in *Dictyostelium discoideum*. *Dev. Biol.* **126**:108-114.
132. Saxe, C.L. III., R. Johnson, P.N. Devreotes, and A.R. Kimmel. 1991. Multiple genes for cell surface cAMP receptors in *Dictyostelium discoideum*. *Dev. Genet.* **12**:6-13.
133. Insall, R. and R.R. Kay. 1990. A specific DIF binding protein in *Dictyostelium*. *EMBO J.* **9**:3323-3328.
134. Broek, D., N. Samily, O. Fasano, A. Fujiyama, F. Tamanoi, J. Northup, and M. Wigler. 1985. Differential activation of yeast adenylate cyclase by wild-type and mutant RAS proteins. *Cell* **41**:763-769.
135. Reymond, C.D., R.H. Gomer, W. Nellen, A. Theibert, P. Devreotes, and R.A. Firtel. 1986. Phenotypic changes induced by a mutated *ras* gene during development of *Dictyostelium* transformants. *Nature* **323**:340-343.
136. van Haastert, P.J.M., F. Kesbeke, C.D. Reymond, R.A. Firtel, E. Luderus, and R. van Driel. 1987. Aberrant transmembrane signal transduction in *Dictyostelium* cells expressing a mutant *ras* gene. *Proc. Natl. Acad. Sci. USA* **84**:4905-4909.
137. Europe-Finner, G.N., M.E.E. Luderus, N.V. Small, R. van Driel, C.D. Reymond, R.A. Firtel, and P.C. Newell. 1988. Mutant *ras* gene induces elevated levels of inositol tris- and hexakisphosphates in *Dictyostelium*. *Cell Sci.* **89**:13-20.
138. van der Kaay, J., R. Draijer, and P.J.M. van Haastert. 1990. Increased conversion of phosphatidylinositol to phosphatidylinositol phosphate in *Dictyostelium* cells expressing a mutated *ras* gene. *Proc. Natl. Acad. Sci. USA* **87**:9197-9201.
139. Datta, S. and R.A. Firtel. 1987. Identification of the sequences controlling cyclic AMP regulation and cell-type specific expression of a prestalk-specific gene in *Dictyostelium discoideum*.. *Mol. Cell. Biol.* **7**:149-159.
140. Pears, C.J. and J.G. Williams. 1987. Identification of a DNA sequence element required for efficient expression of a developmentally regulated and cAMP-inducible gene of *Dictyostelium discoideum*. *EMBO J.* **6**:195-200.
141. Datta, S. and R.A. Firtel 1988. An 80-bp *cis*-acting regulatory region controls cAMP and developmental regulation of a prestalk gene in *Dictyostelium*. *Genes. Dev.* **2**:294-304.
142. Pears, C.J. and J.G. Williams. 1988. Multiple copies of a G-rich element upstream of a cAMP-inducible *Dictyostelium* gene are necessary but not sufficient for efficient gene expression. *Nucleic Acids Res.* **16**:8467-8486.
143. Hjorth, A.L., N.C. Khanna, and R.A. Firtel. 1989. A *trans*-acting factor required for cAMP-induced gene expression in *Dictyostelium* is regulated developmentally and induced by cAMP. *Genes Dev.* **3**:747-759.
144. Hjorth, A.L., C. Pears, J.G. Williams, and R.A. Firtel. 1990. A developmentally regulated *trans*-acting factor recognizes dissimilar G/C rich elements controlling a class of cAMP-inducible *Dictyostelium* genes. *Genes Dev.* **4**:419-432.

145. Pavlovic, J., B. Haribabu, and R.P. Dottin. 1989. Identification of a signal transduction response sequence element necessary for induction of a *Dictyostelium discoideum* gene by extracellular cyclic AMP. *Mol. Cell. Biol.* **9**:4660-4669.
146. Haribabu, B., J. Pavlovic, S.R. Bodduluri, J.F. Doody, B.D. Ortiz, S. Mullings, B. Moon, and R.P. Dottin. 1991. Signal transduction pathways involved in the expression of the uridine diphosphoglucose pyrophosphorylase gene of *Dictyostelium discoideum*. *Dev. Genet.* **12**:35-44.
147. Driscoll, D. and J.G. Williams. 1987. Two divergently transcribed genes of *Dictyostelium discoideum* are cyclic AMP-inducible and co-regulated during development. *Mol. Cell. Biol.* **7**:4482-4489.
148. Schaap, P., T.M. Konijn, and P.J.M. van Haastert. 1984. cAMP pulses coordinate morphogenetic movement during fruiting body formation of *Dictyostelium minutum*. *Proc. Natl. Acad. Sci. USA* **81**:2122-2126.

Advances in Regulation of Cell Growth, Volume 2;
Cell Activation: Genetic Approaches, edited by
James J. Mond, John C. Cambier, and Arthur Weiss.
Raven Press, Ltd., New York.

3

The Role of the c-*fes* Proto-Oncogene Protein-Tyrosine Kinase in Myeloid Differentiation

Robert I. Glazer, Thomas E. Smithgall, Gang Yu[1]
and Flavia Borellini

*Department of Pharmacology, Georgetown University School of Medicine,
Washington, D.C. 20007, and [1]Laboratory of Biological Chemistry, National Cancer
Institute, NIH, Bethesda, MD 20892*

Examining the basis by which the regulation of cell growth and differentiation is modulated by viral oncogenes and their normal cellular counterparts (proto-oncogenes) will allow a better understanding of how these fundamental processes and their attendant signal transduction mechanisms are mediated. Proto-oncogenes are thought to be involved in the control of the normal proliferative and developmental functions of the cell (1-3). Changes in the pattern of expression of proto-oncogenes are sometimes associated with the morphologic changes occurring during differentiation, suggesting that some proto-oncogenes may be directly involved in the differentiation process (4,5). This chapter will describe our studies of the proto-oncogene protein-tyrosine kinase (PTK), $p93^{c\text{-}fes}$, and its role as a mediator of differentiation of myelomonocytic cells.

THE c-*fes* PROTO-ONCOGENE PROTEIN-TYROSINE KINASE

Approximately 50 percent of the viral oncogenes detected thus far produce a PTK activity as their gene product (1). There is an even greater number of PTK genes in normal tissues and a close structural homology exists between many of the domains of the viral oncogene product and the normal cellular enzyme (2). The proto-oncogene PTK's may be classified into two broad categories, those that traverse the plasma membrane and function as receptors for a diverse variety of growth factors, and those that do not have a ligand-binding domain and are distributed entirely intracellularly in the cytoplasm and/or plasma membrane (3,4). With respect to hematopoietic cells, the only example of a ligand-binding PTK is the receptor for macrophage colony-stimulating factor (M-CSF) (5-7). This receptor is encoded by the c-*fms* proto-oncogene, the normal cellular homolog to the transforming gene found in the McDonough strain of feline sarcoma virus (v-*fms*) (8).   The expression of c-*fms* is associated with cells of the monocyte/macrophage lineage (6,9-15), as well as with trophoblastic tissues and breast carcinoma cells (11,16,17).

Examples of subfamilies of non-receptor PTK's are the c-*src* and c-*fes/fps* gene products (18,19).   Although, the c-*fes/fps* PTKs share a high degree of homology within their catalytic (*src* homology region 1) and SH2 (*src* homology region 2) domains with other PTK's such as p60[c-*src*] and p145[c-*abl*] (18), they are distinctly different in their C- and N-terminal sequences (2,20-22) (FIG. 1).

FIG. 1. Regions of structural homology between proto-oncogene tyrosine kinases.

The large family of *src*-related PTK's is widely distributed in all tissues, and many hematopoietic cells contain their own distinct activity such as *hck* (23-25) and c-*fgr* (26-28) which are associated predominantly, but not exclusively with myeloid cells.   In contrast to the c-*src* subfamily, the PTK encoded by the c-*fes/fps* proto-oncogenes is narrowly restricted to cells of the myelomonocytic lineages (29-31).   The c-*fes* and c-*fps* gene products, p93[c-*fes*] and p98[c-*fps*], respectively, are the cellular homologs (32-38) of the transforming gene products expressed by several strains of feline (v-*fes*) and avian (v-*fps*) sarcoma retroviruses (39-41) (**FIG. 2**).

Human Genomic c-fes DNA

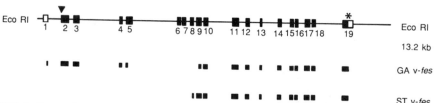

FIG. 2. Comparison of the genomic sequences of human c-*fes* with the Gardner-Arnstein (GA) and Snyder-Theilen (ST) strains of v-*fes*. The human c-*fes* gene contains 19 exons, the first of which is non-coding. Homologous sequences found in the GA and ST strains of feline sarcoma virus are shown as filled boxes. The *inverted triangle* and *asterisk* denote the position of the start and stop codons, respectively.

Unlike their retroviral counterparts, c-*fes*/*fps* is usually non-transforming when expressed in fibroblasts (42-46). However, two recent studies have demonstrated that c-*fes* has a low degree of transforming activity (10,000-fold less than v-*fes*) when expressed in fibroblasts under the control of a strong promoter (47), and that its transforming activity is increased an additional 100-fold by the phosphotyrosine phosphatase inhibitor, sodium vanadate (48).

PROTEIN-TYROSINE KINASES AND MYELOID DIFFERENTIATION

Most previous investigations of proto-oncogene function in myeloid differentiation have focused heavily on genes and gene products which are not PTK's (49-51). Recently, several studies have begun to examine the role of PTK activity in the function of myelomonocytic cells. Uncharacterized PTK activities were elevated in the particulate fraction of monocytes and acute myeloblastic leukemia cells in comparison to lymphocytic cells (52). PTK activities were partially characterized in human neutrophils (53,54) and in promyelocytic cell line HL-60 (53,55-61) and were found to be coupled to the stimulus response of these cells to chemotactic peptide (53) and several differentiating agents (56-62). The first proto-oncogene PTK to be implicated in myeloid differentiation was the c-*src* gene product, pp60$^{c\text{-}src}$, which increased in activity in HL-60 cells and monocytic cell line U937 in response to phorbol ester (63,64) and dimethylsulfoxide (DMSO) (63). However, there is disagreement over whether p60$^{c\text{-}src}$ is activated posttranslationally or transcriptionally since one study reported increased levels of c-*src* kinase (63) while the other study reported that c-*src* mRNA was not increased (64) during differentiation. The purification and characterization of p60$^{c\text{-}src}$ from HL-60 cells treated with DMSO has been described (65), and no correlation was found between its activity and the induction of myeloid differentiation by a variety of differentiating agents (57,59). However, in several of the latter studies from our laboratory, a unique non-denaturing polyacrylamide gel electrophoretic (PAGE) assay (56) was used to detect a membrane-associated PTK whose activity was markedly increased in HL-60 cells by granulocytic and monocytic differentiating agents (56-60) (**FIG. 3**). These results led us to pursue the identity of this PTK and its role in myeloid differentiation.

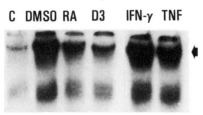

FIG. 3. p93$^{c\text{-}fes}$ Tyrosine kinase activity in cell extracts from HL-60 cells treated with differentiating agents. Cells were treated continuously for 4 days with 1.6% DMSO, 1 μM retinoic acid (RA), 100 nM 1,25-dihydroxyvitamin D_3 (D3), 2,500 units/ml interferon-γ (IFN-γ) or 100 units/ml recombinant tumor necrosis factor-α (TNF). Kinase activity was assayed by the non-denaturing PAGE assay (56). The *arrow* indicates p93$^{c\text{-}fes}$ activity.

Purification of p93$^{c\text{-}fes}$ from Differentiated HL-60 Cells

The plasma membrane-bound PTK associated with myeloid differentiation was purified to homogeneity from DMSO-treated HL-60 cells (65) (**FIG. 4**). Its identity as the c-*fes* gene product, p93$^{c\text{-}fes}$, was established by immunoprecipitation with antibodies against two C-terminal peptides of p93$^{c\text{-}fes}$ (65) (**FIG. 5**), as well as by immunoblotting with a series of antibodies against *E. coli*-expressed proteins encoded by partial human c-*fes* cDNA clones (66) (**FIG. 6**). The PTK activity related to p60$^{c\text{-}src}$ which co-purified with p93$^{c\text{-}fes}$, did not change in response to differentiation (57,59).

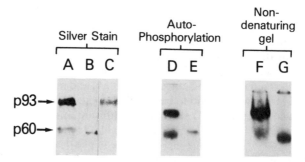

FIG. 4. SDS-PAGE and non-denaturing PAGE of p93$^{c\text{-}fes}$ tyrosine kinase activity following anti-phosphotyrosine and gel filtration HPLC. Lanes *A*, *B*, and *C* represent silver staining after SDS-PAGE of p93$^{c\text{-}fes}$ and p60$^{c\text{-}src}$ following anti-phosphotyrosine HPLC of a partially purified 1% (lane *A*) or 0.1% (lane *B*) Triton X-100 fraction, followed by gel filtration HPLC of the activity in lane *A* to obtain pure p93$^{c\text{-}fes}$ (lane *C*). Lanes *D* and *E* represent the autophosphorylated forms of p93$^{c\text{-}fes}$ and p60$^{c\text{-}src}$ after SDS-PAGE. Lanes *F* and *G* represent the activity of p93$^{c\text{-}fes}$ and p60$^{c\text{-}src}$ by the non-denaturing PAGE assay (65).

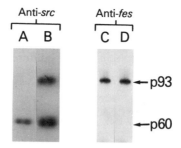

FIG. 5. Immunoprecipitation of p93$^{c\text{-}fes}$ and p60$^{c\text{-}src}$ with anti-p60$^{c\text{-}src}$ and anti-p93$^{c\text{-}fes}$ monoclonal antibodies. The tyrosine kinase purified from the 0.1% (lane *A*) or the 1% (lane *B*) Triton X-100 fractions after anti-phosphotyrosine HPLC was immunoprecipitated with an antibody against p60$^{c\text{-}src}$. The tyrosine kinase activity in the 1% Triton X-100 fraction after anti-phosphotyrosine HPLC was also immunoprecipitated with two different antibodies against C-terminal peptide sequences in p93$^{c\text{-}fes}$ (lanes *C* and *D*). Shown are the autophosphorylated enzymes after SDS-PAGE.

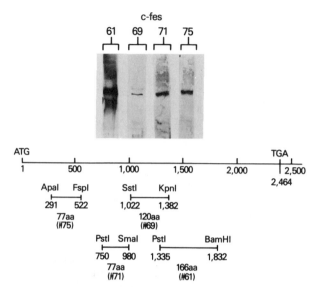

FIG. 6. Western blots of p93^{c-fes} purified from HL-60 cells differentiated with DMSO. Purified p93^{c-fes} was transferred to nitrocellulose after SDS-PAGE and probed with rabbit antisera (numbered 61, 69, 71 and 75) raised against *E. coli*-expressed proteins encoded by partial human c-*fes* cDNA clones (66). The *arrow* denotes p93^{c-fes}.

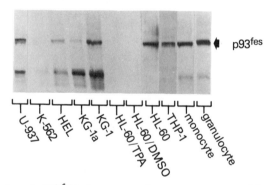

FIG. 7. Identification of p93^{c-fes} in human granulocytes, monocytes and leukemia cell lines by Western blot analysis. **A.** Cell extracts of leukemia cell lines and normal peripheral leukocytes were fractionated by tyrosine-agarose chromatography, and separated by SDS-PAGE. Proteins were transferred to nitrocellulose membranes, and probed for immunoreactive p93^{c-fes} using antiserum 61 (see **FIG. 6.**).

Summary

These studies established conclusively that the differentiation-associated PTK purified from HL-60 cells was identical to the gene product of c-*fes*. In addition, the level and activity of p93$^{c\text{-}fes}$ were associated only with myeloid leukemia cell lines exhibiting a more mature phenotype (HL-60, U-937, HEL, KG-1, THP-1), but were low (KG-1a, HL-60/TPA, HL-60/DMSO) or absent (K562) in poorly differentiated cells (**FIG. 7**).

Transfection Studies

The first direct evidence of a regulatory role of the human c-fes gene in myeloid differentiation was demonstrated by stable transfection of the poorly differentiated myeloblast cell line K562 with the 13.2 kb genomic c-fes sequence (67). The genomic c-fes sequence was ligated into an expression vector dowstream from the SV40 early promoter and used to transfect K562 cells by protoplast fusion. Colonies were selected by their resistance to the neomycin analog, G-418 and several clones were found to have PTK activity as determined by the non-denaturing PAGE assay (**FIG. 8**). Clones WS-1, WS-5 and WS-6 were selected for further studies. p93$^{c\text{-}fes}$ from clone WS-1 possessed the same chromatographic characteristics on tyrosine-agarose as p93$^{c\text{-}fes}$ from DMSO-treated HL-60 cells (**FIG. 9**). The level of c-fes mRNA in the three clonal cell lines was found to correlate with the amount of p93$^{c\text{-}fes}$ activity (**FIG. 10**).

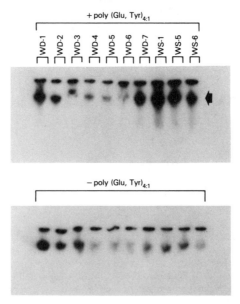

FIG. 8. Non-denaturing gel assay for p93$^{c\text{-}fes}$ tyrosine kinase activity in colonies of K562 cells stably transfected with c-fes. K562 cells were cotransfected with plasmids pECE/fes and pSV2/neo (as a selectable marker) and G418-resistant colonies were screened for p93$^{c\text{-}fes}$ tyrosine kinase activity. The 1.0% Triton X-100 fraction was assayed for tyrosine kinase activity using the non-denaturing PAGE assay in the absence and presence of the synthetic substrate, poly(glutamic acid,tyrosine)$_{4:1}$. The *arrow* denotes p93$^{c\text{-}fes}$ tyrosine kinase activity (67).

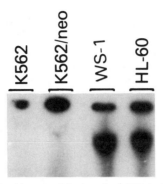

FIG. 9. Comparison of tyrosine kinase activity in colony WS-1 with differentiated HL-60 cells. One percent Triton X-100 extracts were prepared from either wild type K562 (*K562*), pSV2/*neo*-transfected K562 (*K562/neo*), c-fes-transfected K562 (*WS-1*), or DMSO-treated HL-60 (*HL-60*) cells, and $p93^{c\text{-}fes}$ tyrosine kinase was partially purified by tyrosine-agarose chromatography. Eluates were assayed for tyrosine kinase activity using the non-denaturing PAGE assay. The *arrow* denotes $p93^{c\text{-}fes}$ tyrosine kinase activity (67).

Transfected clones of K562 cells expressing the highest levels of $p93^{c\text{-}fes}$ exhibited a significant reduction in their growth rate, responsiveness to the differentiating effects of phorbol ester and expressed characteristics of the mature granulocytic phenotype as exemplified by the reduction of nitroblue tetrazolium and increased erythrophagocytosis (**FIG. 11**). Other markers of myeloid differentiation such as lysozyme activity and surface expression of Mac-1 antigen were also increased (**Table 1**). In contrast to the response of K562 cells, COS-1 cells (SV40-transformed African green monkey kidney cells) transfected with c-fes did not express a myeloid phenotype despite exhibiting $p93^{c\text{-}fes}$ activity that was greater than that present in DMSO-treated HL-60 cells (results not shown).

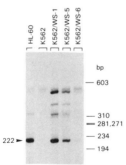

FIG. 10. RNase protection assay of parental and clonal variants of K562 cells stably transfected with pECE/*fes*. Poly-A$^+$ RNA was isolated from wild-type HL-60 (*HL-60*), wild-type K562 (*K562*), WS-1 (*K562/WS-1*), WS-5 (*K562/WS-5*) and WS-6 (*K562/WS-6*) cells, and analysis of c-fes mRNA levels was carried out by RNase protection with a c-fes antisense RNA probe complementary to exon 2. The *arrowhead* denotes the 222 bp RNase-protected c-fes sequence.

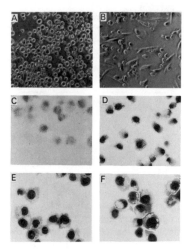

FIG. 11. Photomicrographs of parental K562 cells and K562/*fes* clone WS-1. K562 (*A,C,E*) and WS-1 (*B,D,F*) cells were treated for 2 days with 10^{-7} M TPA (*A,B*), or assayed for their ability to reduce nitroblue tetrazolium (*C,D*) or for their capacity to phagocytize sheep erythrocytes (*E,F*).

TABLE 1. Phenotypic characteristics of K562 cells transfected with c-fes.

Differentiation Marker	HL-60[a]	K562	K562/*neo*	WS-1	WS-5	WS-6
	percent positive					
Phagocytosis	42	0	1	65	56	12
Fc receptors	68	52	48	94	85	81
NBT reduction	77	1	3	64	43	38
Lysozyme activity[b]	n.d.	0	0	4.6	3.6	4.5
Adherence	0	0	0	80	80	65
Response to 100 nM TPA:						
Adherence	75	0	0	50	56	n.d.
Mac-1	88	2	3	70	73	18

[a] HL-60 cells were treated for 4 days with 1.25% Me_2SO.
[b] μg of lysozyme/10^6 cell. Normal leukocytes range from 3.6-8.4.
n.d., not determined.

Summary

These studies demonstrated that the $p93^{c\text{-}fes}$ is sufficient to induce terminal differentiation along the myeloid pathway in poorly differentiated myelocytic leukemia cells which normally do not express this PTK activity.

p93^{c-fes} Expression as a Molecular Marker for Myeloid Leukemia

Since the expression of c-*fes* is highly restricted to hematopoietic cells of the myelomonocytic lineage, p93$^{c\text{-}fes}$ may be a useful diagnostic marker for classifying the phenotype of leukemias. To examine this possibility, the levels of c-*fes* mRNA were determined in several myeloid leukemia cell lines blocked at different stages of differentiation and in cells of lymphoid origin (**FIG. 12**).

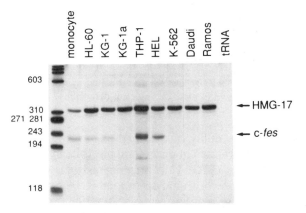

FIG. 12. RNase protection analysis of c-*fes* mRNA levels in human leukemia cell lines and normal peripheral monocytes. Total RNA was isolated from log phase leukemia cells or monocytes, and 20 μg aliquots were analyzed for the presence of c-*fes* and HMG-17 mRNA (an invariant unrelated mRNA) using an RNase protection assay (67,68). Yeast tRNA was added as a negative control to ensure that all unhybridized probe was digested.

Cell lines that were more highly differentiated (HL-60, U-937, THP-1, KG-1, HEL) and capable of responding to differentiation inducers exhibited detectable and variable levels of c-*fes* mRNA, while expression was absent in leukemia cells that were poorly differentiated (K562 and KG-1a), or of B-cell origin (Ramos, Daudi) (**FIG. 12, Table 2**). These data correlated with the levels of p93$^{c\text{-}fes}$ as determined by Western blotting (**FIG. 7**) and indicated that expression of the c-*fes* gene is restricted only to cells that have differentiative potential along the myelomonocytic pathway.

Table 2. c-fes and HMG-17 mRNA Levels in Human Leukemia Cell Lines

Cell line	Cell type	c-fes[a]	HMG-17[a]
HL-60	promyelocytes	0.218	2.451
KG-1	myeloblasts	0.220	1.836
KG-1a	undifferentiated blasts	n.d.	2.208
THP-1	monoblasts	1.106	3.415
HEL	proerythroblasts	0.713	1.788
K-562	undifferentiated blasts	n.d.	1.805
Daudi	B-lymphocyte	n.d.	1.726
Ramos	B-lymphocyte	n.d.	2.409
peripheral monocytes		0.45	0.741

[a] Relative c-fes and HMG-17 mRNA levels were estimated by laser densitometry of the 222 and 317 nucleotide bands shown in **FIG. 12**. *n.d.*, not detectable.

To determine the utility of c-*fes* expression as a diagnostic tool, c-*fes* mRNA levels were determined in peripheral blasts from patients with myeloid and lymphoid leukemias (**FIG. 13**). Significant levels of c-*fes* mRNA were detectable by RNase protection analysis of total RNA from both acute (AML) and chronic (CML) myeloid leukemias. No significant differences were observed in c-*fes* mRNA levels between AML and CML samples, although more variation was observed with AML (3-fold) than with CML (2-fold). RNase protection also revealed c-*fes* expression in two cases of acute lymphocytic leukemia (ALL), although the levels were 3 to 10 times lower than the range of values observed for AML and CML. Surprisingly, patient ALL-1 converted to AML upon relapse, suggesting that the original population of leukemic cells must have contained a small proportion of myeloid precursors. By contrast, no detectable c-*fes* mRNA was detectable in chronic lymphocytic leukemia.

Summary

These data indicate that c-*fes* is an appropriate marker to confirm the myeloid lineage of acute or chronic leukemias, and may be useful in cases of inconclusive histopathologic diagnosis. Moreover, the presence of c-*fes* in cases of ALL may indicate that lineage switching is imminent and that an alternate chemotherapeutic regimen may be in order.

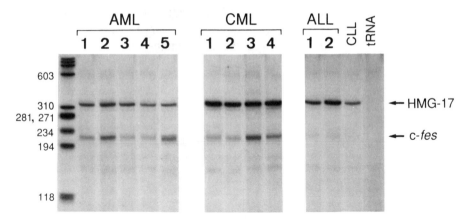

FIG. 13. RNase protection analysis of c-*fes* mRNA levels in primary human leukemia cells. Total RNA was isolated from peripheral blood cells of leukemia patients, and 10 μg aliquots were analyzed for c-*fes* and HMG-17 expression by RNase protection analysis (68).

c-*fes* and Colony-Stimulating Factor Activity

The pioneering studies of Metcalf and others established that colony-stimulating factors (CSF's) function in a concerted manner to produce a regulatory cascade which drives hematopoiesis (69-71) (**FIG. 14**). Conversely, dysregulation of CSF activity at either the receptor or signal transduction level could play an etiological role in myeloproliferative diseases (72). Supporting this hypothesis are studies demonstrating that overexpression of GM-CSF and IL-3 in CSF-dependent cell lines results in factor-independence and tumorigenicity *in vitro* (73-76) and in a myeloproliferative syndrome *in vivo* (77,78). In a similar context, viral oncogene PTK's may produce factor independence *in vitro*. Infection of chick bone marrow cells with v-*fps* (79) or a murine macrophage cell line with v-*fms* (80) resulted in the induction of M-CSF-independent cell growth in a non-autocrine manner.

Evidence of a relationship between c-*fes* expression and CSF's was demonstrated in DMSO-treated HL-60 cells where the extent of expression of p93$^{c\text{-}fes}$ and myeloid differentiation correlated with the level of GM-CSF receptors (66) (**FIG. 15**). From the studies of Gasson et al. (81-83), the GM-CSF receptor level in various myelomonocytic cell lines related to their degree of maturation. Their results are in virtually quantitative agreement with the level of expression of p93$^{c\text{-}fes}$ activity in the same cell lines (66). Of added significance is a recent study where c-*fes* mRNA induction preceded the appearance of c-*fgr* and c-*fms* mRNA's in murine myeloid progenitor cells after treatment with GM-CSF or G-CSF (84) (**FIG. 14**). These investigations suggest that c-*fes* is associated with the cell signalling process mediated via CSF's that are involved very early in the differentiation process.

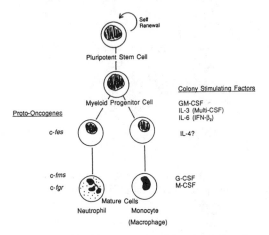

FIG. 14. Proto-oncogene tyrosine kinases and colony-stimulating factors associated with myelopoiesis.

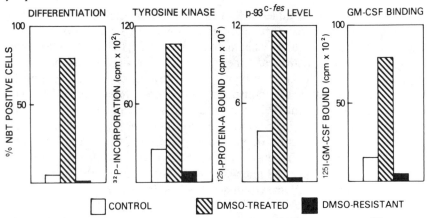

FIG. 15. Correlation of p93^{c-fes} expression and tyrosine kinase activity with [^{125}I]GM-CSF binding in DMSO-treated and resistant HL-60 promyelocytes. HL-60 cells were induced to differentiate by treatment with DMSO, and the proportion of mature granulocytes present was assessed as the percent of cells able to reduce nitroblue tetrazolium (NBT). Cell extracts were prepared from treated, wild-type, and DMSO-resistant HL-60 cells, and tyrosine kinase activity was determined using the non-denaturing PAGE assay. Immunoreactive p93^{c-fes} protein levels were quantitated by Western blotting following tyrosine-agarose chromatography. [^{125}I]GM-CSF binding was assayed using whole cells.

Summary

The onset and elevation of c-*fes* expression is associated with the action of GM-CSF. Although, the GM-CSF receptor has no intrinsic PTK activity (J.C. Gasson and R.I. Glazer, unpublished results), p93$^{c\text{-}fes}$ may be closely associated with the signaling process mediated by GM-CSF.

c-fes PROMOTER REGION

Very little is known about the transcription factors and promoter regions that confer myeloid-specific expression of c-*fes*. Promoter activity within the 13 kb *EcoR* I fragment containing the genomic c-*fes* sequence was first demonstrated in a study where insertion of c-*fes* into an expression vector in an orientation opposite to the SV40 early promoter still resulted in low levels of expression of p93$^{c\text{-}fes}$ kinase activity (30). Further evidence of the myeloid specificity of the c-*fes* promoter region was obtained from two studies involving transgenic animals. Expression of v-*fps* (which lacks 5'- and 3'-untranslated sequences) under the control of the β-2 globin promoter did not result in myeloid tissue specificity (85). In contrast, expression of the 13 kb *EcoR* I genomic c-*fes* sequence resulted in the same myeloid-specific tissue distribution of p93$^{c\text{-}fes}$ that is observed in normal mice (86). These studies indicated that c-*fes* is a self-contained genetic element with all the promoter and enhancer regions required for conferring myeloid tissue specificity (**FIG. 16**). Although the precise promoter region has not been defined, a recent study demonstrated that the promoter sequences immediately upstream to the *Kpn* I site in exon 1 of the c-*fes* sequence did not contain myeloid-specific promoter activity (87). This result suggests that the myeloid-specific promoter/enhancer sequences lie within the remaining 0.5 kb of the 5'-untranslated region (5'-UTR) extending from the *Kpn* I site to the ATG start codon within exon 2 (the first translated exon in the mammalian cellular c-*fes* sequence). Alternatively, a cooperative interaction may exist between the 0.5 kb region downstream from exon 1 and the upstream 5'-UTR region or possibly an enhancer region such as AP-2 in intron 3.

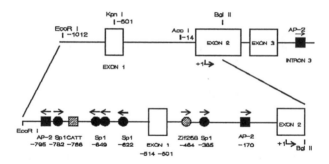

FIG. 16. Genomic organization of the 5'-untranslated region (5'-UTR) of the c-*fes* gene. The c-*fes* sequence contains a 1 kb 5'-UTR extending downstream from the *EcoR I* restriction site to the ATG start codon (+1) in exon 2. Putative binding sites for transcription factors Sp1, Zif268 and AP-2 are indicated, and *arrows* indicate the direction of the consensus sequence on the coding strand (→) or non-coding strand (←). An AP-2 binding site downstream from the 5'-UTR in intron 3 is also indicated.

Transcription factors and phosphorylation

There is increasing evidence that the activity of several transcription factors may be regulated by phosphorylation. Transfection of cells with plasmids containing the catalytic fragment of protein kinase C (88-90), the c-*raf* protein-serine kinase (91) and p60^{v-src} (92) have all resulted in stimulation of transcription through the serum-response element in the c-*fos* enhancer. These studies imply that modulation of transcription by protein-serine and protein-tyrosine kinases may be responsible for the temporal expression of differentiation-specific genes. Preliminary evidence suggests that transfection of K562 cells with the genomic c-*fes* sequence results in activation of the DNA-binding activity of transcription factor Sp1 without changes in its level (93). These results suggest a mechanism by which p93^{c-fes} may activate transcription and thereby initiate a regulatory cascade resulting in the onset of differentiation.

CONCLUSION

Leukemia may be defined as a maturational defect in hematopoietic cells with the retention of their capacity to proliferate. The molecular basis for the neoplasia expressed in stem cells arrested at various stages of maturation may reside, in part, in defects in the signaling process in response to various growth and differentiation factors. It has been established from our studies (56-60,65-68) and from those of other laboratories (29,30) that expression of the c-*fes* gene product, p93^{c-fes}, is closely associated with this signaling process and the ability of myeloid leukemia cells to undergo differentiation. These investigations may provide a useful framework for the design of new therapeutic modalities employing differentiating agents. Of no less significance is the fact that the c-*fes* promoter/enhancer sequences are the first human promoter elements to confer bone marrow tissue specificity for expression of a gene in transgenic mice (86), a finding which may have a significant impact on future gene therapy experiments.

An equally important aspect to this research is the possibility that alterations in the activity or structure of p93$^{c\text{-}fes}$ could represent a developmental step in the formation of myeloproliferative diseases. c-*fes* is highly expressed specifically in myelogenous malignancies (68,94-97). The demonstration that very high levels of c-*fes* expression in fibroblasts can lead to transformation (47,48) suggests that this gene has oncogenic potential. Of relevance perhaps, is the finding that a more distantly related member of the c-*fes* PTK subfamily, the *bcr-abl* fusion protein, can produce a syndrome resembling chronic myelogenous leukemia in mice (98), and may be responsible for this disease in man.

REFERENCES

1. Bishop, J.M. Viral Oncogenes. Cell 42:23-38, 1985
2. Hanks, S.K., Quinn, A.M. and Hunter, T. The protein kinase family: conserved features and deduced phylogeny of the catalytic domains. Science 241:42-52, 1988.
3. Heldin, C-H. and Westermark, B. Growth factors: Mechanism of action and relation to oncogenes. Cell 37:9-20, 1984.
4. Yarden, Y. and Ullrich, A. Molecular analysis of signal transduction by growth factors. Biochemistry 27:3113-3118, 1988.
5. Sherr, C.J., C.W. Rettenmier, R. Sacca, M.F. Roussel, A.T. Look, and E.R. Stanley. The c-*fms* proto-oncogene product is related to the receptor for the mononuclear phagocyte growth factor, CSF-1. Cell 41:665-676 (1985).
6. Woolford, J., Rothwell, V. and Rohrschneider, L., Characterization of the human c-*fms* gene product and its expression in cells of the monocyte-macrophage lineage. Mol. Cell. Biol. 5:3458-346, 1985.
7. Sacca, R., E.R. Stanley, C.J. Sherr, and C.W. Rettenmier, Specific binding of the mononuclear phagocyte colony-stimulating factor CSF-1 to the product of the v-*fms* oncogene. Proc. Natl. Acad. Sci. USA 83:3331-3335 (1986).
8. Sherr, C.J. The *fms* oncogene. Biochim. Biophys. Acta 948:225-243, 1988.
9. Sariban, E., Mitchell, V. and Kufe, D. Expression of the c-*fms* protooncogene during human monocytic differentiation. Nature 316:64-66, 1985.
10. Sariban, E., Mitchell, T., Griffin, J. and Kufe, D.W. Effects of interferon-γ on proto-oncogene expression during induction of human monocytic differentiation. J. Immunol. 138:1954-1958, 1987.
11. Rettenmeier, C.W., Sacca, R., Furman, W.L., Roussel, M.F., Holt, J.T., Nienhuis, A.W., Stanley, E.R. and Sherr, C.J. Expression of the human c-*fms* proto-oncogene product (colony-stimulating factor-1 receptor) on peripheral blood mononuclear cells and choriocarcinoma cells. J. Clin. Invest. 77:1740-1746, 1986.
12. Rambaldi, A., Young, D.C. and Griffin, J.D. Expression of the M-CSF (CSF-1) gene

by human monocytes. Blood 69:1409-1413, 1987.

13. Bicknell, D.C., Williams, D.E. and Broxmeyer, H.E. Correlation between CSF-1 responsiveness and expression of (CSF-1 receptor) c-*fms* in purified murine granulocyte-macrophage progenitor cells (CFU-GM). Exp. Hematol. 16:88-91, 1988.

14. Dubreuil, P., Torres, H., Courcoul, M.A., Birg, F. and Mannoni, P. c-*fms* Expression is a molecular marker of human acute myeloid leukemia. Blood 72:1081-1085, 1988.

15. Ashmun, R.A., Look, A.T., Roberts, W.M., Roussel, M.F., Seremetis, S., Ohtsuka, M. and Sherr, C.J. Monoclonal antibodies to the human CSF-1 receptor (c-*fms* proto-oncogene product) detect epitopes on normal mononuclear phagocytes and on human myeloid leukemic blast cells. Blood 73:827-837, 1989.

16. Horiguchi, J., Sherman, M.L., Sampson-Johannes, A., Weber, B.L. and Kufe, D.W. CSF-1 and c-fms gene expression in human carcinoma cell lines. Biochem. Biophys. Res. Commun. 157:395-401, 1988.

17. Visvader, J. and Verma, I.M. Differential transcription of exon 1 of the human c-*fms* gene in placental trophoblasts and monocytes. Mol. Cell. Biol. 9:1336-1341, 1989.

18. Pawson, T. Non-catalytic domains of cytoplasmic protein-tyrosine kinases: regulatory elements in signal transduction. Oncogene 3:491-495, 1988.

19. Hunter, T., A thousand and one protein kinases. Cell 50:823-829, 1987.

20. Roebroek, A.J.M., Schalken, J.A., Verbeek, J.S., Van den Ouweland, A.M.W., Onnekink, C., Bloemers, H.P.J. and Van de Ven, W.J.M. Structure of the human c-*fes/fps* proto-oncogene. EMBO J. 4:2897-2903, 1985.

21. Roebroek, A.J.M., Schalken, J.A., Bussemakers, M.J.G., van Heerikhuizen, H., Onnekink, C., Debruyne, F.M.J., Bloemers, H.P.J., and Van de Ven, W.J.M. Characterization of human c-*fes/fps* reveals a new transcription unit (fur) in the immediately upstream region of the proto-oncogene. Mol. Biol. Rep. 11:117-125, 1986.

22. Roebroek, A.J.M., Schalken, J.A., Onnekink, C., Bloemers, H.P.J. and Van de Ven, W.J.M. Structure of the feline c-*fes/fps* proto-oncogene: genesis of a retroviral oncogene. J. Virol. 61:2009-2016, 1987.

23. Quintrell, N., Lebo, R., Varmus, H., Bishop, J.M., Pettenati, M.J., Le Beau, M.M., Diaz, M.O. and Rowley, J.D., Identification of a human gene (HCK) that encodes a protein-tyrosine kinase and is expressed in hemopoietic cells. Mol. Cell. Biol. 7:2267-2275, 1987.

24. Ziegler, S.F., Marth, J.D., Lewis, D.B. and Perlmutter, R.M., Novel protein-tyrosine kinase gene (hck) preferentially expressed in cells of hematopoietic origin. Mol. Cell. Biol. 7:2276-2285, 1987.

25. Holtzman, D.A., Cook, W.D. and Dunn, A.R. Isolation and sequence of a cDNA corresponding to a *src*-related gene expressed in murine hematopoietic cells. Proc. Natl. Acad. Sci. USA 84:8325-8329, 1987.

26. Willman C.L., Stewart, C.C., Griffith, J.K., Stewart, S.J. and Tomasi, T.B. Differential expression and regulation of the c-*src* and c-*fgr* protooncogenes in myelomonocytic cells. Proc. Natl. Acad. Sci. USA 84:44809-4484, 1987.

27. Notario, V., Gutkind, J.S., Imaizumi, M., Katamine, S. and Robbins, K.C. Expression of the *fgr* protooncogene product as a function of myelomonocytic cell maturation. J. Cell Biol. 109:3129-3136, 1989.

28. Yi, T.-L. and Wilman, C.L. Cloning of the murine c-*fgr* proto-oncogene cDNA and induction of c-*fgr* expression by proliferation and activation factors in normal bone marrow-derived monocytic cells. Oncogene 4:1081-1087, 1989.

29. Feldman, R.A., Gabrilove, J.L., Tam, J.P., Moore, M.A.S. and Hanafusa, H. Specific expression of the human cellular *fps/fes*-encoded protein NCP92 in normal and leukemic myeloid cells. Proc. Natl. Acad. Sci. USA 82:2379-2385, 1985

30. MacDonald, I, Levy, J, Pawson, T: Expression of the mammalian c-*fes* protein in hematopoietic cells and identification of a distinct *fes*-related protein. Mol. Cell. Biol.

5:2543, 1985

31.   Samarut, J., Mathey-Prevot, B. and Hanafusa, H. Preferential expression of the c-*fps* protein in chicken macrophages and granulocytic cells. Mol. Cell. Biol. 5:1067-1072, 1985.

32.   Barbacid, M., Beemon, K. and Devare, S.G. Origin and functional properties of the major gene product of the Snyder-Theilen strain of feline sarcoma virus. Proc. Natl. Acad. Sci. USA 77:5158-5162, 1980.

33.   Franchini, G., Even, J., Sherr, C.J. and Wong-Staal, F. *onc* sequences (v-*fes*) of Snyder-Theilen feline sarcoma virus are derived from noncontiguous regions of the cat cellular gene (c-*fes*). Nature 290:154-157, 1981.

34.   Franchini, G., Gelmann, E.P., Dalla Favera, R., Gallo, R.C. and Wong-Staal, F. Human gene (c-*fes*) related to the *onc* sequences of Snyder-Theilen feline sarcoma virus. Mol. Cell. Biol. 2:1014-1019, 1982.

35.   Mathey-Prevot, B., Hanafusa, H. and Kawai, S. A cellular protein is immunologically crossreactive with and functionally homologous to the Fujinami sarcoma virus transforming protein. Cell 28:897-906, 1982.

36.   Huang, C.C., Hammond, C. and Bishop, J.M. Nucleotide sequence and topography of chicken c-*fps*. Genesis of a retroviral oncogene encoding a tyrosine-specific protein kinase. J. Mol. Biol. 181:175-186, 1985.

37.   Groffen, J.N., Heisterkamp, N., Shibuya, M., Hanafusa, H. and Stephenson, J.R. Transforming genes of avian (v-*fps*) and mammalian (v-*fes*) retroviruses correspond to a common cellular locus. Virology 125:480-486, 1983.

38.   Pfaff, S.L., Zhou, R.-P., Young, J.C., Hayflick, J. and Duesberg, P.H. Defining the borders of the chicken proto-*fps* gene, a precursor of Fujinami sarcoma virus. Virology 146:307-314, 1985.

39.   Hampe, A., Laprevotte, I., Galibert, F., Fedele, L.A., and Sherr, C.S. Nucleotide sequences of feline retroviral oncogenes (v-*fes*) provide evidence for a family of tyrosine-specific protein kinase genes. Cell 30:775, 1982

40.   Shibuya, M. and Hanafusa, H. Nucleotide sequence of Fujinami sarcoma virus: evolutionary relationship of its transforming gene with transforming genes of other sarcoma viruses. Cell 30:787-794, 1982.

41.   Egan, S.E., Wright, J.A., Jarolim, L., Yanagihara, K., Bassin, R.H. and Greenberg, A.H. Transformation by oncogenes encoding protein kinases induces the metastatic phenotype. Science 238:202-205, 1987.

42.   Sodroski, J.G. and Haseltine, W.A. Transforming potential of a human protooncogene (c-*fes/fps*) locus. Proc. Natl. Acad. Sci. USA 81:3039-3043, 1984.

43.   Foster, D.A., Shibuya, M. and Hanafusa, H. Activation of the transformation potential of the cellular *fps* gene. Cell 42:105-115, 1985.

44.   Verbeek, J.S., Van den Ouweland, A.M.W., Schalken, J.A., Roebroek, A.J.M., Ounekink, C., Bloemers, H.P.J. and Van de Ven, W.J.M. Molecular cloning of the feline c-*fes* proto-oncogene and construction of a chimeric transforming gene. Gene 35:33-43, 1985.

45.   Feldman, R.A., Vass, W.C. and Tambourin, P.E. Human cellular *fes/fps* cDNA rescued via a retroviral shuttle vector encodes myeloid cell NCP92 and has transforming potential. Oncogene Res. 1:441-458, 1987.

46.   Greer, P.A., Meckling-Hansen, K. and Pawson, T. The human c-*fes/fps* gene product expressed ectopically in rat fibroblasts is non-transforming and has restrained protein-tyrosine kinase activity. Mol. Cell. Biol. 8:578-587, 1988.

47.   Feldman, R.A., Lowy, D.R., Vass, W.C. and Velu, T.J. A highly efficient retroviral vector allows detection of the transforming activity of the human c-*fes/fps* proto-oncogene. J. Virol. 63:5469-5474, 1989.

48.   Feldman, R.A., Lowy, D.R. and Vass, W.C. Selective potentiation of c-*fes/fps*

58   *ROLE OF THE c-fes PROTO-ONCOGENE PROTEIN*

transforming activity by a phosphatase inhibitor. Oncogene Res. 5:187-197, 1990.

49. Muller, R. Proto-oncogenes and differentiation. Trends in Biochem. Sci. 11: 129-132, 1986.

50. Rowley P.T. and Skuse, G.R. Oncogene expression in myelopoiesis. Intl. J. Cell Cloning 5:255-266, 1987.

51. Collins, S.J. The HL-60 promyelocytic leukemia cell line: proliferation, differentiation, and cellular oncogene expression. Blood 70:1233-1244, 1987.

52. Punt, C.J.A., Rijksen, G., Vlug, A.M.C., Dekker, A.W. and Staal, G.E.J. Tyrosine protein kinase activity in normal and leukemic human blood cells. Brit. J. Haematol. 73:51-56, 1989.

53. Kraft, A.S. and Berkow, R.L., Tyrosine kinase and phosphotyrosine phosphatase activity in human promyelocytic leukemia cells and human polymorphonuclear leukocytes. Blood 70:356-362, 1987.

54. Berkow, R.L., Dodson, R.W. and Kraft, A.S. Human neutrophils contain distinct cytosolic and particulate tyrosine kinase activities: possible role in neutrophil activation. Biochim. Biophys. Acta 997:292-301, 1989.

55. Frank, D.A. and Sartorelli, A.C. Biochemical Characterization of tyrosine kinase and phosphotyrosine phosphatase activities of HL-60 leukemia cells. Cancer Res. 48:4299-4306, 1988.

56. Glazer, R.I., Yu, G. and Knode, M.C. Analysis of tyrosine kinase activity in cell extracts using nondenaturing polyacrylamide gel electrophoresis. Anal. Biochem. 164:214-220, 1987.

57. Glazer, R.I., Chapekar, M.S., Hartman, K.D. and Knode, M.C., Appearance of membrane-bound tyrosine kinase during differentiation of HL-60 leukemia cells by immune interferon and tumor necrosis factor. Biochem. Biophys. Res. Commun. 140:908-915, 1986.

58. Chapekar, M.S., Hartman, K.D., Knode, M.C. and Glazer, R.I., Synergistic effect of retinoic acid and calcium ionophore A23187 on differentiation, c-*myc* expression and membrane-bound tyrosine kinase activity in human promyelocytic leukemia cell line HL-60. Mol. Pharmacol. 31:140-145, 1987.

59. Glazer, R.I., Chapekar, M.S., Hartman, K.D., Knode, M.C. and Yu, G.: Induction of membrane-bound tyrosine kinase activity in human promyelocytic leukemia cells by differentiating agents. In: Membrane Proteins: Proceedings of the Membrane Protein Symposium. S.S. Goheen (ed.), Bio-Rad Laboratories, Richmond, CA, 1987, pp. 715-728.

60. Yu, G., Grant, S. and Glazer, R.I., Association of p93[c-fes] tyrosine protein kinase with granulocytic/monocytic differentiation and resistance to differentiating agents in HL-60 leukemia cells. Mol. Pharmacol. 33:384-388, 1988.

61. Frank, D.A. and Sartorelli, A.C. Regulation of protein phosphotyrosine content by changes in tyrosine kinase and protein phosphotyrosine phosphatase activities during induced granulocytic and monocytic differentiation of HL-60 leukemia cells. Biochem. Biophys. Res. Commun. 140:440-447, 1986.

62. Frank, D.A. and Sartorelli, A.C. Alterations in tyrosine phosphorylation during the granulocytic maturation of HL-60 leukemia cells. Cancer Res. 48:52-58, 1988.

63. Gee, C.E., Griffin, J., Sastre, L., Miller, L.J., Springer, T.A., Piwnica-Worms, H. and Roberts, T.M. Differentiation of myeloid cells is accompanied by increased levels of pp60[c-src] protein and kinase activity. Proc. Natl. Acad. Sci. USA 83:5131-5135, 1986.

64. Barnekow, A. and Gessler, M., Activation of the pp60[c-src] kinase during differentiation of monomyelocytic cells in vitro. The EMBO J. 5:701-705, 1986.

65. Yu, G. and Glazer, R.I. Purification and characterization of p93[fes] and p60[src]-related tyrosine protein kinase activities in differentiated HL-60 leukemia cells. J. Biol. Chem. 262:17543-17548, 1987

66. Smithgall, T.E., Yu, G. and Glazer, R.I., Identification of the differentiation-associated p93 tyrosine protein kinase of HL-60 leukemia cells as the product of the human c-*fes* locus and its expression in myelomonocytic cells. J. Biol. Chem. 263:15050-15055, 1988.

67. Yu, G., Smithgall, T.E. and Glazer, R.I., K562 leukemia cells transfected with the human c-*fes* gene acquire the ability to undergo myeloid differentiation. J. Biol. Chem. 264:10276-10281, 1989.

68. Smithgall, T.E., Johnston, J.B., Bustin, M. and Glazer, R.I. Elevated expression of the c-*fes* proto-oncogene in adult myeloid leukemias in the absence of gene amplification. J. Natl. Cancer Inst., in press.

69. Nicola, N.A. Hemopoietic growth factors and their interactions with specific receptors. J. Cell. Physiol. Suppl. 5:9-14, 1987.

70. Miyajima, A., Miyatake, S., Schreurs, J., De Vries, J., Arai, N., Yokota, T. and Arai, K.-I. Coordinate regulation of immune and inflammatory responses by T-cell derived lymphokines. FASEB J. 2:2462-2473, 1988.

71. Metcalf, D. The molecular control of cell division, differentiation commitment and maturation in haemopoietic cells. Nature 339:27-30, 1989.

72. Browder, T.M., Dunbar, C.E. and Nienhuis, A.W. Private and public autocrine loops in neoplastic cells. Cancer Cells 1:9-17, 1989.

73. Lang, R.A., Metcalf, D., Gough, N.M., Dunn, A.R. and Gonda, T.J. Expression of a hemopoietic growth factor cDNA in a factor-dependent cell line results in autonomous growth and tumorigenicity. Cell 43:531-542, 1985.

74. Laker, C., Stocking, C., Bergholz, U., Hess, N., De Lamarter, J.F. and Ostertag, W. Autocrine stimulation after transfer of the granulocyte/macrophage colony-stimulating factor gene and autonomous growth are distinct but interdependent steps in the oncogenic pathway. Proc. Natl. Acad. Sci. USA 84:8458-8462, 1987.

75. Wong, P.M.C., Chung, S.-W. and Nienhuis, A.W. Retroviral transfer and expression of the interleukin-3 gene in hematopoietic cells. Genes and Develop. 1:358-365, 1987.

76. Chang, J.M., Metcalf, D., Lang, R.A., Gonda, T.J. and Johnson, G.R. Nonneoplastic hematopoietic myeloproliferative syndrome induced by dysregulated multi-CSF (IL-3) expression. Blood 73:1487-1497, 1989.

77. Johnson, G.R., Gonda, T.J., Metcalf, D., Hariharan, I.K. and Cory, S. A lethal myeloproliferative syndrome in mice transplanted with bone marrow cells infected with a retrovirus expressing granulocyte/macrophage colony stimulating factor. EMBO J. 8:441-448, 1989.

78. Wong, P.M.C., Chung, S.-W., Dunbar, C.E., Bodine, D.M., Ruscetti, S. and Nienhuis, A.W. Retrovirus-mediated transfer and expression of the interleukin-3 gene in mouse hematopoietic cells result in a myeloproliferative disorder. Mol. Cell. Biol. 9:798-808, 1989.

79. Carmier, J.F. and Samarut, J. Chicken myeloid stem cells infected by retroviruses carrying the v-*fps* oncogene do not require exogenous growth factors to differentiate in vitro. Cell 44:159-165, 1986.

80. Wheeler, E.F., Rettenmeier, C.W., Look, T.A.T. and Sherr, C.J. The v-*fms* oncogene induces factor independence and tumorigenicity in CSF-1 dependent macrophage cell line. Nature 324:377-380, 1986.

81. Gasson, J.C., Kaufman, S.E., Weisbart, R.H., Tomonaga, M. and Golde, D.W. High-affinity binding of granulocyte-macrophage colony-stimulating factor to normal and leukemic human myeloid cells. Proc. Natl. Acad. Sci. USA 83:669-673, 1986.

82. Tomonaga, M., Golde, D.W. and Gasson, J.C., Biosynthetic (Recombinant) human granulocyte-macrophage colony-stimulating factor:effect on normal bone marrow and leukemia cells. Blood 67:31-36, 1986.

83. DiPersio, J., Billing, P., Kaufman, S., Eghtesady, P., Williams, R.R. and Gasson, J.C., Characterization of the human granulocyte-macrophage colony-stimulating factor

receptor. J. Biol. Chem. 263:1834-1841, 1988.

84. Liebermann, D.A. and Hoffman-Liebermann, B., Proto-oncogene expression and dissection of the myeloid growth to differentiation developmental cascade. Oncogene 4:583-592, 1989.

85. Yee, S.-P. Y., Mock, D., Maltby, V., Silver, M., Rossant, J., Bernstein, A. and Pawson, T. Cardiac and neurological abnormalities in v-*fps* transgenic mice. Proc. Natl. Acad. Sci. U.S.A. 86:5873-5877, 1989.

86. Greer, P., Maltby, V., Rossant, J., Bernstein, A. and Pawson, T. Myeloid expression of the human c-*fps/fes* proto- oncogene in transgenic mice. Mol. Cell. Biol. 10:2521-2527, 1990.

87. Alcalay, M., Antolini, F., Van de Ven, W.J., Lanfrancone, L., Grignani, F. and Pelicci, P.G. Characterization of human and mouse c-*fes* cDNA clones and identification of the 5' end of the gene. Oncogene 5:267-275, 1990.

88. Hata, A., Akita, Y., Konno, Y., Suzuki, K. and Ohno, S. Direct evidence that the kinase activity of protein kinase C is involved in transcriptional activation through a TPA-responsive element. FEBS Lett. 252:144-146, 1989.

89. Muramatsu , M.-A., Kaibuchi, K. and Arai, K.-I. A protein kinase C cDNA without the regulatory domain is active after transfection *in vivo* in the absence of phorbol ester. Mol. Cell. Biol. 9:831-836, 1989.

90. Kaibuchi, K., Fukumoto, Y., Oku, N. and Takai, Y. Molecular genetic analysis of the regulatory and catalytic analysis of the regulatory domains of protein kinase C. J. Biol. Chem. 264:13489-13496, 1989.

91. Kaibuchi, K., Fukumoto, Y., Oku, N., Hori, Y., Yamamoto, T., Toyoshima, K. and Takai, Y. Activation of the serum response element and 12-*O*-tetradecanoylphorbol-13-acetate response element by the activated c-*raf*-1 protein in a manner independent of protein kinase C. J. Biol. Chem. 264:20855-20858, 1989.

92. Fujii, M., Shalloway, D. and Verma, I.M. Gene regulation by tyrosine kinases: *src* protein activates various promoters, including c-*fos*. Mol. Cell. Biol. 9:2493-2499, 1989.

93. Glazer, R.I., Borellini, F., Aquino, A., Yu, G. and Josephs, S.: Increased expression of SP1 and NFκB transcriptional enhancer elements in K562 cells transfected with the myeloid-specific c-*fes* tyrosine kinase gene. U.C.L.A. Symposium on Molecular and Cellular Biology, "Transcriptional Control of Cell Growth," January 27 to February 3, 1990, Keystone, Colorado. J. Cell. Biochem. 14B:181, 1990.390.

94. Ferrari, S, Torelli, U, Selleri, L, Donelli, A, Venturelli, D, Moretti, L and Torelli, G: Expression of human c-*fes* onc-gene occurs at detectable levels in myeloid but not in lymphoid cell populations. Br. J. of Haematol. 59:21, 1985

95. Emilia, G., Donelli, A., Ferrari, S., Torelli, U., Selleri, L., Zucchini, P., Moretti, L., Venturelli, D., Ceccherelli, G. and Torelli, G. Cellular levels of mRNA from c-*myc*, c-*myb* and c-*fes* onc-genes in normal myeloid and erythroid precursors of human bone marrow: an *in situ* hybridization study. Brit. J. Haematol. 62:287-292, 1986.

96. Slamon, D.J., de Kernion, J.B., Verma, I.M. and Cline, M.J. Expression of cellular oncogenes in human malignancies. Science 224:256-262, 1984.

97. Mavilio, F, Sposi, NM, Petrini, M, Bottero, L, Marinucci, M, De Rossi, G, Amadori, S, Mandelli, F and Peschle, C. Expression of cellular oncogenes in primary cells from human acute leukemias. Proc. Natl. Acad. Sci. USA 83:4394-4398, 1986.

98. Daley, G.Q., Van Etten, R.A. and Baltimore, D. Induction of chronic myelogenous leukemia in mice by the P210$^{bcr/abl}$ gene of the Philadelphia chromosome. Science 247:824-830, 1990.

Advances in Regulation of Cell Growth, Volume 2;
Cell Activation: Genetic Approaches, edited by
James J. Mond, John C. Cambier, and Arthur Weiss.
Raven Press, Ltd., New York © 1991.

4

Cyclic AMP-Dependent Transactivation of Gene Transcription Mediated by the CREB Phosphoprotein

Terry E. Meyer, Ph.D. and Joel F. Habener, M.D.

Laboratory of Molecular Endocrinology, Massachusetts General Hospital, Howard Hughes Medical Institutes, Harvard Medical School, Boston, Massachusetts 02114

The cAMP-response element binding protein, CREB, belongs to a group of transcription factors that, in response to stimuli of second messenger pathways, activate the expression of specific genes (1-4). In general, an intra- or extracellular signal such as binding of a ligand to its cytoplasmic or plasma membrane bound receptor triggers a cascade of transient biochemical events that modulate the final physiological responses (Fig. 1). The mechanisms which ensure that these responses direct 'normal' developmental, cell-specific pathways are many and complex (1-11). Many of the signal transduction pathways regulate gene activity through protein phosphorylation. Two of these pathways ultimately control gene transcription by the activation of the cAMP-dependent protein kinase A (PKA) and the diacylglycerol-dependent protein kinase C (PKC) (12,13). It is likely that additional pathways such as those that activate receptor tyrosine kinases and calcium-dependent calmodulin kinases also either directly or indirectly control the transcription of genes (14-17). At least some of the proteins that are subsequently phosphorylated by protein kinase A and/or protein kinase C or other kinases function as transcription factors by binding to specific DNA enhancer or suppressor sequences located in the control regions of their target genes. In the present article, we focus discussion on one of these factors, the cyclic AMP-responsive CREB protein, with particular emphasis on the diverse mechanisms that have evolved to control the expression and activation of this key factor located at the end of the cAMP-mediated protein kinase A pathway involved in the regulation of gene transcription.

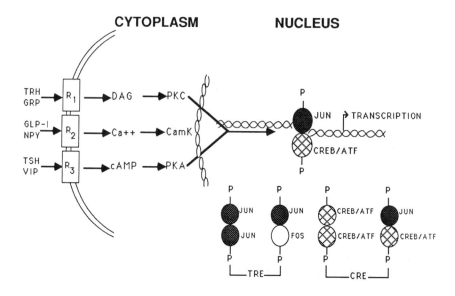

FIG. 1. Diagram showing three cell-surface receptor signal transduction pathways involved in the activation of a superfamily of nuclear transcription factors. Peptide hormone ligands, for example,   thyrotropin releasing hormone (TRH), gastrin-releasing peptide (GRP), thyroid stimulating hormone (TSH), and vasoactive intestinal peptide (VIP) interact with receptors coupled to either the diacylglycerol (DAG)/protein kinase C (PKC) or cAMP/protein kinase A (PKA) pathways. Other signal transduction pathways may involve regulation by calcium ions that stimulate calcium calmodulin kinases (CamK) resulting in the phosphorylation of transcription factors. It appears that in neuronal and pancreatic endoneurodermal endocrine cells that CREB/ATFs may be regulated by phosphorylation by both cAMP-dependent protein kinase A and calcium sensitive calmodulin-linked protein kinases (our unpublished data). The protein kinases phosphorylate members of the CREB/ATF, Jun/AP-1 and Fos/Fra families of DNA-binding proteins to modulate DNA-binding affinities and/or transcriptional activation. The various proteins bind as dimers determined by a poorly understood code that is not promiscuous inasmuch as only certain homodimer or heterodimer combinations are permissible. For example, c-Jun will heterodimerize with CRE-BP1/ATF-2, but not with CREB-327/341 or ATF-1. In addition recognition by the dimerized proteins of a CRE or a TPA-response element (TRE) is somewhat specific. CREB-327 homodimers bind to CREs but not to TREs, and Jun/Fos heterodimers bind just the reverse, whereas Jun homodimers will bind to both CREs and TREs.

HORMONAL REGULATION OF GENE EXPRESSION AT THE LEVEL OF TRANSCRIPTION

The expression of many cellular genes in higher eukaryotes is controlled by hormones. In the case of the steroid and thyroid hormones, substances that readily pass through the plasma membrane and enter cells, the DNA-binding, transcription activators are the hormone receptors themselves (18). Binding of the hormone ligand to its receptor results in either translocation of the receptor to the nucleus where it binds to the target DNA element and activates transcription (steroid hormones) and/or activates prebound receptors (T3, retinoic acid receptors).

The poorly diffusable peptide hormones, on the other hand, activate gene transcription indirectly by binding to receptors which, in turn, trigger the formation of intracellular second messengers such as cAMP, inositol phosphates, diacyl glycerol, etc. (13,19), and initiate the second messenger pathways.

Comparisons and studies of the transcriptional control regions of the genes for the hormones proenkephalin (20), vasoactive intestinal polypeptide (21,22), somatostatin (23), and alpha chorionic gonadotropin (hCG) (24,25), and of many other genes (26-29) that are activated by cAMP have revealed the presence of a common palindromic octamer of the general motif, 5'-TGACGTCA-3', referred to as a cAMP-responsive element (CRE). Transfection studies of many of these genes confirmed that cAMP acts via CREs. Similarly, the heptamer, 5'-TGAGTCA-3' (TRE), has been shown to mediate protein kinase C, phorbol-ester gene activation (30,31). The affinity and specificity with which the relevant transcription factors bind to CREs and TREs is dependent on the five or six bases surrounding these enhancers (24,25). By using assays designed to study DNA-protein interactions, the cell-specific expression and the biochemical properties of the proteins that bind to CREs and TREs have been partially characterized.

It should be emphasized that the final transcriptional response mediated by any given transcription factor, *i.e.*, DNA-binding protein, is certain to come about by a convergence of the actions of several distinct signal transduction pathways. That is, multiple phosphorylations, and/or other modifications of the proteins are effected by the combined actions of several different protein kinases, phosphatases, and other enzymes. For example, a convergence of the cAMP-dependent protein kinase A and diacylglycerol-dependent protein kinase C pathways has been described in the transcriptional activation of genes directed by proteins that bind to a TRE (32).

In the case of TREs, a key link was made between oncogenesis and transcription factors when Vogt *et al.* (33) reported a convincing homology for the viral v-Jun oncogene from chicken fibrosarcomas, and the DNA-binding domain of the yeast transcription factor GCN4. At that time, GCN4 was known to activate transcription by binding to the sequence 5'-ATGA(C/G)TCAT-3'. Subsequent studies of the cellular phorbol ester-responsive transcription factor AP1, a protein complex that was eventually shown to contain a heterodimer between c-Jun and Fos, established the connection between protein kinase C stimulation and transcriptional activation mediated through TREs (34). As discussed by Smeal *et al.*(35), and Turner and Tjian (36) these observations were confirmed upon characterization of recombinant Jun and Fos proteins expressed from cloned cDNAs.

Concomitant studies of the regulatory regions from adenovirus genes (37) and the cellular gene for somatostatin (38) revealed the presence of a 43-45 kDa CRE-binding protein(s), CREB/ATF. As will be discussed later, there are a large number of DNA-binding proteins in the molecular weight range of approximately 38 to 47 KDa that comprise a superfamily of transcription factors known collectively as CREB/ATFs. ATF is the abbreviation for activating transcription factor, one of several proteins that activate the transcription of adenovirus genes and are structurally similar to CREB. One of these proteins, rat CREB protein, was purified by sequence-specific DNA affinity chromatography and partially characterized *in vitro* (38). By DNase I protection analysis, the purified CREB exhibited specific binding for the rat somatostatin CRE. In addition, protein kinase A phosphorylated the purified CREB, confirming a role for this protein in the cAMP-mediated signal transduction pathway.

CLONING OF CREB327/341 cDNAs

The cloning of cDNAs for CREB (39-41) has provided the opportunity to study the control mechanisms by which this protein mediates transcriptional transactivation via the protein kinase A pathway. In the first reported isolation of a CREB cDNA, Hoeffler *et al.* (39) utilized the CRE-binding specificity of CREB to screen a human placental cDNA expression library after the method of Singh *et al.* (42). Additional CREB cDNAs have been cloned by polymerase chain reaction (PCR) amplification of cDNAs from rat (41,43) and human (40,43) tissue mRNAs. Comparison of the predicted amino acid sequences of these clones reveals the presence of two primary CREB proteins, CREB327 and CREB341, expressed in both rats and humans. These proteins are actually isoforms, derived from a single gene, and differ only by the presence or absence of a 14 residue peptide that is the result of alternative splicing of a 42 bp exon (40,44). The high degree of sequence homology between rat and human CREB327/341 (only three conservative amino acid substitutions), is suggestive that much of the protein structure of CREB is critical to its function.

THE bZIP FAMILY OF DNA-BINDING PHOSPHOPROTEINS

As shown in Figure 2, the secondary structure of CREB is bipartite. The carboxy-terminal 66 amino acids are necessary and sufficient for protein dimerization and DNA-binding, while the remainder of the protein participates in the transcriptional transactivation of target genes (45-48). The DNA-binding region consists of a dense concentration of positively charged, basic amino acids. Adjacent to this basic domain, is a series of four leucines in a heptad repeat, referred to as a leucine zipper. Together, the basic domain and leucine zipper are characteristic of a recently defined set of transcription factors, the "leucine zipper" or bZip proteins (49,50). Other members of this family of proteins and their homology in the bZip domain are listed in Table 1, and include C/EBP, ATF-2, and the Jun and Fos proteins.

The leucine zipper and basic domain coordinate dimerization of these proteins and govern the DNA binding specificity. Residues of the leucine repeat twist to form an alpha-helix. Intertwining of such helices from two 'compatible' bZip proteins generates a coiled-coil, a zipper, that serves to stabilize the proteins as dimers. Both homo- and heterodimers may be formed, depending on some, as yet unclear, structural code that restricts which proteins participate. CREB, for example, normally binds as a homodimer to its cognate CRE. On the other hand, c-Jun binds to TREs weakly as a homodimer, and actually binds preferentially as a c-Jun/Fos heterodimer, part of the AP1 complex (36,51-55).

By forming dimers, the bZip proteins bring to bear a pair of DNA-binding basic domains. Each basic domain forms a helix-bend-helix motif in which the

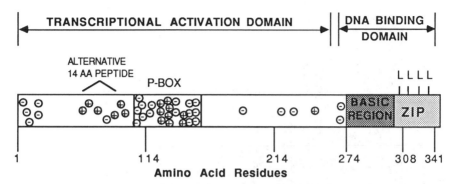

FIG. 2. Diagramatic structure of cloned CREB341 protein. An additional isoform of CREB, CREB327, differs from CREB341 by the splicing-out of a 42 base pair exon encoding 14 amino acids. The phosphorylation box (P-Box), is phosphorylated by multiple protein kinases, initiating the transcriptional transactivation function of CREB.

two amphipathic alpha-helices with positively charged faces of lysine and arginine residues are broken by an asparagine cap (49). The bend between the helices apparently allows the basic domain to wrap around the major groove of the DNA helix, bringing to register the positively-charged protein with the negatively-charged phosphate backbone of the DNA. Vinson *et al.* (49) have proposed a "scissors-grip" model to describe this form of DNA-protein interaction. Of considerable interest has been the finding that the basic domains of the bZip proteins appear to exist as random structures in the absence of DNA. When the bZip proteins come into contact with DNA of certain defined nucleotide sequences the basic domains then assume the configuration of α-helices (56-60). This property of structural plasticity of the DNA-binding domains dictated by the nucleotide sequence of the cognate DNA element is likely to be responsible for the observations that a given DNA-binding protein such as C/EBP can recognize and bind to several different elements comprised of distinctly different nucleotide sequences. Although the "scissors-grip" model is based on the characterization of C/EBP (49) and GCN4 (51), it promises to be generally instructive for studies of other members of this highly homologous bZip family of proteins.

PHOSPHORYLATION REGULATES THE TRANSCRIPTIONAL ACTIVATION OF CREB

The amino-terminal residues upstream of the CREB bZip region constitute the transcriptional activation domain. More specifically, there is a stretch of about 60 residues that include potential sites for phosphorylation by several protein kinases (Fig. 3). Consistent with CREB's role in the cAMP pathway, there is a consensus RRPSY peptide site for phosphorylation by protein kinase A at serine residue 119 in CREB327, and serine 133 in CREB341 (referred to hereafter as serine-119/133). The protein kinase A (PKA) site is flanked by consensus sites (serines and threonines) for phosphorylations by protein kinase C (PKC), casein kinase II (CK II), and glycogen synthase kinase-3 (GSK -3).

TABLE 1. Homology of the Basic Regions and Leucine Zippers of the bZip Protein Super-Family*.

Protein	Species	Reference	Basic Region	Leucine Zipper
			• •.•N.•. •C.•••	L. L. L. L. Y L
CREB327/341	Human, Rat	39-41	AEEAARKREVRLMKNREAAARECRRKKKEYVKCLENRVAVLENQNKTLIEELKALKDLYCHKSD	
GCN4	Yeast	51	VPESSDPAALKRARNTEAARSRARKLQRMKQLEDKVEELLSKNYHLENEVARLKKLVGER	
CREBP1,ATF2	Human,Mouse	61-64	NEDPDEKRRKFLERNRAAASRCRQKRKVWVQSLEKKAEDLSSLNGQLQSEVTLLRNEVAQLKQ	
c-Jun	Human,Mouse	65	ESQERIKAERKRMRNRIAASKCRKRKLERIARLEEKVKTLKAQNSELASTANMLREQVAQLKQ	
Jun-B	Human,Mouse	65,66	EDQERIKVERKRLRNRIAASKCRKRKLERIARLEDKVKTLKAENAGLSSAAGLLREQVAQLKQ	
Jun-D	Mouse	65	DTQERIKAERKRLRNRIAASKCRKRKLERISRLEEKVKTLKSQNTELASTASLLREQVAQLKQ	
dJRA	Drosophila	67	EAQEKIKLERKRQRNRVAASKCRKRKLERISKLEDRVKVLKGENVDLASIVKNLKDHVAQLKQ	
c-Fos	Human,Mouse	68	SPEEEEKRRIRRERNKMAAAKCRNRRRELTDTLQAETDQLEDEKSALQTEIANLLKEKEKLEF	
fra-1	Human,Rat	68	SPEEEERRVRRERNKLAAAKCRNRRKELTDFLQAETDKLEDEKSGLQREIEELQKQKERLEL	
fra-2	Human	68	SPEEEKRRIRRERNKLAAAKCRNRRRELTEKLQAETEELEEKSGLQKEIAELQKEKEKLEF	
fosB	Mouse	68	TPEEEEKRRVRRERNKLAAAKCRNRRRELTDRLQAETDQLEEEKAELESEIAELQKEKERLEF	
dFRA	Drosophila	68	TPEEEQKRAVRRERNKQAAARCRKRRVDQTNELTEEVEQLEKRGESMRKEIEVLTNSKNQLEY	
ATF-1	Human	62	TDDPQLKREIRLMKNRE.ARECRRKKKEYVKCLENRVAVLENQNKTLTEELKTLKDLYSNKSV	
ATF-3	Human	62	APEEDERKKRRERNKIAAAKCRNKKKEKTECLQKESEKLESVNAELKAQIEELKNEKQHLIY	
TREB5	Human	69	HLSPEEKALRRKLKNRVAAQTARDRKKARMSELEQQVVDLEEENQKLLENQLLREKTHGLVV	
c-Myc	Human	51	RSSDTEENVKRRTHNVLERQR.RNELKRSFFALRDQIPELENNEKAPKVVILKKATAYILSVQ	
L-Myc	Human	51	PVSSDTEDVTKRKNHNFLERKRRNDLRSRFLALRDQVPTLASCSKAPKVVILSKALEYLQALV	
NF-IL6	Human	70	VDKHSDEYKIRRERNNIAVRKSRDKAKQRNVETQQKVLELTAENERLQKKVEQLSRELSTLRN	
C/EBP	Human	70	VDKNSNEYRVRRERNNIAVRKSRDKAKQRNVETQQKVLELTSDNDRLRKRVEQLSRELDTLRG	
HBP-1	Wheat	71	WDERELKKQKRKLSNRESARRSRLRKQAECEELGQRAEAALKSENSSLRIELDRIKKEYEELLS	
TGA1b	Tobacco	72	NNDEDEKKRARLVRNRESAQLSRQRKKHYVEELEDKVRIMHSTIQDLNAKVAYIIAENATLKT	
O2	Maize	73	KMPTEERVRRKKESNRESARRSRYRKAAHLKELEDQVAQLKAENSCLRRIAALNQKYNDANV	

*The DNA-binding (basic region) and dimerization (leucine zipper) domain of human CREB327 and CREB341 proteins belong to the bZip super-family of transcription factors. The species indicated are not all-inclusive, but signify those species for which bZip sequences have been reported. This sample of bZip proteins exhibits a striking sequence conservation, especially when the diversity of species represented is considered. Note those amino acid positions that have been marked by a bullet (•), asparagine (N), cysteine (C), tyrosine (Y), or leucine (L) for particularly conserved charge or polarity. Somewhere in this arrangement of conserved and not so conserved residues lies the code, only partially understood, that governs the DNA binding-sequence specificity.

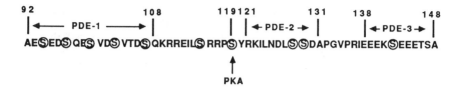

FIG. 3. The panel shows the amino acid sequence of the CREB327 phosphorylation box (P-Box), indicating the single serine-119 that is phosphorylated by cAMP-dependent protein kinase A (PKA). Phosphorylation by PKA is essential, but not sufficient for the full transcriptional activity of CREB; additional phosphorylations by glycogen synthase kinase-3, casein kinase II, and/or protein kinase C are required to fully activate CREB transcriptionally. These additional phosphorylations occur in the PDE-1 and PDE-2 subdomains, and are triggered by the initial phosphorylation of serine-119 by protein kinase A. The high acidity of the PDE-3 subdomain implicate it as part of the transcriptional activation domain of CREB, however, experimental data does not show the PDE-3 subdomain to be required for transcriptional activation.

It is notable that both the casein kinases and glycogen synthase kinases are known to be "processive kinases" inasmuch as their capacity to phosphorylate serines or threonines is markedly facilitated by phosphorylations of amino acid residues located three to four residues away from the target phosphorylation sites. This "phosphorylation box" (P-box), nested in a dense cluster of negatively charged amino acids, appears to be the key to the transcriptional activity of CREB. Computer modelling of CREB327 suggests that phosphorylation of the P-box would allosterically drive this region from a mostly random coil to a highly structured series of amphipathic alpha helices with negatively charged hydrophilic faces. Furthermore, the modelling predicts that this allosteric effect would be potentiated by the insertion of the alternatively spliced 14 amino acids that comprise the difference between CREB327 and CREB341 (our unpublished observations). Ptashne (4) has proposed that this kind of negatively charged amphipathic helix interacts with RNA polymerase II to stimulate transcription. In fact, transcription studies of the yeast GAL4 and GCN4 protein counterparts to the CREB P-box, have demonstrated a precedence for this means of activating RNA polymerase II (74,75). Studies by Horikoshi *et al.* (76) of purified ATF/CREB have actually demonstrated a direct interaction with the TATA binding factor TFIID. Studies of the promoter of the rat somatostatin gene in which the CRE is located only 9 bps upstream of the TATA box also provides evidence for the involvement of CREB interactions with components of the basal transcriptional complex (77).

More recently, functional studies of recombinant CREB and synthetic CREB polypeptides support the notion that phosphorylation of the P-box confers and/or modifies alpha-helix formation (48, Yun *et al.*, unpublished data) and transcriptional transactivation (22,32,45,48,78). Protein kinase A does indeed phosphorylate CREB at the single predicted site, but fails to phosphorylate site-specific mutants of CREB. Yun (unpublished data) and Yamamoto *et al.* (48) have shown by circular dichroism that the structure of the P-box of CREB is predominantly alpha helical and Yamamoto *et al.* (48) have reported that the 14 amino acid insert in CREB341 resulting from the alternatively spliced exon is also alpha helical.

Phosphorylation is necessary, but not sufficient to render CREB fully active in its capacity to transactivate gene transcription. In an elegant series of experiments on the CREB P-box, Lee *et al.* (45) demonstrated that the transcriptional activation of CREB, in addition to phosphorylation by protein kinase A, requires phosphorylation of a sequence located ten residues amino terminal to serine-119/133 (PDE-1 sequence, Fig. 3) and the presence of a sequence of eleven residues located carboxy terminal to serine-119/133 (PDE-2 sequence). Also demonstrated was the ability of purified casein kinase II to phosphorylate serine residues in the amino terminal PDE-1 part of the P-box. Casein kinase II is expressed ubiquitously, but an exact role for this enzyme has yet to be described (45). Interestingly, casein kinase II has been shown to phosphorylate proteins involved in cellular proliferation, including Myc (79) and the E7 protein of human papilloma virus (80). Phosphorylation of CREB by glycogen synthase kinase-3 is presently under investigation. As mentioned earlier, both the casein kinases and glycogen synthase kinase-3 are known to be processive kinases inasmuch as phosphorylation of specific serines and threonines is markedly facilitated by prior phosphorylations of adjacent residues (81,82). Of course the counterpart to the transcriptional activation of CREB by phosphorylation will likely be inactivation by phosphatases. The potentially important role of phosphatases as counter regulators of protein kinases in gene transcription is only now becoming a focus of investigations.

CHARACTERIZATION OF THE CREB GENE AND ITS EXPRESSION

In contrast with the array of reports concerning CREB protein function, relatively little information has been published about the control of *creb* gene expression. Consequently, we have isolated for study several human genomic *creb* clones from cosmid and lambda phage libraries (40,44,83) (Fig. 4). DNA sequencing of subclones of the cosmids reveal that *creb* consists of at least 10 exons (40). The exons are separated by large introns, spread over at least 30 kb of DNA, as determined by polymerase chain reaction and Southern blot analyses. Alternative splicing of one of the exons generates the CREB327 and CREB341 protein isoforms as predicted earlier by sequencing of cDNAs. Further alternative RNA splicing, found only in testes and not in other rat tissues, appears to result in truncated CREB proteins that lack a putative nuclear translocation signal and the bZip domain (43). These truncated CREB proteins fail to transactivate transcription through the usual binding to CRE target sequences, and may serve to down-regulate CREB-mediated transcription. Additional alternatively spliced CREB RNAs have been detected in a tissue type-specific distribution (Waeber, unpublished observations). The functions of these RNAs, however, has not yet been determined. One speculation is that certain of the RNAs may encode "co-activators," bridges between CREB and other transcription factors and TATA-box binding proteins (TFIID) as has been postulated for activation by Sp1 (84). Additional functions for such alternatively spliced transcripts have been reviewed by Maniatis (85), and include the possibility that alternative splicing may serve as a simple on/off switch to regulate gene expression through the use of mRNAs encoding full length versus truncated proteins. It is notable that the multiexonic structure of the *creb* gene is

mRNA

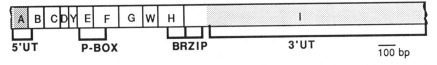

Gene

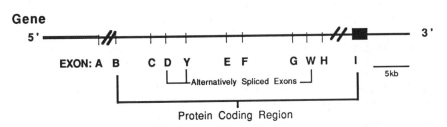

Protein Coding Region

FIG. 4. Diagram of the CREB327/341 gene (bottom) and a messenger RNA (top). The gene consists of multiple small exons, several of which are alternatively spliced (exons D, Y, and W) in a tissue type-specific manner. It appears likely that yet additional alternatively spliced exons exist for this CREB gene. Exons E and F comprise the phosphorylated transcriptional transactivation domain (P-Box). The DNA-binding domain, consisting of the basic region (BR) and leucine zipper (ZIP), is encoded in parts of exons H and I.

not shared by several of the other genes encoding members of the bZip protein family. For example, the genes for *c-jun, jun-B* and C/EBP are intronless.

Knowledge of the human *creb* intron and exon sequences allowed us to investigate the chromosomal location of *creb* (40). Human *creb* exon-specific oligonucleotides were used to PCR clone genomic DNA from a battery of hamster-human and rat-human somatic cell hybrids, each hybrid representing different human chromosomes. Southern blots of the PCR products were then hybridized with oligonucleotide probes made to human *creb* intron sequences. Because the human *creb* intron sequences differ from those of other species, we were able to differentially hybridize the PCR samples, and have tentatively assigned the *creb327/341* locus to human chromosome 2 (40). Another study has more precisely localized the *creb* gene to the long arm of chromosome 2 (86). Considering the critical role that CREB plays in signal transduction, localization of the *creb* locus should prove useful in studies of diseases resulting from abnormalities in human chromosome 2.

Most recently, we have set about characterization of the human *creb* gene promoter and its 5'-flanking region (83). A 1.2 kb clone containing an exon for part of the 5'-untranslated tract and the promoter region has been sequenced. By inspection, the DNA sequence is highly enriched in guanines and cytosines, and contains canonical binding sites for the transcription factors AP2 (87,88) , C/EBP (50), Sp1 (89), GCF (90), and CREB. Transfection studies are in progress to determine which of these sites, if any, participate in regulating transcription from the *creb* gene. A few stretches of the 5'-flanking/promoter sequence resemble TATA and CCAAT sites, but none of them fit the canonical sequences for these sites. The lack of true TATA and CCAAT sites are consistent with our primer extension data which indicate the presence of multiple transcription start sites for the *creb* gene. These features are reminiscent of many of the house-keeping genes that have recently been described (91-94). Typically these genes are expressed ubiquitously in cells of diverse phenotypes. In fact many of the genes found to have the GC-rich promoter sequences devoid of TATA and CCAAT boxes encode components of signal transduction pathways, for example, plasma membrane bound receptors, protein kinases, GTP-binding proteins and cell adhesion molecules. Characteristically, CREB has been detected in all cells and tissues tested thus far. Whereas the CREB protein is easily detected by either antibodies or DNA probes, the CREB mRNA is particularly difficult to detect by standard Northern blotting. This suggests that the mRNA may be expressed at a relatively low, basal level and/or is rapidly degraded, and the protein must be relatively stable. A constant supply of stable CREB protein would be consistent with the observation that cAMP activation of the transcriptional transactivation functions of CREB, mediated through the protein kinase A pathway, occurs rapidly, reaching a peak within 30 minutes of stimulation, even when cycloheximide is used to block new protein synthesis (95,96). Thus, post translational modifications of CREB by way of phosphorylations play a major and critical role in the regulation of its transactivational functions.

Some recent, interesting results by Waeber *et al.* (43) provide the first evidence that expression of the *creb* gene is not at a constitutive, basal level, but is modulated in certain tissues. Multiple tissues and cells were examined by histohybridization with anti-CREB antibodies, and by *in situ* mRNA hybridization with *creb* oligonucleotides. Most of the samples contained levels of CREB mRNA (low) and CREB protein (reasonably detectable) consistent with the Northern blotting and antibody studies mentioned earlier. In the rat testes, however, CREB mRNA and protein levels exhibited striking increases at specific developmental stages, in specific cell types. For example, the intensity from a one-day autoradiographic exposure of *in situ* hybridized CREB mRNA from adult rat testes is comparable to that for a 6-week exposure of a similarly probed rat brain section. This burst of CREB expression is also interesting in that it exhibits developmental regulation. Certainly we are eager to understand the control mechanisms of CREB expression in the testes, and to identify what role this enormous burst of CREB synthesis must serve in spermatogenesis.

REGULATION OF SECOND MESSENGER PATHWAYS OCCURS AT MULTIPLE LEVELS

The turning on and off of genes in a cell-specific, timely manner requires careful regulation at the molecular level. To meet this demand for a finely tuned system, cells have evolved intricate second messenger pathways. In the protein kinase A pathway, for example, adenylyl cyclase is stimulated to catalyze the synthesis of cAMP, which then triggers the release of the active catalytic subunits of protein kinase A. Already at this stage in the pathway, two primary mechanisms are invoked to down-regulate the system back to a basal state; phosphodiesterase cleaves and inactivates the cAMP, and the stimulation of adenylyl cyclase by ligands is halted by "uncoupling" of the receptor from regulatory GTP-binding proteins or by degradation of the ligand or receptor/ligand complex.

Further down the pathway, protein kinase A probably phosphorylates multiple transcription factors including CREB and other CREB/ATF-like proteins. CRE-BP1 and ATF-1 have consensus serines for phosphorylations by protein kinase A, although phosphorylation of these sites has not yet been directly demonstrated. By activating a variety of transcription factors, protein kinase A causes a branching out of the second messenger pathway, and triggers the expression of the genes needed to deliver the appropriate physiological responses. The physiological responses are delivered in a cell- and differentiation stage-specific manner as the result of a complex interplay among transcription factors and their target DNA enhancer/repressor elements.

Transcription Is Regulated by Protein-Protein Interactions

One level of complexity in gene regulation stems from the biochemical constraints placed on interactions among the transcription factors. As mentioned earlier for the bZip family of proteins, the leucine zippers control homo- and heterodimerization, thereby partially controlling which target DNA elements the dimers transactivate. Jun-Fos heterodimers transactivate TREs better than do Jun-Jun homodimers, and CREB preferentially binds as a homodimer to transactivate CREs. Also, the bZip protein ATF-2 (also called CRE-BP1) has been shown to heterodimerize with Jun in a complex that binds more tightly to CREs than to TREs (97), and ATF-2 heterodimerizes with ATF-3 to bind CREs (62). Because increasing the number of these possible dimer combinations may further broaden the variety of CRE- and TRE-like elements which can be transactivated, it is interesting to note that more than ten different human cDNAs have already been discovered for such bZip proteins. From these observations, the control of gene expression depends, in part, on which of these factors is present in a particular cell type, and on whether or not the gene being activated contains any of the DNA enhancer elements for these factors.

Protein-protein interactions other than leucine zipper-mediated dimerization also serve to regulate transcription factors. A recent publication by Busch and Sassone-Corsi (98) suggests that a third protein, ABP (auxiliary binding protein), interacts with the bZip region of Fos/Jun complexes during transactivation. This type of quaternary structure probably provides a different function than that described for the "coactivator" factor proposed for some Sp1 transcription complexes (84). Abate *et al.* (99) recently reported another protein, as yet unnamed, which contributes to Fos/Jun transactivation. This protein chemically reduces Fos and Jun and stimulates DNA-binding activity, presumably by modification of a particular pair of cysteine residues. We (Yun and Habener, unpublished data) have observed a similar effect of reduction on the binding of CREB proteins to CREs *in vitro*. Interestingly, CREB, Fos, Jun and many other members of the bZip family of proteins exhibit a positional conservation of either the candidate cysteine or a serine in their DNA-binding domains (Table 1). One intriguing interpretation for the conservation of either a cysteine or serine in the same relative position in the DNA-binding domains of the bZip proteins is that these residues can be modified to create a negative charge that may in turn modify (inhibit) binding of the proteins to their respective DNA elements. Cysteine may be oxidized to sulfinic acid SO_2- or sulfonic acid $SO_3=$, perhaps by changes in intracellular redox potentials mediated by actions of signalling pathways on the transport of hydrogen ions. Serine is a favored residue for phosphorylation, that, similar to the oxidation of cysteine, converts a neutral amino acid, serine, to one with a high negative charge, phosphoserine. The formation of a negative charge in a critical region of the DNA-binding domain may inhibit close juxtapositioning of the protein to the negatively charged phosphate backbone of the DNA.

A third set of protein-protein interactions adds yet another level of control to gene expression. In this case, interaction of the transcription factors occurs in such a way as to confer "cooperativity", "synergism", or "squelching" during transactivation. Cooperative interactions of transcription factors occurs when binding of one factor facilitates binding of additional factors. Synergism results when two or more transcription factors bind to several enhancers along the control region of one gene, and their combined transcriptional activation is greater than the simple sum of their activities were they to function independently (4). In the human alpha chorionic gonadotropin gene, for example, a regulatory element (URE) upstream of the CREs synergizes with the CREs—by itself, the URE contributes little basal and no cAMP-stimulated CREB activity, but in tandem with the CREs, the URE gives 10-20-fold greater cAMP-response than do the CREs alone (32).

Whereas cooperativity and synergism usually produce up-regulation of transcription, squelching contributes to down-regulation of activity. Ptashne (4) describes squelching, essentially, as an over-abundant release of a factor(s) such that a competition is set up, and the amount of the inactive form of the affected transcription protein is disproportionately greater than that of the active form, effectively blocking transcription. For example, the squelching effect has been demonstrated during the over-expression of recombinant CREB in transfection assays such that increased CREB transcription activity should have been detected but wasn't (Hoeffler *et al.*, unpublished observations). In that instance, the limiting component of the system was the amount of endogenous protein kinase A; co-transfection of a protein kinase A expression plasmid along with the CREB expression plasmid restored the expected increase in CREB-mediated transcriptional transactivation.

Regulation of Transcription Factors by Post-Translational Modifications

The activity of many transcription factors is regulated by post-translational modifications. Such modifications include O-linked and N-linked glycosylation, phosphorylation/dephosphorylation, and oxidation/reduction of cysteine/cystine residues. Regulation of the transcriptional activity of CREB likely involves all of these. As mentioned earlier, the oxidation/reduction state of CREB may affect binding to CREs. The modification of CREB by glycosylation at serine and asparagine residues is predicted by inspection of the protein sequence for consensus sites, but remains formally unproven.

The phosphorylation and dephosphorylation of CREB and other transcription factors is an especially interesting area of research because this form of modification occurs in a reversible fashion to directly control transcriptional activity. As discussed earlier, CREB has consensus sites for and/or has been directly shown to be phosphorylated by protein kinaseA, protein kinase C, glycogen synthase kinase-3, and casein kinase II. Presumably, phosphorylation of CREB by these kinases activates transcription. Besides activating transcription, phosphorylation has also been shown to affect both specificity and affinity of binding of proteins to the CRE or TRE elements. In studies of c-Jun, Hunter *et al.* (100) reported that activation of protein kinase C leads to dephosphorylation (of phosphorylation by glycogen synthase kinase-3) of sites adjacent to the DNA binding domain, thereby enhancing binding of Jun/Fos heterodimers. Merino *et al.* (101) examined the binding of HeLa cell proteins to CRE elements in the adenovirus early region III and IV promoters, and demonstrated that phosphorylation and dephosphorylation affected both transcriptional activity and binding affinity. One conclusion from the study was that phosphorylation of CRE-binding proteins, including the CREB-related CRE-BP1, contribute to the regulation of cellular responses to the adenovirus E1a protein.

Autoregulation of Transcription Factor Expression

While post-translational modification and control of protein-protein interactions provide cells with the means for relatively quick modulation of transcription factors in response to stimuli, a more direct way to regulate transcription factors is to govern their synthesis. It is notable, then, that one of the mechanisms for regulating the expression of transcription factors involves autoregulation. For example, the pituitary-specific transcription factor Pit-1, also referred to as GHF1, exhibits both positive and negative autoregulation (102,103). Positive autoregulation also occurs for the expression of the *c-jun* proto-oncogene, but in this case, the activating factor actually consists of the Jun/Fos heterodimer (66,104). On the other hand, the Jun/Fos complex provides negative autoregulation of the *fos* gene (105,106). Jun-B, another member of the Jun family, functions as a negative transactivator of *c-jun* gene expression (66). Figure 5 illustrates this complex interplay of transcription factors, and Figure 6 depicts how the *creb* gene may also be autoregulated through the recently identified CRE in its promoter.

SUMMARY

Early studies of cAMP-dependent transactivation of gene transcription established a connection between PKA phosphorylation of a protein, CREB/ATF, and CRE enhancer-mediated gene expression. By taking advantage of the observation that specific mutants of the CRE sequence fail to mediate cAMP stimulation, it was possible to devise a strategy to screen cDNA expression libraries and isolate clones that exhibited binding specificity for CREs and not for mutant CREs. As one might predict from the role of CREB in the cAMP pathway, the first reported CREB amino acid sequence deduced from the cloned cDNA possesses a site for phosphorylation by protein kinase A. The cDNA clone encodes a protein of 327 amino acids, CREB327. Later, CREB327 was found to be an isoform of the 341 amino acid CREB341, the result of alternatively spliced RNAs transcribed from the CREB327/341 gene, located on human chromosome 2. Both CREB327 and CREB341 require phosphorylation by cyclic AMP-dependent protein kinase A of a single serine in the "P-box," as well as additional phosphorylations by one or more of the kinases glycogen synthase kinase-3, protein kinase C, or casein kinase II to render CREB transcriptionally active. That is, phosphorylation of CREB by protein kinase A is necessary but not sufficient to generate the transactivation functions. The structures of the CREB327/341 proteins are bipartite in that transcriptional activity is driven by the amino terminal 4/5 of the protein, and the CRE-binding is governed by the carboxy-terminal basic region and leucine zipper, bZip domain.

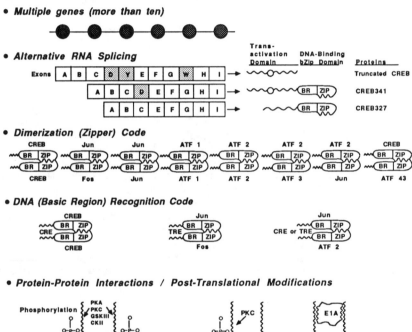

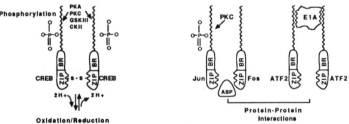

FIG. 5. Schematic representation of the multiple levels of gene expression at which diversification of CREB functions takes place. From top to bottom: At least ten distinct CREB/ATF genes have been identified. The CREB 327/341 gene is comprised of multiple small exons that are alternatively spliced in or out during the processing of the premRNA into the mature mRNAs. The alternative splicing patterns differ in a tissue type-specific manner leading to the formation of a diversity of structurally different CREB's. The dimerization (zipper, ZIP) domains dictate a code for homo- or heterodimerization among the different members of the bZip proteins. The basic region (BR) domain determines DNA element binding specificities, *e.g.*, whether a particular dimer will bind to a CRE or to a TRE. Protein-protein interactions and/or post-translational modifications, particularly by phosphorylations, can modulate binding and/or transcriptional transactivation activities.

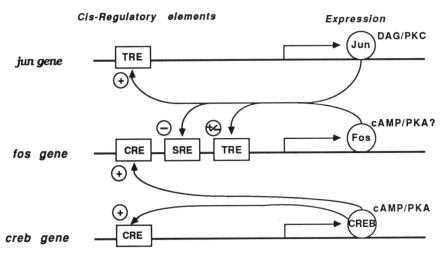

FIG. 6. Transcriptional autoregulation of the *jun, fos,* and *creb* genes. The *c-jun* gene is positively autoregulated by heterodimers of the Jun and Fos (AP-1) protein complex through interaction with a tumor-promoting agent (phorbol esters)-element,TRE. The regulation of the *fos* gene is highly complex, involving interactions of Jun, Fos, and CREB proteins with several regulatory DNA elements. *Fos* gene transcription is up-regulated by interactions of CREB, or CREB/ATF-like proteins, with a cyclic AMP response element (CRE). In addition, Jun-Fos heterodimers interact in a complex manner with both a TRE and a serum-response element (SRE) to either up- or down-regulate *fos* gene transcription. The *creb* gene promoter is likewise complex and is currently being analyzed; among many other elements it appears to have a CRE—predictive of positive autoregulation.

The bZip domain of CREB places it in a super-family of DNA-binding transcription factors, the bZip proteins, which have remarkably similar and conserved residues in their basic regions and leucine zippers. These proteins bind to their target DNA enhancers as either homo- or heterodimers, in select combinations that are dictated by the constituent amino acids in the leucine zippers. Whether the dimer binds to CREs, TREs, or other enhancer elements is determined by both the basic regions and the leucine zippers, as well as the contextual sequence of the enhancers. Recent studies of several newly identified transcription factors have shown that some bZip proteins bind with specificity to naturally occurring variants of the canonical CRE and TRE, thus expanding the scope of genes which can be transactivated by second messenger pathways. Also, the CREB327/341 proteins are likely not to be solely responsible for protein kinase A-mediated gene expression; both of the recently cloned CRE-BP1 and ATF-1 proteins, for example, bind to CREs and possess consensus protein kinase A phosphorylation sites.

Future studies of CREB and of transcription factors with structures similar to that of CREB will be aimed at sorting out the complex system which has evolved to regulate gene transcription via second messenger pathways—pathways that activate protein kinases and phosphorylations of specific transcription factors which in turn, transactivate CREs, TREs and related enhancers. Southwestern blots of proteins probed by CRE-containing oligonucleotides and hybridizations of genomic DNA fragments by Southern blots suggest that many CRE-binding proteins remain to be cloned, potentially adding new insights to the study of CRE-mediated gene transcription. Finally, it is worth noting that CREB, similar in structure and function to the proto-oncoproteins c-Jun, c-Fos, c-Myc, and their oncoprotein counterparts, has been cloned and characterized only as a native, endogenous cellular transcription factor. Historically, viral and cellular oncogenes appear to have evolved from mutations of 'normal' cellular genes (protooncogenes) which encode key factors in signal transduction pathways, *e.g.*, receptors, protein kinases, GTP-binding proteins, and transcription factors. Therefore, it seems likely that an oncoprotein form of CREB exists and remains to be discovered.

ACKNOWLEDGMENTS

We thank the members of our laboratory, both past and present, for their contributions to the work cited herein. Thanks also, to Townley Budde for much of the artwork and typing, and to Gerard Waeber, Mario Vallejo, and David Ron for reviewing the material presented. The studies were supported in part by USPHS grants DK25532 and DK30457.

REFERENCES

1.   Herschman HR. Extracellular signals, transcriptional responses and cellular specificity. *Trends Biochem Sci* 1989;14:455-58.
2.   Mitchell PJ, Tjian R. Transcriptional regulation in mammalian cells by sequence-specific DNA binding proteins. *Science* 1989;245:371-78.
3.   Druker BJ, Mamon HJ, Roberts TM. Oncogenes, growth factors, and signal transduction. *N Engl J Med* 1989;321:1383-91.
4.   Ptashne M. How eukaryotic transcriptional activators work. *Nature* 1988; 335: 683-89.
5.   Ptashne M, Gann AAF. Activators and targets. *Nature* 1990;346:329-31.
6.   Lewin B. Commitment and activation at pol II promoters: a tail of protein-protein interactions. *Cell* 1990; 61:1161-64.
7.   Levine M, Manley JL. Transcriptional repression of eukaryotic promoters. *Cell* 1989; 59:405-408.
8.   Struhl K. Helix-turn-helix, zinc-finger, and leucine-zipper motifs for eukaryotic transcriptional regulatory proteins. *Trends Biochem Sci* 1989;14:137-140.
9.   Dynan WS. Modularity in promoters and enhancers. *Cell* 1989;58:1-4.
10.   Ross J. The turnover of messenger RNA. *Sci Am* 1989; 260:48-55.
11.   Sharp PA. RNA splicing and genes. *JAMA* 1988;260:3035-41.
12.   Roesler WJ, Vanderbark GR, Hanson RW. Cyclic AMP and the induction of eukaryotic gene expression. *J Biol Chem* 1988;263:9063-66
13.   Nishizuka Y. The role of protein kinase C in cell surface signal transduction and tumor promotion. *Nature* 1984;308:693-98.
14.   Gold MR, Law DA, DeFranco AL. Stimulation of protein tyrosine phosphorylation by the B-lymphocyte antigen receptor. *Nature* 1990;345:810-13.

15. Northwood IC, Davis RJ. Signal transduction by the epidermal growth factor receptor after functional desensitization of the receptor tyrosine protein kinase activity. *Proc Natl Acad Sci USA* 1990;87:6107-11.

16. Babcock AE, Norenberg MD, Norenberg LO, Neary JT. Calcium/calmodulin-dependent protein kinase activity in primary astrocyte cultures. *Glia* 1989;2:112-118.

17. Wang JK, Walaas SI, Greengard P. Protein phosphorylation in nerve terminals: comparison of calcium/calmodulin-dependent and calcium/diacylglycerol-dependent systems. *J Neurosci* 1988;8:281-88.

18. Beato M. Gene regulation by steroid hormones. *Cell* 1989:56:335-44.

19. Roesler WJ, Vanderbark GR, Hanson RW. Cyclic AMP and the induction of eukaryotic gene expression. *J Biol Chem* 1988;263:9063-66.

20. Comb M, Birnberg NC, Seasholtz A, Herbert E, Goodman HM. A cyclic AMP- and phorbol ester-inducible DNA element. *Nature* 1986;323:353-56.

21. Tsukada T, Fink JS, Mandel G, Goodman RH. Identification of a region in the human vasoactive intestinal polypeptide gene responsible for regulation by cyclic AMP. *J Biol Chem* 1987;262:8743-47

22. Fink JS, Verhave M, Kasper S, Tsukada T, Mandel G, Goodman RH. The CGTCA sequence motif is essential for biological activity of the vasoactive intestinal peptide gene cAMP-regulated enhancer. *Proc Natl Acad Sci USA* 1988; 85: 6662-66.

23. Montminy MR, Sevarino KA, Wagner JA, Mandel G, Goodman RH. Identification of a cyclic-AMP-responsive element within the rat somatostatin gene. *Proc Natl Acad Sci USA* 1986;83:6682-86.

24. Deutsch PJ, Hoeffler JP, Jameson JL, Lin JC, Habener JF. Structural determinants for transcriptional activation by cAMP-responsive DNA elements. *J Biol Chem* 1988; 263:18466-72.

25. Deutsch PJ, Hoeffler JP, Jameson JL, Habener JF. Cyclic AMP and phorbol ester-stimulated transcription mediated by similar DNA elements that bind distinct proteins. *Proc Natl Acad Sci USA* 1988;85:7922-26.

26. Nagamine Y, Reich E. Gene expression and cAMP. *Proc Natl Acad Sci USA* 1985;82:4606-10.

27. Lee CQ, Miller HA, Schlichter D, Dong JN, Wicks WD. Evidence for a cAMP-dependent nuclear factor capable of interacting with a specific region of a eukaryotic gene. *Proc Natl Acad Sci USA* 1988;85:4223-27.

28. Sassone-Corsi P. Cyclic AMP induction of early adenovirus promoters involves sequences required for E1A trans-activation. *Proc Natl Acad Sci USA* 1988; 85:7192-96.

29. Bokar JA, Roesler WJ, Vandenbark GR, Kaetzel DM, Hanson RW, Nilson JH. Characterization of the cAMP responsive elements from the genes for the alpha-subunit of glycoprotein hormones and phosphoenolpyruvate carboxykinase (GTP). *J Biol Chem* 1988; 263:19740-47.

30. Angel P, Imagawa M, Chiu R, et al. Phorbol ester-inducible genes contain a common cis element recognized by a TPA-modulated trans-acting factor. *Cell* 1987; 49:729-39.

31. Lee W, Mitchell P, Tjian R. Purified transcription factor AP-1 interacts with TPA-inducible enhancer elements. *Cell* 1987;49:741-52.

32. Hoeffler JP, Deutsch PJ, Lin J, Habener JF. Distinct adenosine 3',5'-monophosphate and phorbol ester-responsive signal transduction pathways

converge at the level of transcriptional activation by the interactions of DNA-binding proteins. *Mol Endocrinol* 1989;3:868-80.

33. Vogt PK, Bos TJ, Doolittle . Homology between the DNA- binding domain of the GCN4 regulatory protein of yeast and the carboxyl-terminal region of a protein coded for by the oncogene Jun. *Proc Natl Acad Sci USA* 1987;84:3316-19.

34. Curran T, Franza BR Jr. Fos and Jun: the AP-1 connection. *Cell* 1988;55: 95-97.

35. Smeal T, Angel P, Meek J, Karin M. Different requirements for formation of Jun:Jun and Jun:Fos complexes. *Genes Dev* 1989;3:2091-2100.

36. Turner R, Tjian R. Leucine repeats and an adjacent DNA binding domain mediate the formation of functional cFos-cJun heterodimers. *Science* 1989; 243: 1689-94.

37. Lin YS, Green MR. Interaction of a common cellular transcription factor, ATF, with regulatory elements in both E1a- and cyclic AMP-inducible promoters. *Proc Natl Acad Sci USA* 1988;85:3396-3400.

38. Montminy MR, Bilezikjian LM. Binding of a nuclear protein to the cyclic-AMP response element of the somatostatin gene. *Nature* 1987;328:175-78.

39. Hoeffler JP, Meyer TE, Yun Y, Jameson JL, Habener JF. Cyclic- AMP responsive DNA-binding protein: structure based on a cloned placental cDNA. *Science* 1988;242:1430-33.

40. Hoeffler JP, Meyer TE, Waeber GW, Habener JF. Multiple adenosine 3', 5'-cyclic monophosphate response element DNA-binding proteins generated by gene diversification and alternative exon splicing. *Mol Endocrinol* 1990;4:920-30.

41. Gonzalez GA, Yamamoto KK, Fischer WH, Karr D, Menzel P, Biggs WIII, Vale WW, Montminy MR. A cluster of phosphorylation sites on the cyclic AMP-regulated nuclear factor CREB predicted by its sequence. *Nature* 1989;337:749-52.

42. Singh H, LeBowitz JH, Baldrin AS Jr, Sharp PA. Molecular cloning of an enhancer binding protein: isolation by screening of an expression library with a recognition site DNA. *Cell* 1988;52:415-23.

43. Waeber GW, Meyer TE, LeSieur M, Hermann H, Gérard N, Habener JF. Developmental stage-specific expression of the cyclic AMP response element binding protein CREB during spermatogenesis involves alternative exon splicing. (Submitted).

44. Waeber GW, Meyer TE, Hoeffler JP, Habener JF. Diversification of cyclic AMP-responsive enhancer-binding proteins generated by alternative exon splicing. *Trans Assoc Am Physicians* (in press).

45. Lee CQ, Yun Y, Hoeffler JP, Habener JF. Cyclic-AMP-responsive transcriptional activation of CREB-327 involves interdependent phosphorylated subdomains. *EMBO J* 1990;9:4455-65

46. Yun Y, Dumoulin M, Habener JF. DNA-binding and dimerization domains of cyclic AMP-responsive protein CREB reside in the carboxyl-terminal 66 amino acids. *Mol Endocrinol* 1990;4:931-39.

47. Dwarki VJ, Montminy M, Verma IM. Both the basic region and the 'leucine zipper' domain of the cyclic AMP responsive element binding protein are essential for transcriptional activation. *EMBO J* 1990;9:225-32.

48. Yamamoto KK, Gonzalez GA, Menzel P, Rivier J, Montminy MR. Characterization of a bipartite activator domain in transcription factor CREB. *Cell* 1990; 60:611-17.

49. Vinson CR, Sigler PB, McKnight SL. Scissors-grip model for DNA recognition by a family of leucine zipper proteins. *Science* 1989;246:911-22.

50. Landschultz WH, Johnson PF, McKnight SL. The leucine zipper: a hypothetical structure common to a new classs of DNA binding proteins. *Science* 1988;240:1759-64.

51. Kouzarides T, Ziff E. The role of the leucine zipper in the Fos-Jun interaction. *Nature* 1988;336:646-51.

52. Nakabeppu Y, Ryder K, Nathans D. DNA binding activities of three murine Jun proteins: stimulation by Fos. *Cell* 1988;55:907-15.

53. Schuermann M, Neuberg M, Hunter JB, Jenuwein T, Ryseck R-P, Bravo R, Müller R. The leucine repeat motif in Fos protein mediates complex formation with Jun/AP1 and is required for transformation. *Cell* 1989;56:507-16.

54. Nakabeppu Y, Nathans D. The basic region of Fos mediates specific DNA binding. *EMBO J* 1989;8:3833-41.

55. Cohen DR, Curran T. Analysis of dimerization and DNA binding functions in Fos and Jun by domain-swapping: involvement of residues outside the leucine zipper/basic region. *Oncogene* 1990;3:929-39.

56. Talonian RV, McKnight CJ, Kim PS. Sequence-specific DNA binding by a short peptide dimer. *Science* 1990; 249:769-71.

57. Shuman JD, Vinson CR, McKnight SL. Evidence of changes in protease sensitivity and subunit exchange rate in DNA binding by C/EBP. *Science* 1990;249:771-74.

58. O'Neil KT, Hoess RH, DeGrado WF. Design of DNA-binding peptides based on the leucine zipper motif. *Science* 1990;774-78.

59. Patel L, Abate C, Curran T. Altered protein conformation on DNA binding by Fos and Jun. *Nature* 1990;347:572-75.

60. Weiss MA, Ellenberger T, Wobbe CR, Lee JP, Harrison SC, Struhl K. Folding transition in the DNA-binding domain of GCN4 on specific binding to DNA. *Nature* 1990;347:575-78.

61. Maekawa T, Sakura H, Kanei-Ishii C, *et al.* Leucine zipper structure of the protein CRE-BP1 binding to the cyclic AMP response element in the brain. *EMBO J* 1989;8:2023-28.

62. Hai T, Liu F, Coukos WJ, Green MR. Transcription factor ATF cDNA clones: an extensive family of leucine zipper proteins able to selectively form DNA-binding heterodimers. *Genes Dev* 1989; 3:2083-90.

63. Kara CJ, Liou H-C, Ivashkiv LB, Glimcher LH. A cDNA for a human cyclic AMP response element-binding protein which is distinct from CREB and expressed preferentially in brain. *Mol Cell Biol* 1990;10:1347-57.

64. Ivashkiv LB, Liou H-C, Kara CJ, Lamph WW, Verma IM, Glimcher LH. mXBP/CRE-BP2 and c-Jun form a complex which binds to the cyclic AMP, but not to the 12-0-tetradecanoylphorbol-13-acetate, response element. *Mol Cell Biol* 1990;10:1609-21.

65. Ryder K, Lanahan A, Perez-Albuerne E, Nathans D. Jun D: a third member of the Jun gene family. *Proc Natl Acad Sci USA* 1989;86:1500-03.

66. Schütte J, Viallet J, Nau M, Segal S, Fedorko J, Minna J. Jun-B inhibits and c-Fos stimulates the transforming and trans-activating activities of c-Jun. *Cell* 1989;59:987-97.

67. Perkins KK, Admon A, Patel N, Tjian R. The Drosophila Fos-related AP-1 protein is a developmentally regulated transcription fac tor. *Genes Dev* 1990;4:822-34.

68. Matsui M, Tokuhara M, Konuma Y, Nomura N, Ishizaki R. Isolation of human Fos-related genes and their expression during monocyte-macrophage differentiation. *Oncogene* 1990;5:249-55.

69. Yoshimura T, Fujisawa J, Yoshida M. Multiple cDNA clones encoding nuclear proteins that bind to the tax-dependent enhancer of HTLV-1: all contain a leucine zipper structure and basic amino acid domain. *EMBO J* 1990;9:2537-42.

70. Akira S, Isshiki H, Sugita T, et al. A nuclear factor for IL-6 expression (NF-IL6) is a member of a C/EBP family. *EMBO J* 1990;9:1897-1906.

71. Tabata T, Takase H, Takayama S. A protein that binds to a cis-acting element of wheat histone genes has a leucine zipper motif. *Science* 1989;245:965-67.

72. Katagiri F, Lam E, Chua N-H. Two tobacco DNA-binding proteins with homology to the nuclear factor CREB. *Nature* 1989;340:727-30.

73. Schmidt RJ, Burr FA, Aukerman MJ, Burr B. Maize regulatory gene opaque-2 encodes a protein with a "leucine-zipper" motif that binds to zein DNA. *Proc Natl Acad Sci USA* 1990;87:46-50.

74. Ma J, Ptashne M. Deletion analysis of GAL4 defines two transcriptional activating segments. *Cell* 1987; 48:847-53.

75. Hope IA, Mahadevan S, Struhl K. Structural and functional characterization of the short acidic transcriptional activation region of yeast GCN4 protein. *Nature* 1988; 333:635-40.

76. Horikoshi M, Hai T, Lin YS, Green MR, and Roeder RG. Transcription Factor ATF interacts with the TATA factor to facilitate establishment of a preinitiation complex. *Cell* 1988;54:1033-42.

77. Andrisani OM, Pot DA, Zhu Z, Dixon JE. Three sequence-specific DNA-protein complexes are formed with the same promoter element essential for expression of the rat somatostatin gene. *Mol Cell Biol* 1988;8:1947-56.

78. Gonzalez GA, Montminy MR. Cyclic AMP stimulates somatostatin gene transcription by phosphorylation of CREB at serine 133. *Cell* 1989;59:675-80.

79. Luscher B, Kuezel EA, Krebs EG, Eisenmann RN. Myc oncoproteins are phosphorylated by casein kinase II. *EMBO J* 1989;8:1111-19.

80. Firzlaff JM, Galloway DA, Eisenmann RN, Luscher B. The E7 protein of human papillomavirus type 16 is phosphorylated by casein kinase II. *The New Biologist* 1989;1:44-53.

81. Flotow H, Roach PJ. Synergistic phosphorylation of rabbit muscle glycogen synthase by cyclic AMP-dependent protein kinase and casein kinase I. *J Biol Chem* 1989; 264:2961-68.

82. Roach PJ. Control of glycogen synthase by hierarchal protein phosphorylation. *FASEB J* 1990; 4:2961-68.

83. Meyer TE, Waeber G, Beckmann W, Lin J, Habener JF. Cloning of the human CREB gene and characterization of its promoter and 5'flanking region. (in preparation).

84. Pugh BF, Tjian R. Mechanism of transcriptional activation by Sp1: evidence for coactivators. *Cell* 1990;61:1187-97.

85. Maniatis T. Mechanisms of alternative pre-mRNA splicing. *Science* 1991;251:33-34.

86. Taylor AK, Klisak I, Mohandas T, Sparkes RS, Li C, Gaymor R, Susis AJ. Assignment of the human gene for CREB1 to chromosome 2q32.3-q34. *Genomics* 1990;7:416-21.

87. Imagawa M, Chiu R, Karin M. Transcription factor AP-2 mediates induction by two different signal-transduction pathways: protein kinase C and cAMP. *Cell* 1987;51:251-60.

88. Williams T, Admon A, Lüscher B, Tjian R. Cloning and expression of AP-2, a cell-type-specific transcription factor that activates inducible enhancer elements. *Genes Dev* 1988;2:1557-69.

89. Kadonaga JT, Carner KR, Masiarz FR, Tjian R. Isolation of cDNA encoding transcription factor Sp1 and functional analysis of the DNA binding domain. *Cell* 1987;51:1079-90.

90. Kageyama R, Pastan I. Molecular cloning and characterization of a human DNA binding factor that represses transcription. *Cell* 1989;59:815-25.

91. Blake MC, Jambou RC, Swick AG, Kahn JW, Azizkhan JC. Transcriptional initiation is controlled by upstream GC-box interactions in a TATAA-less promoter. *Mol Cell Biol* 1990;10:6632-41.

92. Boyer TG, Maquat LE. Minimal sequence and factor requirements for the initiation of transcription from an atypical, TATATAA box-containing housekeeping promoter. *J Biol Chem* 1990;265:20524-32.

93. Ishii S, Xu YH, Stratton RH, Roe BA, Merlino GT, Pastan I. Characterization and sequence of the promoter region of the human epidermal growth factor receptor gene. *Proc Natl Acad Sci USA* 1985;82:4920-24.

94. Tilley WD, Marcelli M, McPhaul MJ. Expression of the human androgen receptor gene utilizes a common promoter in diverse human tissues and cell lines. *J Biol Chem* 1990;265:13776-81.

95. Lewis EJ, Harrington CA, Chikaraish DM.Transcriptional regulation of the tyrosine hydroxylase gene by glucocorticoid and cyclic AMP. *Proc Natl Acad Sci USA* 1987;84:3550-54.

96. Sasaki K, Cripe TP, Koch SR, Andreone TL, Petersen DD, Beale EG, Granner DK. Multihormonal regulation of phosphoenolpyruvate carboxykinase gene transcription. *J Biol Chem* 1984;259:15242-51.

97. Macgregor PF, Abate C, Curran T. Direct cloning of leucine zipper proteins: Jun binds coperatively to the CRE with CRE-BP1. *Oncogene* 1990;4:451-58.

98. Busch SJ, Sassone-Corsi P. Fos, Jun and CREB basic-domain peptides have intrinsic DNA-binding activity enhanced by a novel stabilizing factor. *Oncogene* 1990;5:1549-56.

99. Abate C, Patel L, Rauscher FJIII, Curran T. Redox regulation of Fos and Jun DNA-binding activity in vitro. *Science* 1990; 249:1157-61.

100. Hunter TI, Boyle B, Lindberg R, Jaehner D, Middlemus D, Pines J. Signal transducing protein kinases and their targets. *J Cell Biochem* 1990; [suppl] 14E: (Abstract) 139.

101. Merino A, Buckbinder L, Mermelstein FH, Reinberg D. Phosphorylation of cellular proteins regulates their binding to the cAMP response element. *J Biol Chem* 1989;264:21266-76.

102. Chen R, Ingraham HA, Treacy MN, Albert VR, Wilson L, Rosenfeld MG. Autoregulation of pit-1 gene expression mediated by two cis-active promoter elements. *Nature* 1990;346:583-86.

103. McCormick A, Brady H, Theill LE, Karin M. Regulation of the pituitary-specific homeobox gene GHF1 by cell-autonomous and environmental cues. *Nature* 1990;345:829-32.

104. Angel P, Hatteri K, Smeal T, Karin M. The Jun proto-oncogene is positively autoregulated by its product, Jun/AP-1. *Cell* 1988;55:875-85.

105. König H, Ponta H. Rahmsdorf N, Büscher M, Schönthal A, Rahmsdorf HJ, Herrlich P. Autoregulation of fos: the dyad symmetry element as the major target of repression. *EMBO J* 1989;8:2559-66.

106. Schönthal A, Büscher M, Angel P, *et al.* The Fos and Jun/AP-1 proteins are involved in the downregulation of Fos transcription. *Oncogene* 1989;4:629-36.

Advances in Regulation of Cell Growth, Volume 2;
Cell Activation: Genetic Approaches, edited by
James J. Mond, John C. Cambier, and Arthur Weiss.
Raven Press, Ltd., New York © 1991.

5

Induction of Gene Expression in Response to Mitogens, Tumor Promoters and Neurotransmitters

Harvey R. Herschman

*Department of Biological Chemistry, and Laboratory of Biomedical and Environmental
Sciences, UCLA School of Medicine, Los Angeles, California 90024*

For a number of years my laboratory has studied the mitogen-induced biochemical, molecular and physiological events that occur when resting, G_0 cells are stimulated by mitogens to re-enter the cell cycle. We have used the murine embryo Swiss 3T3 cell line to study the causal events in the G_0 to G_1 transition, employing both genetic and molecular strategies to identify events involved in this mitogenic response. In our first approach we have isolated mitogen-nonproliferative mutants of 3T3 cells, to identify causal steps in the pathways of the mitogenic responses to epidermal growth factor and tetradecanoyl phorbol acetate. In our second strategy, we have cloned genes whose message levels are elevated as part of the early response to mitogenic stimulation. To our suprise, we have found that many of these same primary response genes are expressed in a number of differerent cell types, in response to a variety of extracellular ligands, as a part of a wide variety of biological responses. In this context, the mitogenic response becomes a special case of a common problem confronting biologists interested in oncology, developmental biology, endocrinology, neurobiology and other areas in which specific cellular responses to extracellular ligands are mediated by changes in gene expression. In this chapter I will (i) review our isolation and characterization of mitogen-nonproliferative 3T3 mutants, (ii) describe the cloning of mitogen-inducible primary response genes from 3T3 cells, (iii) discuss the nature of the gene products encoded by the primary response genes, (iv) demonstrate the diversity of cellular responses accompanied by induction of these genes, and (v) provide several alternative explanations for the manner in which a common set of primary response genes could mediate both cell-specific and ligand-specific biological responses.

SWISS 3T3 CELLS ARE A USEFUL MODEL
FOR THE STUDY OF MITOGENESIS

Some cells in the body, such as platelets, red blood cells, central nervous system neurons and fused muscle cells are post-mitotic; they cannot divide. In contrast, many cells in a multicellular organism are in a growth-arrested, G_o state. These arrested cells can respond to appropriate, cell-specific mitogens by leaving the growth-arrested state and reentering the cell cycle, to synthesize DNA and divide. For example, following partial hepatectomy, hepatocytes can divide and regenerate the liver. During pregnancy, under appropriate hormonal stimulation, mammary cells can undergo both hypertrophy and hyperplasia. Antigen-responsive cells undergo a clonal, antigen-driven expansion before maturation to antibody producing cells. The study of the mitogenic response *in vivo* is, however, difficult, for a number of reasons. Target tissues are heterogeneous with regard to cellular composition. Injected "mitogens" may in fact be effective *in vivo* not by acting directly on the cell-type under consideration, but by initiating an endocrine or paracrine interaction, with the target tissue as a distal responder. Cells in a target tissue *in vivo* may be asynchronously distributed around the cell cycle. Finally, large quantities of often difficult-to-obtain mitogens are required for *in vivo* studies. For these reasons, most of the cellular and molecular studies of the mitogenic response have been carried out with cell culture model systems.

Swiss 3T3 cells are a clonal, immortal murine embryo cell line (1). 3T3 cells can grow only when they are attached to an appropriate substrate. They grow to a density proportional to the amount of serum in the medium, then exit the cell cycle and arrest in the G_o state. The growth-arrested state is reversible; if additional serum is added to the medium density-arrested cells will, as a population, exit the resting state and enter the cell cycle. In addition to serum, a variety of well defined chemical agents can stimulate density-arrested 3T3 cells to divide. These mitogens include both polypeptide growth factors such as epidermal growth factor (EGF), fibroblast growth factor (FGF) and platelet derived growth factor (PDGF), and the macrocyclic tumor promoter tetradecanoyl phorbol acetate (TPA).

EGF-SPECIFIC NONPROLIFERATIVE MUTANTS OF 3T3 CELLS CAN BE ISOLATED

The mitogenic response is accompanied by an extraordinary array of biochemical, physiological and molecular alterations. Changes in ion and small molecule transport, phosphorylation of a variety of substrates, and gene expression occur rapidly after the binding of mitogens to quiescent cells. Which of these events are causal -- necessary for the initiation of DNA synthesis and subsequent cell division -- and which of them are correlative, but not essential for the ensuing cell division? Several years ago we initiated a genetic approach to this issue. If we could isolate mitogen-specific, nonproliferative mutants of 3T3 cells, we reasoned, the identification of the biochemical and molecular lesions in such mutants should reveal the causal, essential steps in the mitogenic pathway for the particular mitogen in question.

We developed a procedure, described in detail elsewhere (2), to effectively isolate mitogen-specific nonproliferative mutants of 3T3 cells. Using this selection procedure, we isolated EGF-nonproliferative mutants of Swiss 3T3 cells (3-5). Our EGF-nonproliferative mutants share a common phenotype; they are unable to bind EGF (3-5), and do not produce either an antigen cross-reactive with antisera to the EGF receptor (6) or mRNA for the EGF receptor (6). The basis for their EGF-nonproliferative phenotype appears to be a defect(s) in the synthesis of message for the EGF receptor, rendering the cells deficient in production of EGF receptor. When a cloned EGF receptor gene is introduced into the EGF-nonproliferative mutant 3T3-NR6, the resulting transfectant is capable of mounting a mitogenic response to EGF (7). The EGF-receptorless 3T3 mutants have been of great use in elucidating the roles of tyrosine kinase activation and ligand-induced receptor internalization in the EGF-induced mitogenic response (7,8).

TPA-SPECIFIC NONPROLIFERATIVE MUTANTS OF 3T3 CELLS CAN BE ISOLATED

A large body of work in recent years has lead to the conclusion that TPA exerts most, if not all, of its biological effects by acting as a pharmacologic analogue of diacylglycerol to activate protein kinase C (PKC). PKC is a serine, threonine protein kinase that requires calcium and phospholipid for activity. In the presence of these agents diacylglycerol, released from lipid precursors in response to ligand-stimulated phospholipase C, can activate the PKC molecule. TPA, a hydrophobic molecule, can penetrate the cell membrane and bypass phospholipase activation, directly binding to and activating PKC. Because TPA is effective at very low concentrations, and is much more slowly metabolized than is diacylglycerol, it is a potent pharmacologic agonist of PKC-mediated events.

We have also used our selection protocol (2) to isolate TPA-nonproliferative mutants of 3T3 cells (9). The TPA-nonproliferative mutant 3T3-TNR9, while unable to mount a mitogenic response to TPA, still contains PKC that can be activated by TPA (10). Moreover, 3T3-TNR9 cells treated with TPA still demonstrate phosphorylation of both the EGF receptor and the prominent 80 kilodalton fibroblast substrate that occurs in wild-type 3T3 cells (10), suggesting that the initial activation of PKC in these TPA-nonproliferative cells occurs in a fashion similar to that found in the parental cell line. However, the TPA-induced phosphorylation of the ribosomal S6 protein is substantially reduced in 3T3-TNR9 cells (11), suggesting that a/the molecular defect in these cells lies in a protein kinase cascade initiated by PKC. Recent studies have suggested that the 3T3-TNR9 cell may have a defect in the ability to regulate the expression of PKC as well (12).

Perhaps the simplest explanation for the inability of TPA-nonproliferative mutants to mount a mitogenic response to TPA would be a block in the PKC-mediated induction of gene expression that occurs in response to TPA administration. TPA, like other mitogens, is known to stimulate the accumulation of mRNA for ornithine decarboxylase (ODC), the rate limiting step in polyamine biosynthesis. When 3T3-TNR9 cells are exposed to TPA the induction of ODC mRNA accumulation is similar to that seen in response to a competent mitogen such as serum (Figure 1). Moreover, the induction of message for "major excreted protein" (MEP), another TPA-inducible protein (13), also occurs in TPA-treated 3T3-TNR9 cells (14). These data suggest that at least some transcriptional responses are not impaired in the TPA-nonproliferative mutant. While immunoprecipitation experiments have demonstrated that TPA can also induce the synthesis of MEP protein in 3T3-TNR9 cells (Herschman, unpublished), TPA cannot induce an elevation of ODC enzymatic activity in these cells (15), despite the appropriate elevation of ODC message (Figure 1). In contrast, ODC enzymatic activity is induced in 3T3-TNR9 cells in response to a competent mitogen such as serum (Ref. 15 and Figure 1B). Some TPA-specific step in either ODC protein synthesis or a post-translational modification of the ODC protein necessary for enzymatic activity must be defective in 3T3-TNR9 cells. It will be of great interest to examine the synthesis of ODC protein and its phosphorylation in the 3T3-TNR9 mutant. The physiological and biochemical properties of the EGF- and TPA-nonproliferative mutants have been reviewed in detail previously (16).

PRIMARY RESPONSE GENES INDUCED BY TPA CAN BE CLONED FROM 3T3 CELLS

It seemed likely to us that, when cells are induced by mitogens to divide, new programs of gene expression must be initiated. Stiles and his associates had demonstrated that treatment of quiescent 3T3 cells with platelet-derived

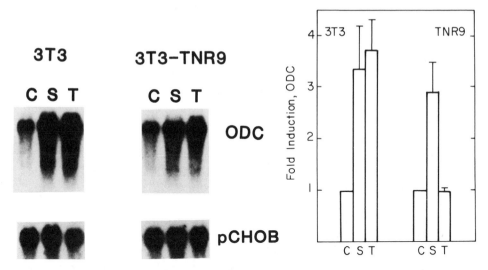

FIG. 1.   Ornithine decarboxylase mRNA accumulation, but not ODC enzymatic activity, can be induced by TPA in the TPA-nonproliferative Swiss 3T3 mutant 3T3-TNR9. Panel A. Quiescent 3T3 and 3T3-TNR9 cultures were treated with TPA (100 ng/ml) or elevated serum (ten percent). After three hours the cells were harvested, mRNA was prepared, and the level of ODC mRNA was measured by electrophoresis, transfer to nitrocellulose filters, and hybridization with a murine ODC cDNA probe. Each lane was loaded with 20 µg of total RNA. pCHOB is a probe for a constitutively expressed message, used to normalize for variations in RNA loading. Panel B. Quiescent 3T3 and 3T3-TNR9 cells were treated with TPA or elevated serum. After three hours cell extracts were prepared and assayed for ODC enzymatic activity (15). Error bars show S.E.M. of five experiments. C; control untreated cells. S; serum-treated cells. T; TPA-treated cells.

growth factor stimulated the accumulation of messages for several new genes (17). Transient expression of the *c-fos* proto-oncogene in response to serum and other mitogens has been known for several years (18). We decided, therefore, to clone cDNAs for genes whose expression is induced in response to a mitogenic stimulus of quiescent, nondividing 3T3 cells.

What mitogen should we use in these studies? We selected TPA as the mitogen of choice for several reasons. (i) TPA is thought to stimulate only a single second messenger pathway, the PKC pathway. If other mitogens such as EGF or FGF work through pathways mediated by other second messengers, perhaps we would find a subset of genes induced by TPA but not by other mitogens. (ii) TPA is the most powerful tumor promoter known, both in experimental animal systems and in cell culture models. Perhaps we would clone genes that are also involved in the process of tumor promotion, another active area of interest in our laboratory (Refs. 19-22; reviewed in Ref. 23). (iii) Finally, the 3T3-TNR9 mutant retains PKC activity (10), and at least some transcriptional responses to TPA (Figure 1 and Ref. 14). Perhaps, after cloning cDNAs for TPA-induced messages in 3T3 cells, we could identify genes that are differentially expressed in the wild-type and TPA-nonproliferative cell lines; genes which, by virutue of their failure to induce in the 3T3-TNR9 cell in response to TPA, might be identified as potential causal players in the TPA-induced mitogenic response.

At what time after TPA administration should we isolate message for subsequent cloning? Gene expression in response to extracellular ligands (growth factors, hormones, tumor promoters, neurotransmitters) has been divided into two physiologically distinct categories. "Primary response" genes are genes whose transcriptional responses can occur in the absence of any protein synthesis. The transcriptional machinery necessary for their expression is already present, but needs to be activated as a result of some post-translational modification resulting from ligand-receptor interaction. Transcription of "secondary response" genes following ligand-receptor interaction requires the synthesis of an intervening protein (24). The accumulation of message for primary response genes subsequent to ligand binding will, therefore, occur in the presence of an inhibitor of protein synthesis. In contrast, the appearance of message for a secondary response gene after ligand binding will not occur if an inhibitor of protein synthesis is present. We decided to concentrate initially on the earliest transcriptional events occurring in response to TPA-induced mitogenesis, and to clone primary response genes induced as part of the early events initiated by TPA in quiescent 3T3 cells.

Density-arrested 3T3 cells, grown in five percent fetal calf serum, were treated with TPA plus cycloheximide (CHX). After three hours, cells were harvested and polyA+ RNA was prepared from the cell extract. This RNA was used as template to prepare double stranded cDNA which, after appropriate linker ligation, was cloned into a lambda gt10 cloning vector. We then used a differential screening procedure to examine, from a library of about 150,000 phage, approximately 50,000 phage . Duplicate nitrocellulose lifts of plates were screened with two probes. The first single-stranded ^{32}P cDNA probe -- the "+ probe" -- was prepared from message isolated from 3T3 cells treated with both TPA and CHX. The second single-stranded ^{32}P cDNA probe -- the "- probe" -- was prepared from mRNA isolated from cells treated with cycloheximide alone. Thus, phage plaques that hybridized on the duplicate filter lifts to both cDNA probes represented cDNAs for messages expressed in both TPA treated and untreated cells. Plaques that hybridized preferentially to the + probe contained cDNAs for messages whose level was elevated in the cells treated with TPA, relative to untreated cells. Plaques were picked, placed in ordered arrays, and rescreened a second time with the + and - probes. Cross-hybridization studies and examination of message sizes on northern analysis reduced a population of 35 such clones to a collection of seven cDNAs that clearly distinguished seven distinct messages whose levels were increased following TPA treatment (25). We refer to these genes as TPA Induced Sequences, or TIS genes.

THE TIS GENES ARE INDUCED RAPIDLY AND TRANSIENTLY BY TPA, AND ARE SUPERINDUCED IN THE PRESENCE OF CYCLOHEXIMIDE

The levels of all the TIS gene messages (TIS1, 7, 8, 10, 11, 21 and 28), when characterized by northern analysis, are low in quiescent, density-arrested 3T3 cells. Following stimulation with TPA, however, a rapid accumulation of these messages occurs (Figure 2, Figure 3). The message levels for the TIS genes reach maximal values at various times for the different TIS genes, but in general peak between 30 and 90 minutes. The elevated message levels are transient, declining rapidly after the peak values are reached, returning to near baseline values by three to four hours. The TIS genes are induced rapidly and transiently in 3T3 cells in response to TPA.

If TPA induction is carried out in the presence of cycloheximide a

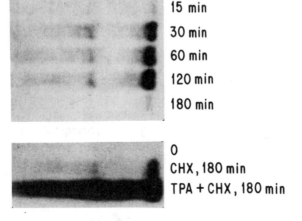

FIG. 2.   TIS8 mRNA accumulates rapidly and transiently in response to TPA, and is superinduced in the presence of cycloheximide. Left panel. Quiescent 3T3 cells were treated with TPA (50 ng/ml) for the times indicated. Cells were harvested, RNA was prepared, and TIS8 mRNA was analyzed by northern blotting. Right panel. 3T3 cells were incubated in the presence of either cycloheximide (10 µg/ml) alone, or TPA plus cycloheximide for three hours.

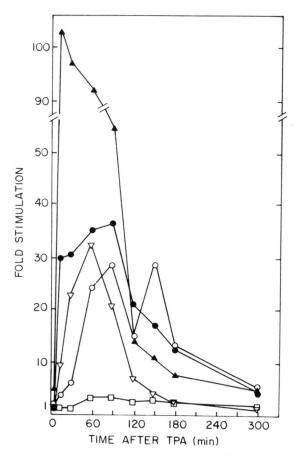

FIG. 3.   TIS gene mRNAs all accumulate rapidly and transiently in response to TPA. Quiescent cultures of Swiss 3T3 cells were treated for the times indicated with TPA (50 ng/ml) RNA was prepared and transferred to nitrocellulose by slot-blotting. Autoradiographs of slot blots were analyzed by densitometric scanning. The levels of TIS mRNA expression were normalized to the zero time controls. TIS1 (●), TIS7(□), TIS8 (▽), TIS10 (○), TIS11 (▲). Data are from Reference 25.

remarkable "superinduction" of all the TIS gene messages occurs. An example of this phenomenon is shown in Figure 2. At three hours, when the level of the TIS8 mRNA has returned to near basal levels, a large superinduction of the TIS8 message is observed in cells exposed concomitantly to both TPA and CHX. The superinduction phenomenon occurs for all the TIS genes (25).

ALL THE TIS GENES ARE INDUCED
IN THE TPA-NONPROLIFERATIVE MUTANTS

One reason why we chose TPA as the mitogen in our cloning experiments was the possiblility that our TPA-nonproliferative mutants might be defective in their ability to express one or more of the TIS genes after TPA exposure. Such a gene might, therefore, play a causal role in the mitogenic response to TPA. However, when 3T3-TNR2 or 3T3-TNR9 cells were treated with TPA their ability to express the TIS mRNAs, when compared to 3T3 cells, was unimpaired (26). When normalized to serum induction, there is no decrement in the TPA-induced expression of TIS genes in the TPA-nonproliferative mutants. An example of the ability of these TPA-nonproliferative 3T3 mutants to express the TIS8 gene is shown in Figure 4.

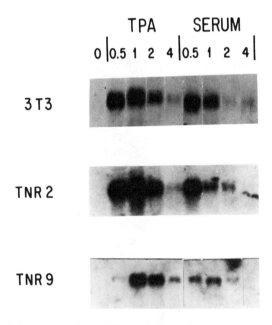

FIG. 4. The TIS8 gene is inducible by TPA and serum in both Swiss 3T3 cells and the TPA-nonproliferative mutants 3T3-TNR9 and 3T3-TNR2. Quiescent cultures were treated with TPA (50 ng/ml) or fetal calf serum (10 percent) for the times (in hours) shown. RNAs were isolated and analyzed for TIS8 mRNA by northern blot analysis. Each lane had 10 μg of total RNA. Data are from Reference 26.

THE TIS GENES ARE INDUCED BY MULTIPLE, INDEPENDENT PATHWAYS IN 3T3 CELLS

We used TPA as the mitogen of choice for the isolation of TIS genes in part because we thought that mitogenesis mediated by a PKC-activated pathway might result in the expression of some primary response genes not stimulated by growth factors such as EGF or FGF. However, TIS gene expression is also rapidly and transiently induced in 3T3 cells by both EGF and FGF (Figure 5).

It is possible that the EGF and FGF induction of TIS gene expression is, in fact, occuring through a PKC-mediated pathway. To examine this question more carefully, we eliminated the bulk of PKC from 3T3 cells by prolonged

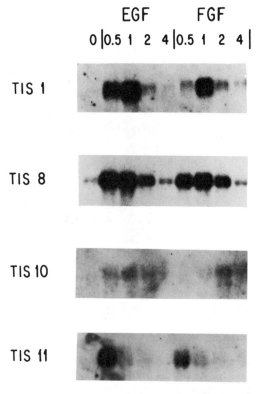

FIG. 5.   TIS genes are induced in 3T3 cells by EGF and FGF. Quiescent 3T3 cultures were treated with either EGF (20 ng/ml) or FGF (10 ng/ml) for the times (in hours) indicated. RNAs were isolated and analyzed by northern blot analysis. Each lane was loaded with 10 μg of total RNA. Data are from Reference 26.

NO
PRETREATMENT

TPA
PRETREATMENT

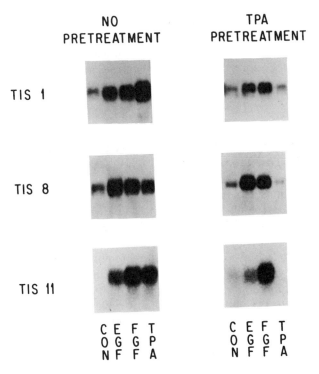

TIS 1

TIS 8

TIS 11

C E F T   C E F T
O G G P   O G G P
N F F A   N F F A

FIG. 6.   TIS genes can still be induced by EGF or FGF after protein kinase C has been down regulated. Quiescent 3T3 cultures were either treated for 24 hours with TPA (50 ng/ml) or were left untreated. Cells were then exposed to EGF (100 ng/ml), FGF (10 ng/ml), or additional TPA (50 ng/ml), or left without further treatment (CON). RNAs were isolated one hour later, and analyzed by northern blot analysis. Each lane was loaded with 10 µg of total RNA. Data are from Reference 26.

incubation with TPA, then challenged the cells with either TPA, EGF or FGF and measured the ability of the cells to express the TIS genes (Figure 6). A second challenge with TPA was, as expected, unable to ellicit TIS gene expression, since the PKC level had (presumably) been reduced by TPA-mediated down regulation. If EGF and/or FGF were exerting their primary effects on TIS gene message levels via a PKC-mediated pathway, we would expect no substantial expression of these genes in cells in which PKC had been eliminated. Conversely, if EGF or FGF are working through a distinct, nonPKC-mediated pathway, then down regulation of PKC should have little or no effect on the ability of EGF or FGF to induce TIS gene expression. Prior pretreatment with TPA had only a relatively small effect on the induction of several TIS genes by EGF or FGF, suggesting that both PKC-dependent and PKC-independent pathways for TIS gene induction by extracellular ligands exist in 3T3 cells (Figure 6). These data, therefore, suggest that the TIS genes can be induced by multiple, independent pathways in 3T3 cells. The TIS genes can also be induced in 3T3 cells by serum, PDGF, and bombesin, suggesting that the various second messenger pathways leading to mitogenic responses in these cells converge on this common set of genes.

THE VARIOUS TIS GENES EACH HAVE A UNIQUE DEVELOPMENTAL TISSUE DISTRIBUTION PATTERN

While the various TIS genes are expressed with somewhat distinct kinetics in 3T3 cells, they are all rapidly and transiently induced in response to a common set of agents. If the co-ordinate induction of these genes in 3T3 cells in response to a variety of mitogens is indicative of common induction mechanisms, then study of the regulatory regions governing expression of several of these genes might be redundant. To initially pursue this issue, we examined the expression of the TIS genes in the organs of neonatal and adult mice (27). Each TIS gene has a unique developmental tissue distribution pattern. This early set of experiments suggested that each TIS gene must have unique regulatory elements governing its expression, mRNA processing and/or mRNA stability; regulatory elements that contribute to its unique expression pattern in the intact animal.

THE TIS GENES ARE ALSO RAPIDLY AND TRANSIENTLY INDUCED BY MITOGENS FOR NEOCORTICAL RAT BRAIN ASTROCYTES

While the murine 3T3 cell system is an excellent model to study the mitogenic response, it frequently suffers from the criticism that it is not a "normal" cell population. Cultured astrocytes from neonatal rats are relatively easy to prepare, well characterized, quite consistent from one preparation to the next, and able to mount either a mitogenic response to polypeptide growth factors such as EGF and FGF or to "stellate", a differentiation response, when exposed to agents that increase cAMP levels (for a review of the characteristics of these cells see Ref. 28). When secondary cultures of rat neocortical astrocytes were challenged with TPA, EGF or FGF, all the TIS genes (TIS1, 7, 8, 10, 11, 21 and 28) were rapidly and transiently induced, with time courses similar to those observed previously in 3T3 cells (Figure 7 and Ref. 29). The general features of TIS gene expression accompanying the mitogenic responses in 3T3 cells also occur with this well characterized population of normal cells in culture.

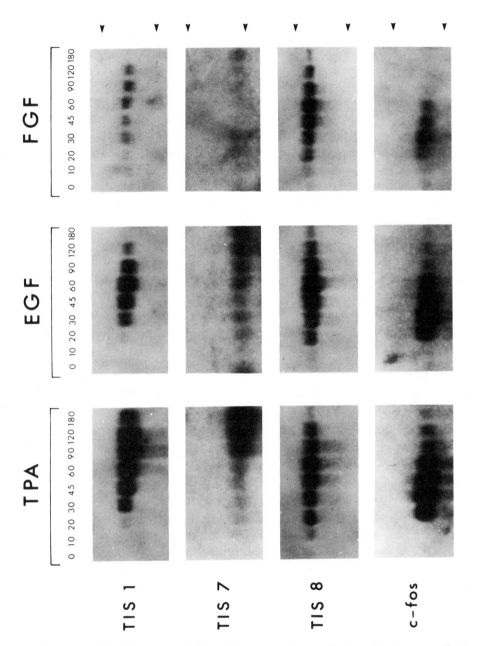

FIG. 7. TIS mRNAs are induced in secondary cultures of rat neocortical astrocytes in response to the mitogens TPA, EGF or FGF. Secondary cultures of rat neocortical astrocytes, were treated with TPA (100 ng/ml), EGF (25 ng/ml) or FGF (100 ng/ml) for the times (in minutes) indicated. RNA was isolated and analyzed by northern blot analysis. Each lane was loaded with 8 μg of total RNA. Data are from Reference 29.

INDUCTION OF THE TIS GENES IN ASTROCYTES BY COMBINATIONS OF MITOGENS ALSO DEMONSTRATES THAT TIS GENES ARE INDUCED BY MULTIPLE, INDEPENDENT PATHWAYS

Recall that EGF and FGF can induce TIS gene mRNA accumulation in 3T3 cells previously treated with TPA, to down regulate their PKC content (26). These experiments lead to the conclusion that the TIS genes can be induced by multiple independent pathways. If this is indeed the case, it is possible that these pathways might be additive or even synergistic for TIS gene induction. To examine this question, we exposed secondary cultures of rat astrocytes to combinations of mitogens, and measured the accumulation of several TIS gene messages. When EGF was presented to astrocyte cultures along with a maximally inducing level of FGF, there was no increase in TIS gene message accumulation above that induced by the EGF alone. In contrast, when TPA was presented to astrocyte cultures along with a maximally inducing level of EGF, the induction of the various TIS genes examined was at least as great as the sums of the individual TPA and EGF inductions observed in parallel cultures (30). These data suggest that the EGF and FGF pathways saturate some common, limiting step in their induction pathways. In contrast, the TPA and polypeptide mediated induction pathways are able to augment one another, suggesting that they utilize at least partially independent and additive mechanisms of induction. Our data suggest that, in both 3T3 cells and in secondary rat astrocyte cultures, there exist multiple, independent, additive pathways for TIS gene induction.

THE TIS GENES ENCODE TRANSCRIPTION FACTORS, PRESUMPTIVE CYTOKINES, AND PROTEINS OF AS YET UNKNOWN FUNCTION

One of our major objectives, after isolating the original cDNA clones for the TIS genes, was to identify the protein products they encode. Mitogenic responses to serum and several growth factors have been known for several years to induce a rapid and transient transcriptional response of the *c-fos* protooncogene. Moreover, this response is superinducible by cycloheximide. We therefore anticipated that one of the TIS cDNAs might be a partial clone of *c-fos*. When the initial seven TIS cDNAs were screened with a *v-fos* probe we found this to be the case; the TIS28 cDNA was a partial *c-fos* clone (25).

The laboratories of Vikas Sukhatme, Rodrigio Bravo and Daniel Nathans have isolated primary response genes following serum stimulation of 3T3 cells, in the presence of cycloheximide. Sukhatme et al. (31,32) have described one such gene, which they named *egr-1*, for "early growth response" gene 1. When we sequenced a partial cDNA clone of TIS8 (33) we found it to be identical to the *egr-1* sequence. The *egr-1*/TIS8 gene codes for a zinc-finger containing protein which has recently been shown to be a transcription factor that recognizes a sequence-specific promoter element (34,35).

When we sequenced the TIS1 cDNA (Varnum, unpublished), we found that it had extensive homology to the estrogen receptor and the other members of the ligand-dependent transcription factor family that includes receptors for glucocorticoids, retinoinc acid, vitamin D, thyroid hormone, and a variety of other ligands. At about this time Hazel et al. (36) reported the sequence of the serum-induced gene they called *nur77*. Our sequence for TIS1 and the reported sequence of *nur77* are identical. The TIS1/*nur77* gene codes for a mitogen-inducible member of this superfamily; a presumptive ligand-dependent transcription factor.

Primary response genes have also been cloned from other biological systems. PC12 cells are a rat pheochromocytoma cell line that responds to NGF by ceasing cell division and differentiating into cells that resemble sympathetic neurons morphologically, biochemically and electrophysiologically; (for a review of the effects of NGF on PC12 cells, see Ref. 37). The rat cDNA homologues of TIS8/*egr-1* (38) and TIS1/*nur77* (39) were cloned from PC12 cells exposed to nerve growth factor (NGF) in the presence of cycloheximide.

Unlike the sequences for TIS1 and TIS8, the sequence of the TIS7 cDNA does not predict an open reading frame that suggests a transcription factor (33). The rat homologue of the TIS7 cDNA was cloned from NGF treated PC12 cells by Tirone and Shooter (40). There are only thirteen amino acid differences, nearly all of which are conservative substitutions, in the predicted 449 amino acid proteins from rat and mouse. Tirone and Shooter (40) suggest that *PC4*, the rat homologue of TIS7, shows some sequence similarity to rat interferon gamma. While it is attractive to speculate that the TIS7 protein product might be a cytokine-like molecule that can participate in autocrine or paracrine circuits in response to growth factor stimulation of cells, such conclusions must await expression of the protein product of this gene and analysis of its biological activity. It should be pointed out, however, that the sequence of several other mitogen inducible genes suggest that they might be cytokines. For example, the serum-inducible primary response gene *KC/N51* is the murine homologue of the human cytokine Melanoma Growth-Stimulating Activity, or MGSA (reviewed in Ref. 41).

The sequence of the TIS11 cDNA predicts an open reading frame of 183 amino acids (42). Extensive sequence homology searching did not find any previously reported sequence with substantial similarity to the TIS11 open reading frame. However, there is an interesting repeat present in the predicted amino acid sequence of the TIS11 protein:

...SSR**YKTELC**RTYSESGRCRYGAKCQFAHGLGELRQANRHPK**YKTELC**...

Many of the mitogen-induced primary response genes are members of multiple-gene families (41). We suspected that the repeat present in the TIS11 protein might represent a protein motif found in other members of a family of proteins with related functions. By using direct screening of cDNA libraries with degenerate oligonucleotides we have identified a second cDNA, TIS11b, which has the YKTELC sequence repeated in the same position. Moreover, the TIS11 and TIS11b open reading frames have 47 identical amino acids in this region over a 67 amino acid sequence (Varnum et al., submitted for publication). The rat homologue, cMG1, of TIS11b has recently been reported (43). Using polymerase chain reaction amplification from genomic DNA with degenerate oligonucleotides spanning the conserved region, we have identified a third member of the TIS11 family, TIS11d, which has a similar conserved sequence. The TIS11 cDNA appears to be the first member isolated of a family of genes that encode proteins with this highly conserved sequence of 67 amino acids. We are currently preparing antibodies to these proteins, to determine where they are found in the cell. We suggest that their highly conserved sequence motif is likely to be indicative of related and important functions in cells.

The sequence of the TIS21 cDNA is now completed. Like the sequence for TIS11, no strong sequence similarities with any known proteins were identified by comparing the predicted TIS21 open reading frame with protein sequences currently available in the data banks. Antisera to fusion proteins are currently being prepared.

The TIS10 gene codes for a message of approximately four kilobases. To date, a contiguous 2.9 kilobase sequence has been compiled for this cDNA. No extensive open reading frame has been identified in this sequence. We are currently extending our sequence in the five prime direction.

The sequence data from our laboratory and the laboratories of other groups that have cloned primary response genes suggest that the TIS genes encode transcription factors (TIS1, TIS8), potential cytokines (TIS7), and proteins whose functions are not as yet known (TIS11, TIS21). A summary of the presumptive functions of the TIS genes, the names given to them as either serum or NGF induced primary response genes, and the references to the identification of their open reading frames by sequence analysis of cDNAs are given in Table 1. Current estimates of the number of primary response genes induced by mitogens are in the range of 100-200 (44).

THE TIS GENES CAN BE INDUCED IN RAT ASTROCYTES BY NEUROTRANSMITTERS

Our laboratory originally cloned the TIS cDNAs as part of a program to identify causal events in the TPA mitogenic pathway. However, these genes are also induced in response to a variety of other mitogens, both in 3T3 cells

TABLE 1. *The TIS gene products*

Name	Potential function	Cloned from serum-treated 3T3 cells	Cloned from NGF-treated PC12 cells
TIS1	Ligand-dependent transcription factor, member of the glucocorticoid receptor superfamily	*nur77* (36), *N10* (61)	*NGFIB* (39)
TIS7	Possible cytokine	---	*PC4* (40)
TIS8	Zinc-finger containing transcription factor	*egr-1* (32), *zif/268* (62), *krox24* (63)	*NGFIA* (38)
TIS10	Not known; sequence not completed	---	---
TIS11	Not known; contains internal repeat	---	---
TIS11b	Not known; contains internal repeat		
TIS11d	Not known; contains internal repeat		
TIS21	Not known; sequence completed	---	---
TIS28	Partial *c-fos* clone; transcriptional modulator	---	---

and in cultured astrocytes. Moreover, many of these same genes can be induced in a differentiation response to NGF and -- as described above -- have been cloned as NGF-inducible genes (Table 1). It appears, therefore, that TIS gene expression may occur in response to ligands other than mitogens.

The secondary rat astrocyte populations described above are able to mount distinct biological responses to a wide variety of ligands. They are known to contain both alpha and beta adrenergic receptors, as well as muscarinic cholinergic receptors, in addition to their growth factor receptors. To determine whether the TIS gene response repertoire extended to agents such as neurotransmitters, in addition to growth factors and differentiation agents, we examined the ability of adrenergic and cholinergic agonists to induce TIS gene expression in cultured rat neocortical astrocytes (45).

The cholinergic agonist carbachol elevates diacylglycerol levels in cultured astrocytes. Since TPA is a potent inducer of the TIS genes, it seemed likely that carbachol might also induce the expression of the TIS genes. Message for the TIS genes accumulates rapidly and transiently when secondary astrocyte cultures are treated with carbachol (Figure 8). Carbachol induction of the TIS genes was enhanced by the presence of lithium, suggesting that this response is mediated by an inositol phosphate pathway, since lithium inhibits inositol-1-phosphatase (45). Carbachol induction of the TIS genes in these secondary rat astrocyte cultures is blocked by co-administration of atropine sulphate (45), consistent with the proposed muscarinic cholinergic pathway of TIS gene activation.

Adrenergic agonists are able to induce changes in astrocyte metabolism and morphology. Norepinephrine is a potent inducer of TIS gene expression (Figure 9). Moreover, both the pure beta-adrenergic agonist isoproterenol and the alpha-adrenergic agonist phenylephrine can induce TIS gene message accumulation, suggesting that both adrenergic pathways are operative in these cells. The norepinephrine induction of the TIS1 gene could be substantially reduced by either the beta blocker propranolol or the alpha-1 antagonist prazosin (Figure 10). The combination of these two antagonists blocked norepinephrine induction of TIS1 expression by greater than 90 percent, suggesting that adrenergic stimulation of this transcriptional response is mediated almost entirely by beta and alpha-1 receptor mediated pathways. Consistent with this hypothesis, the alpha-2 antagonist yohimbine had no significant effect on norepinephrine-induced TIS1 message accumulation.

TIS gene induction in astrocytes can be observed in response to protein kinase C mediated pathways, cAMP-mediated pathways such as the adrenergic induction responses discussed above, and pathways thought not to involve either of these second messengers; e.g., EGF or FGF mediated TIS gene induction. Moreover, the induction of the TIS genes by many agents can be enhanced in astrocytes by peripheral benzodiazepines (30), suggesting that additional hormonal modulations of TIS gene expression occur in normal astrocyte physiology and metabolism. Astrocytes in brain are known to demonstrate substantial regional heterogeneity in receptor distribution and inositol phosphate metabolism. The wide variety of ligands, second messenger systems, and potential astrocyte subtypes that participate in TIS gene expression suggest that these genes may play major roles in the complex interactions among glial cells, as well as in the interactions between glial cells and neurons.

THERE ARE SEVERAL POTENTIAL EXPLANATIONS FOR THE MANNER IN WHICH CELL-SPECIFIC AND LIGAND-SPECIFIC BIOLOGICAL RESPONSES TO EXTRACELLULAR LIGANDS OCCUR

Mitogens such as serum, EGF, FGF, PDGF and TPA stimulate resting cells to enter the cell cycle. In contrast, NGF stimulates PC12 cells to exit the cell cycle and differentiate. Both of these responses to extracellular ligands induce

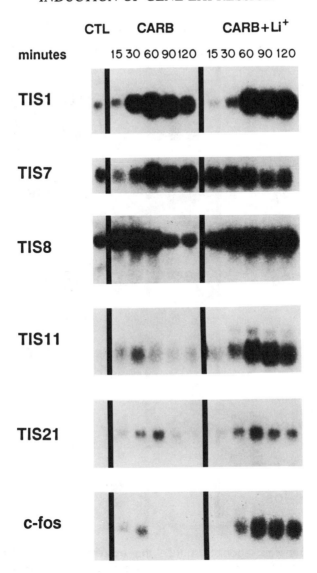

FIG. 8.   Carbachol induces expression of the TIS genes in secondary rat neocortical astrocyte cultures. Cultures were exposed to carbachol (CARB, l00 μM) in the presence or absence of lithium chloride (Li$^+$, 5mM). When present, lithium was added fifteen minutes before carbachol. "CTL" indicates control, untreated cells. RNA was isolated at the times (in minutes) shown, and analyzed for TIS mRNA levels by northern blotting. Each lane was loaded with 6 μg of total RNA. Data are from Reference 45.

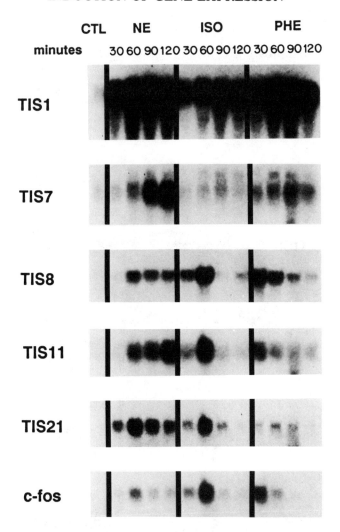

FIG. 9.   Adrenergic agonists induce the expression of the TIS genes in secondary rat neocortical astrocyte cultures. Cultures were exposed to norepinephrine (NE, 6.25 μM), isoproterenol (ISO, 6.25 μM) or phenylephrine (PHE, 6.25 μM). RNA was isolated at the times (in minutes) shown, and analyzed for TIS mRNA levels by northern blotting. Each lane was loaded with 6 μg of total RNA. Data are from Reference 45.

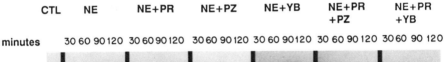

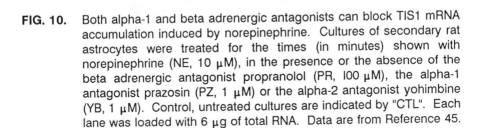

TIS1

FIG. 10. Both alpha-1 and beta adrenergic antagonists can block TIS1 mRNA accumulation induced by norepinephrine. Cultures of secondary rat astrocytes were treated for the times (in minutes) shown with norepinephrine (NE, 10 µM), in the presence or the absence of the beta adrenergic antagonist propranolol (PR, 100 µM), the alpha-1 antagonist prazosin (PZ, 1 µM) or the alpha-2 antagonist yohimbine (YB, 1 µM). Control, untreated cultures are indicated by "CTL". Each lane was loaded with 6 µg of total RNA. Data are from Reference 45.

the rapid and transient expression of the *c-fos* gene. Adrenergic and cholinergic stimulation of astrocytes also induced a rapid and transient *c-fos* response (45). Moreover, a wide variety of additional ligands, interacting with receptors on many different cell types to stimulate an enormous range of biological responses, all induce the expression of the *c-fos* gene. These include -- but are not restricted to -- vitamin D stimulated macrophage differentiation, thyrotropin activation of thyroid cells, depolarization of neurons, and granulocyte-macrophage colony stimulating factor driven activation of neutrophils (46).

Those genes cloned as primary response cDNAs following NGF induction of differentiation of PC12 cells -- *NGFIA*, *NGFIB* and *PC4* -- have also been identified as serum and/or TPA inducible genes in 3T3 cells (Table 1). The *c-fos* and *c-jun* genes are also rapidly and transiently induced both in mitogenically responding 3T3 cells and in PC12 cells induced to differentiate in response to NGF (reviewed in Ref. 41). If distinct extracellular ligands, binding to a variety of receptors on many different cell types, induce the same transcriptional responses, then how do cells achieve either (i) ligand-specific responses in a common cell, or (ii) cell-type specific responses to a common ligand? I have previously suggested (46) several possible mechanisms by which such specificity may be obtained.

Do ligand-specific primary response genes exist? Since serum is a complex mixture of mitogens, it is unlikely that serum-specific primary response genes would be identified. However, for TPA-induced mitogenesis a specific second messenger pathway is activated. Nevertheless, we were unable to identify any TPA-specific primary response genes accompanying TPA mitogenic stimulation of 3T3 cells (26). Moreover, we examined the induction of the TIS genes in PC12 cells following administration of NGF, EGF and TPA and

INDUCTION OF GENE EXPRESSION

CONT   TPA   NGF

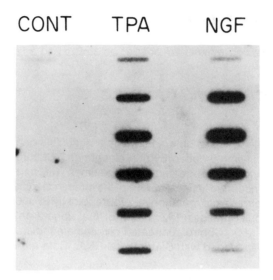

FIG. 11. TIS1 mRNA accumulation can be induced in PC12 pheochromocytoma cells in response to both TPA and NGF. RNA was isolated from PC12 cells at various times after TPA (50 ng/ml) or NGF (50 ng/ml) addition. The time points shown are 15, 30, 60, 90 and 180 minutes after addition of TPA or NGF. RNA was analyzed by slot-blotting. Each slot contains 10 μg of total RNA. Data are from Reference 47.

found that these agents could all induce TIS1/*NGFIB* (Figure 11), TIS7/*PC4*, TIS8/*NGFIA*, TIS11 and TIS21 (47). Thus, while it is certainly the case that these genes are induced by NGF, they are also induced in PC12 cells by other ligands -- ligands which do not induce many of the biological responses observed with NGF. To date, no ligand-specific primary response gene has been described.

Qualitatively different expression of primary response genes may be induced by distinct ligands. In our studies of mitogenic stimulation of TIS genes in astrocytes we found that the various TIS genes can be differentially induced by alternative growth factors (30). Thus the TIS1 gene is induced much more effectively by TPA than by EGF or FGF, while the TIS8 gene is induced with approximately equal intensity by all three of these ligands (Figure 12). In collaborative experiments with the laboratory of Ralph Bradshaw, we have observed a similar result in PC12 cells; carbachol is a potent inducer of both TIS1 and TIS8 message accumulation, while FGF is a potent inducer of TIS8 but not TIS1 gene expression (Altin et al., submitted). Bartel et al. (48) demonstrated that depolarization and growth factor stimulation have distinctly different abilities to induce the TIS1/*NGFIB* and TIS8/*NGFIA* genes in PC12 cells. These and similar data suggest that, within a single cell type, distinct

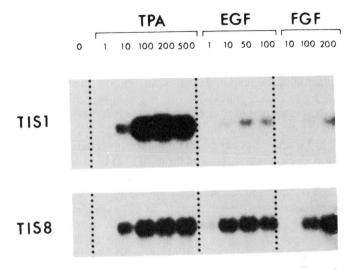

FIG. 12.   The TIS1 and TIS8 genes are induced to different extents in secondary rat astrocyte cultures in response to different ligands. Secondary rat neocortical astrocyte cultures were treated for one hour with increasing concentrations of TPA, EGF or FGF. The numbers represent concentrations of ligands, in ng/ml. RNA was analyzed by northern blotting. Each lane was loaded with 8 μg of total RNA. Data are from Reference 30.

extracellular ligands may stimulate differing quantitative combinations of second messenger signals. The promoter regions of the TIS genes may be differentially stimulated by alternative second messenger pathways. If we consider just the possibilities raised by the differential induction of the TIS1 and TIS8 transcription factors in response to alternative ligands in a common cell (30,40; Altin et al., submitted) the possibilities for ligand-specific responses, without the existence of ligand-specific primary response genes, becomes apparent. It is not difficult to devise a number of potential scenarios by which the presence of differing ratios of these two transcription factors could result in distinct "downstream" gene activation events and consequent alternative biological responses.

Post-translation modification of the structure and function of primary response genes. The *c-fos*/TIS28 gene product FOS forms heterodimers with the JUN gene family proteins. These heterodimers are the components of the AP-1 transcription factor family (41). Post-translational modification of these, and other, primary response gene products should surely modify their functional properties. Morgan and Curran (49,50) showed that the *c-fos* gene could be induced to approximately equal levels by growth factors and by depolarization. However, the FOS protein produced following NGF induction was much more extensively phosphorylated than was the FOS protein that accumulated in response to depolarization (49,50). This differential post-translational modification of primary response gene products following induction by distinct ligands might well be a major mechanism by which ligand-specific biological responses can occur, without the induction of ligand-specific primary response genes.

Cell-specific restriction of the expression of primary response genes. When we examined the expression of the TIS genes in PC12 cells (47), we found that the TIS1, 7, 8, 11, and 21 genes could be induced by NGF, EGF, TPA and K^+ depolarization. However, accumulation of TIS10 mRNA could not be detected in response to any ligand in PC12 cells. These data suggested that the expression of the TIS10 gene might be "restricted" or "extinguished" in a cell-specific manner during development. To eliminate the possibility that our inability to detect TIS10 message in rat PC12 cells is due to a lack of cross hybridization capability with this partial murine cDNA clone, we have performed parallel experiments with Rat1 cells. PC12 cells induced with either TPA or EGF, in the presence of cycloheximide to superinduce the TIS genes and thus provide maximum sensitivity, do not express any detectable TIS10 message (Figure 13). The probe used in the experiment is able to

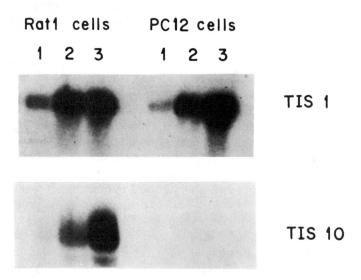

FIG. 13. The TIS10 gene cannot be induced in PC12 cells, even under superinducing conditions. Rat1 or PC12 cells were exposed to cycloheximide (10 µg/ml; lanes 1), cycloheximide plus TPA (50 ng/ml; lanes 2), or cycloheximide plus EGF (5 ng/ml; lanes 3) for three hours. RNA was analyzed by northern blotting. Each lane was loaded with 10 µg of RNA. Data are from Reference 52.

detect rat TIS10 message, since it hybridizes to RNA prepared from TPA- or EGF-induced Rat1 cells. The RNA preparation from the PC12 cells is intact and the two ligands have generated appropriate intracellular signals to activate TIS genes, since the TIS1/*NGFIB* gene is induced robustly in PC12 cells in this experiment. We conclude that the TIS10 gene cannot be expressed in PC12 cells, in response to any inducer; the TIS10 gene is silent in PC12 cells.

The inability of PC12 cells to express the TIS10 gene suggests another, developmentally regulated, basis for cellular specificity of responses to extracellular ligands. In addition to the differential induction of TIS genes by alternative ligands, it appears that the qualitative elimination of expression of subsets of primary response genes may be an additional mechanism by which cell-type specific responses to alternative ligands may occur. To extend these studies further we turned to a cell system distinct from both the 3T3 "fibroblast" and PC12 "neuronal" systems we had analyzed previously. The 32Dclone3 cell is a murine myeloid cell line that requires either IL-3 or granulocyte macrophage colony stimulating factor (GM-CSF) to grow (51). When 32Dclone3 cells were brought to quiescence by withdrawal of IL-3/GM-CSF, then stimulated with either TPA or GM-CSF, the TIS7, 8, 10, and 11 message levels increased (52). However, no TIS1 message could be detected in response to either GM-CSF or TPA in 32Dclone3 cells (Figure 14). To extend this assay to maximum sensitivity, 32Dclone3 cells were treated with either TPA or GM-CSF in the presence of cycloheximide, to superinduce the expression of the TIS genes. Even under superinducing conditions, no TIS1 message could be detected in these cells (Figure 15). The TIS1 probe used in this latter experiment could detect TIS1 message in RNA, present on the same filter, prepared from 3T3 cells treated with TPA. The RNA preparations from the 32Dclone3 cells were intact, and the second messenger pathways used by these agents had been activated, since substantial induction of the TIS11 gene could be seen in the same experiment. It appears that, in the myeloid-derived 32Dclone3 cells, the expression of the TIS1 gene, which encodes a presumptive transcription factor (Table 1), is restricted. The pattern of restriction of TIS gene expression for 3T3, PC12 and 32Dclone3 cells is shown in Table 2.

We then extended the analysis of restricted expression of primary response genes by examining the induction of a number of different genes in a variety of cell lines. We used probes for *c-jun* and *junB*, in addition to the TIS gene probes. Cells were treated with TPA, to bypass the issues of distinct receptors on the various cell types and differential coupling of receptor-ligand systems to various second messenger systems in different cell types. Activation of only the protein kinase C system should occur in the TPA-treated cell lines. To maximize our ability to detect signals, cycloheximide was included in the inductions of all these cell lines with TPA. Thus to escape detection by northern analysis the primary response gene in question must not be

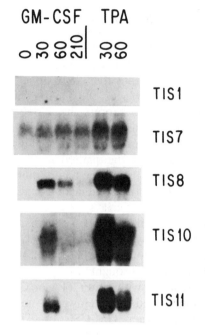

FIG. 14. The TIS1 gene cannot be induced in 32Dclone3 cells. 32Dclone3 cells were deprived of GM-CSF overnight. TPA (50 ng/ml) or GM-CSF (1 nM) was added for the times (in minutes) shown. RNA was analyzed by northern blotting. Each lane was loaded with 10 µg of RNA. Data are from Reference 52.

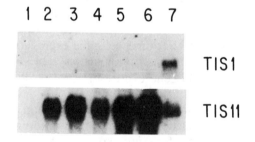

FIG. 15. The TIS1 gene cannot be induced in 32Dclone3 cells, even under superinducing conditions. Lanes 1 contain RNA from control, untreated 32Dclone3 cultures deprived of GM-CSF overnight. Lanes 2, 3, 4, 5, and 6 contain RNAs from similar cultures that had been treated with GM-CSF (1 nM) for 30 minutes, TPA (50 ng/ml) for 30 minutes, cycloheximide (10 µg/ml) for three hours, GM-CSF plus cycloheximide for three hours, or TPA plus cycloheximide for three hours. Lane 7 contains RNA from 3T3 cells treated with TPA. Each lane was loaded with 10 µg of total RNA. Data are from Reference 52.

Table 2. *Inducibility of TIS genes*

Gene	Inducibility in 3T3 cells			Inducibility in PC12 cells			Inducibility in 32Dc13 cells	
	TPA	EGF	FGF	TPA	EGF	NGF	TPA	GM-CSF
TIS1	+	+	+	+	+	+	-	-
TIS7	+	+	+	+	+	+	+	+
TIS8	+	+	+	+	+	+	+	+
TIS10	+	+	+	-	-	-	+	+
TIS11	+	+	+	+	+	+	+	+

susceptible to cycloheximide-superinduced TPA induction. Representative data are shown in Figure 16. From these data we can conclude that distinct patterns of induction of primary response genes can be observed within cell lines. In this limited survey we see that the *junB* gene is the most ubiquitously expressed, while TIS1 appears to show restriction in other cells, in addition to 32Dclone3 cells. Moreover we find that some cells can be restricted in their ability to express more than one of this small sample of primary response genes analyzed. The four genes whose induction is characterized in this study are all transcription factors or modulators of transcription. It is clear that the restricted combinatorial utilization of groups of interacting transcription factors can develop an enormous repertoire of ligand-specific responses, while still providing a mechanism for establishing cell-type specific constraints on the ability of cells to respond to a common ligand.

The experiments described above on cell-type specific restriction were all performed with cell lines. Recall that, when the induction of TIS genes was examined in secondary rat astrocyte cultures, all the TIS genes cloned from 3T3 cells could be expressed in response to TPA, EGF, FGF and adrenergic and muscarinic agonists. It is possible that the restricted expression of the TIS genes observed in the cultured cell lines (Table 2 and Figure 16) is due to alterations that have occurred as these cells were either established in culture or carried as long-term cultures. To address this question we analyzed the expression of the TIS genes in purified human neutrophils. These cells are a post-mitotic cell population of myeloid lineage that can respond to GM-CSF in a variety of ways (53-55). Neutrophils purified from human blood, when treated with either TPA or GM-CSF, expressed both the TIS8 and TIS11 genes. However, like the murine myeloid cell line 32Dclone3, human neutrophils exposed to either TPA or GM-CSF could not express the TIS1 gene (52). From these data we conclude that cell-type specific restriction of the ability to express primary response genes is likely to be a developmentally regulated process by which cell-type specificity of biological responses to extracellular stimuli is determined.

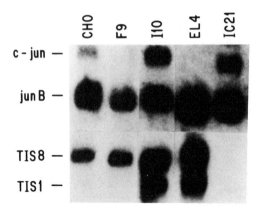

FIG. 16. Restriction of primary response gene induction is a common phenomenon in cultured cell lines. Cultures of the cell lines indicated in the figure were treated with TPA (100 ng/ml) and cycloheximide (10 µg/ml) for three hours. RNA was isolated and analyzed by northern blotting. Each lane contains 10 µg of total RNA. Individual filters containing the panel of RNAs were hybridized to the various cDNA probes. The illustration is a composite photo; the relative positions of the autoradiographic signals is not a reflection of the size of the messages for the various primary response gene mRNAs. CHO; chinese hamster ovary cells. F9; murine embryo carcinoma cells. I10; murine testicular tumor cells. EL4; murine lymophocytic cells. IC21; murine macrophage tumor cells.

CONCLUSIONS AND FUTURE DIRECTIONS

Activation of pre-existing, latent transcription factors such as CREB and NFκB must be responsible for the ligand-dependent transcription of the primary response genes. Cis-acting response elements for these and other constitutively expressed transcription factors have been identified in the 5' promoter regions of several of the TIS genes (41). It is clear (i) that expression of the TIS genes and the other primary response genes can be induced by multiple second messenger pathways and (ii) that different TIS genes can be induced to differing degrees by alternative ligands in the same cell. We, and others, are studying the basis for the quantitative differences in expression of primary response genes following treatment with alternative ligands, by isolating the promoter elements of the TIS genes and characterizing the relative inducibility of fusion genes constructed with TIS gene promoters and reporter genes such as luciferase and chloramphenicol acetyl transferase. Quantitative studies of the expression of fusion gene products should lead to an understanding of the potential for expression of these various promoter/enhancer regions.

When we began our studies, we anticipated that we would find both cell-type specific and ligand-specific primary response genes. To our surprise, we found that all the primary response genes induced following TPA treatment in 3T3 cells were also induced in response to other mitogens. Moreover, each gene we identified as a primary response gene for mitogen induction in 3T3 cells could also be induced in other cell types, in response to alternative ligands. A number of the same genes we cloned as TPA-inducible sequences were also cloned, by other workers, as NGF-inducible PC12 genes. We conclude that it is unlikely that truly ligand or cell-type specific primary response genes exist. Instead, it appears that distinct combinatorial patterns of utilization of a pool of primary response genes, and alternative modifications of their protein products, are likely to account both for ligand-specific responses to alternative ligands in a common target cell and for distinct, cell-type specific responses to a common ligand. One of the major challenges for the future lies in determining how unique combinations of these gene products result in specific biological responses. Much of our future work will center on these two issues of ligand and cell-type specificity.

We have only begun to identify the members of the pool of primary response genes inducible by extracellular ligands. Each laboratory that has identified primary response genes, whether following TPA, serum, lectin or NGF treatment, has reported a large number of distinct cDNAs remaining to be characterized. The size of the ligand inducible primary response gene pool is estimated at 100 to 200 genes (44). Given these conclusions, it is surprising that several of the genes, such as TIS1 and TIS8, have been cloned a number of times, in several different laboratories. These may represent the more "promiscuous" of the primary response genes; genes whose products are required for more general aspects of cellular responses to extracellular signals. It is possible that the remaining primary response genes identified in the various screening procedures will show somewhat more restricted cell type expression and/or greater differential induction by alternative ligands. We are currently attempting to develop procedures to search for primary response genes whose expression is more highly restricted to specific cell types.

We have emphasized cell-type specific "restriction" of primary response gene expression as a possible basis for lineage-specific differences in biological responses to a common ligand. Restriction might occur as a result of several alternative mechanisms. "Restricted" cells, in which expression of a particular TIS gene cannot be induced by any ligand, might result either from the absence of a trans-acting factor required for induction, or from the presence of a trans-acting factor that acts as a repressor. Alternatively, cis-acting alterations in gene or chromatin structure, such as methylation or chromosomal packing, might restrict gene expression despite the presence of appropriate second messenger activation pathways and the relevant transcription factors. We are currently investigating these possibilities by somatic cell hybridization, transfection analysis of reporter gene constructs, and genomic sequencing/*in vivo* footprinting (56,57).

Many primary response gene products are transcription factors. It is likely that the expression of these new transcription factors results in a "wave" of expression of secondary response genes. We suggest that "restricted" expression of secondary response genes is an additional mechanism by which cell-type specificity occurs. Thus, for example, we suggest that induction of the TIS8 transcription factor will cause the expression of overlapping, but not identical, sets of secondary response genes in 3T3 and PC12 cells. Some TIS8 responsive genes will be induced in both cell types when TIS8 protein is expressed. Other potentially responsive TIS8 target genes will be "restricted" to PC12 cells, and not expressed in 3T3 cells. Still other TIS8 target genes will be inducible in 3T3 cells, but not in PC12 cells. The search for secondary response genes in different target tissues, and the mechanisms by which alternative secondary response genes are expressed in distinct tissues when a common primary response transcription factor is induced, will be of great interest.

What might be the role of those primary response gene products that are not transcription factors? The context in which their cDNAs have been cloned and, for the most part, studied has centered around clonal lines in cell culture. It is possible that many of the primary response gene products, particularly those that are secreted, are involved in autocrine loops that regulate the immediate cellular response under investigation. Alternatively, these secreted products may serve, *in vivo*, as paracrine agents that modulate the behavior of surrounding cells. In the clonal cell culture systems primarily used to date to study these responses, the induction of such genes might appear gratuitous. However, their roles in regulating cellular cascades of behavior may be of great importance. Once these gene products are produced in quantity, it should be possible to identify receptors for such molecules, and study their effects on cellular phenotype.

While it is now clear that primary response genes are induced by alternative ligands in many cellular contexts, it is not so clear that their expression is essential for the various biological responses. The question of the causal role of a primary response gene in mitogenesis has been addressed only for the FOS gene product. Both antisense oligonucleotide (58,59) and antibody injection (60) experiments have suggested that the induced expression of FOS protein is essential for mitogen-induced DNA synthesis and cell division. However, the details of this result differ among the various experiments. Moreover, FOS protein expression has not been shown to be requisite in other biological responses in which its induction is observed. The induction of the other primary response genes has not yet been shown to be required for a biological response. Antibody injection experiments, antisense experiments, and other studies will be necessary before we can conclude that the induction of expression of these genes is essential for the many biological responses they precede.

REFERENCES

1. Todaro GJ, Green H. Quantitative studies of the growth of mouse embryo cells in culture and their development into established cell lines. *J Cell Biol* 1963;17:299-313.
2. Herschman HR. Isolation of mitogen-specific nonproliferative variant cell lines. In: Barnes D, Sirbasku D, eds. *Peptide Growth Factors: Methods in Enzymology*, vol 147, part B. New York: Academic Press, 1987; pp 355-369.
3. Pruss RM, Herschman HR. Variants of 3T3 cells lacking mitogenic response to epidermal growth factor. *Proc Natl Acad Sci USA* 1977;74:2790-2794.
4. Terwilliger E, Herschman HR. 3T3 variants unable to bind epidermal growth factor cannot complement in co-culture. *Biochem Biophys Res Commun* 1984;118:60-64.
5. Terwilliger E, Herschman HR. Dominant and recessive mitogen-nonproliferative variants of 3T3 cells. *J Cell Physiol* 1985;123:321-325.
6. Schneider CA, Lim RW, Terwilliger E, Herschman HR. Epidermal growth factor nonresponsive 3T3 variants do not contain epidermal growth factor receptor-related antigens or messenger RNA. *Proc Natl Acad Sci USA* 1987;83:333-336.
7. Wells A, Welsh JB, Lazar CS, Wiley HS, Gill GN, Rosenfeld MG. Ligand-induced transformation by a noninternalizing epidermal growth factor receptor. *Science* 1990;247:962-964.
8. Jiang L-W, Schindler M. Nucleocytoplasmic transport is enhanced concomitant with nuclear accumulation of epidermal growth factor (EGF) binding activity in both 3T3-1 and EGF receptor reconstituted NR-6 fibroblasts. *J Cell Biol* 1990;110:559-568.
9. Butler-Gralla E, Herschman HR. Variants of 3T3 cells lacking mitogenic response to the tumor promoter tetradecanoyl-phorbol-acetate. *J Cell Physiol* 1987;107:59-68.
10. Bishop R, Martinez R, Weber MJ, Blackshear PJ, Beatty S, Lim R, Herschman HR. Protein phosphorylation in a TPA non-proliferative variant of 3T3 cells. *Mol Cell Biol* 1985;5:2231-2237.
11. Erikson RL, Alcorta D, Bedard P-A, et al. Molecular analyses of gene products associated with the response of cells to mitogenic stimulation. *Cold Spring Harbor Symp Quant Biol* 1988;53:143-151.
12. Biemann, H-PN, Erikson RL. Abnormal protein kinase C down regulation and reduced substrate levels in non-phorbol ester-responsive 3T3-TNR9 cells. *Mol Cell Biol* 1990;10:2122-2132.

13. Gottesman MM, Sobel ME. Tumor promoters and Kirsten sarcoma virus increase synthesis of a secreted glycoprotein by regulating levels of translatable mRNA. *Cell* 1980;19:449-455.

14. Herschman HR. Mitogen-specific nonresponsive mutants and the initiation of cellular proliferation. *BioEssays* 1987;6:270-274.

15. Butler-Gralla E, Herschman HR. Glucose uptake and ornithine carboxylase activity in tetradecanoyl phorbol acetate non-proliferative variants. *J Cell Physiol* 1983;114:317-320.

16. Herschman HR. Mitogen specific nonproliferative variants of Swiss 3T3 cells. In: Colburn NH, Moses HL, Stanbridge EJ, eds. *Growth Factors, Tumor Promoters, and Cancer Genes*, vol 58. New York: Alan R Liss, Inc, 1984; pp 395-310.

17. Cochran BH, Reffel AC, Stiles CD. Molecular cloning of gene sequences regulated by platelet-derived growth factor. *Cell* 1983;33:939-947.

18. Greenberg ME, Ziff EB. Stimulation of 3T3 cells induces transcription of the c-fos proto-oncogene. *Nature (London)* 1984;311:433-438.

19. Herschman HR, Brankow DW. Ultraviolet irradiation transforms C3H10T½ cells to a unique, suppressible phenotype. *Science* 1986;234;1385-1388.

20. Herschman HR, Brankow DW. Colony size, cell density, and nature of the tumor promoter are critical variables in expression of a transformed phenotype (focus formation) in co-cultures of UV-TDTx and C3H10T½ cells. *Carcinogenesis* 1987;8:993-998.

21. Chen AC, Herschman HR. Tumorigenic methylcholanthrene transformants of C3H10T½ cells have a common nucleotide alteration in the c-Ki-*ras*. *Proc Natl Acad Sci USA* 1989;86:1608-1611.

22. Chen AC, Brankow DW, Herschman HR. A reassessment of methylcholanthrene transformation in the C3H10T½ cell culture system. *Carcinogenesis* 1990;11:817-822.

23. Herschman HR, Brankow D. Suppression and expression of the transformed phenotype in two-stage C3H10T½ transformants. In: Colburn N, ed. *Genes and Signal Transduction in Multistage Carcinogenesis*. New York: Marcel Dekker, Inc, 1988, pp 69-90.

24. Yamamoto KR, Alberts BM. Steroid receptors: elements for modulation of eukaryotic transcription. *Annu Rev Biochem* 1976;45:721-746.

25. Lim RW, Varnum BC, Herschman HR. Cloning of tetradecanoyl phorbol ester induced "primary response" sequences and their expression in density-arrested Swiss 3T3 cells and a TPA nonproliferative variant. *Oncogene* 1987;1:263-270.

26. Lim RW, Varnum BC, O'Brien TG, Herschman HR. Induction of tumor promoter inducible genes in murine 3T3 cell lines and TPA non-

proliferative 3T3 variants can occur through both protein kinase C dependent and independent pathways. *Mol Cell Biol* 1989;9:1790-1793.

27. Tippetts MT, Varnum BC, Lim RW, Herschman HR. Tumor promoter inducible genes are differentially expressed in the developing mouse. *Mol Cell Biol* 1988;8:4570-4572.

28. Arenander AT, Lim R, Varnum B, Cole R, Herschman HR, de Vellis J. Astrocyte response to growth factors and hormones: early nuclear events. In: Reier R, Bunge R, Seif F, eds. *Current Issues in Neural Regeneration Research*. New York: Alan R Liss, Inc, 1988, pp 257-269.

29. Arenander AT, Lim RW, Varnum BC, Cole R, de Vellis J, Herschman HR. TIS gene expression in cultured rat astrocytes: induction by mitogen and stellation agents. *J Neurosci Res* 1989;23:247-256.

30. Arenander AT, Lim RW, Varnum BC, Cole R, de Vellis J, Herschman HR. TIS gene expression in cultured rat astrocytes: multiple pathways of induction by mitogens. *J Neurosci Res* 1989;23:257-265.

31. Sukhatme VP, Kartha S, Toback FG, Taub R, Hoover RG, Tsai-Morris C-H. A novel early growth response gene rapidly induced by fibroblast, epithelial cell and lymphocyte mitogens. *Oncogene Res* 1987;1:343-355.

32. Sukhatme VP, Cao X, Chang LC, et al. A zinc finger-encoding gene coregulated with *c-fos* during growth and differentiation, and after cellular depolarization. *Cell* 1988;53:37-43.

33. Varnum BC, Lim RW, Herschman. Characterization of TIS7, a gene induced in Swiss 3T3 cells by the tumor promoter tetradecanoyl phorbol acetate. *Oncogene* 1989;4:1263-1265.

34. Christy B, Nathans D. DNA binding site of the growth factor-inducible protein Zif268. *Proc Natl Acad Sci USA* 1989;86:8737-8741.

35. Lemaire P, Vesque C, Schmitt J, Stunnenberg H, Frank R, Charnay P. The serum-inducible mouse gene *Krox-24* encodes a sequence-specific transcriptional activator. *Mol Cell Biol* 1990;10:3456-3467.

36. Hazel TG, Nathans D, Lau LF. A gene inducible by serum growth factors encodes a member of the steroid and thyroid hormone receptor superfamily. *Proc Natl Acad Sci USA* 1988;85:8444-8448.

37. Greene LA, Tischler AS. PC12 pheochromocytoma cultures in neurobiological research. In: Federoff S, Hertz L, eds. *Advances in Cellular Neurobiology*, vol 3. New York: Academic Press, Inc, 1982, pp 373-414.

38. Milbrandt J. A nerve growth factor-induced gene encodes a possible transcriptional regulatory factor. *Science* 1987;238:797-799.

39. Milbrandt J. Nerve growth factor induces a gene homologous to the glucocorticoid receptor gene. *Neuron* 1988;1:183-188.

40. Tirone E, Shooter EM. Early gene regulation by nerve growth factor in PC12 cells: induction of an interferon-regulated gene. *Proc Natl Acad Sci USA* 1989;86:2088-2092.

41. Herschman HR. Primary response genes induced by growth factors and tumor promoters. In: Richards CH, ed. *Annual Reviews of Biochemistry*, vol 60. Palo Alto: Annual Reviews, Inc, In Press.

42. Varnum BC, Lim RW, Sukhatme VP, Herschman HR. Nucleotide sequence of a cDNA encoding TIS11, a message induced in Swiss 3T3 cells by the tumor promoter tetradecanoyl phorbol acetate. *Oncogene* 1989;4:119-120.

43. Gomperts M, Pascall JC, Brown KD. The nucleotide sequence of a cDNA encoding an EGF-inducible gene indicates the existence of a new family of mitogen-induced genes. *Oncogene* 1990;5:1081-1083.

44. Almendral JM, Sommer D, MacDonald-Bravo H, Burckhardt J, Perera J, Bravo R. Complexity of the early genetic response to growth factors in mouse fibroblasts. *Mol Cell Biol* 1988;8:2140-2148.

45. Arenander AT, de Vellis J, Herschman HR. Induction of *c-fos* and TIS genes in cultured rat astrocytes by neurotransmitters. *J Neurosci Res* 1989;24:107-114.

46. Herschman HR. Extracellular signals, transcriptional responses, and cellular specificity. *Trends in Biochemical Sciences* 1989;14:455-458.

47. Kujubu DA, Lim RW, Varnum BC, Herschman HR. Induction of transiently expressed genes in PC-12 pheochromocytoma cells. *Oncogene* 1987;1:257-262.

48. Bartel DP, Sheng M, Lau LF, Greenberg ME. Growth factors and membrane depolarization activate distinct programs of early response gene expression: dissociation of *fos* and *jun* induction. *Genes and Devel* 1989;3:304-313.

49. Morgan JI, Curran T. The role of ion flux in the control of *c-fos* expression. *Nature* 1986;322:552-555.

50. Curran T, Morgan JI. Barium modulates *c-fos* expression and post-translational modification. *Proc Natl Acad Sci USA* 1986;83:8521-8524.

51. Greenberger JS, Sakakeeny MA, Humphries RK, Eaves CJ, Eckner RJ. Demonstration of permanent factor-dependent multipotential (erythroid/neutrophil/basophil) hematopoietic progenitor cell lines. *Proc Natl Acad Sci USA* 1983;80:2931-2935.

52. Varnum BC, Lim RW, Kujubu D, et al. Granulocyte-macrophage colony-stimulating factor and tetradecanoyl phorbol acetate induce a distinct, restricted subset of primary-response TIS genes in both proliferating and terminally differentiated myeloid cells. *Mol Cell Biol* 1989;9:3580-3583.

53. Fleischmann J, Golde DW, Weisbart RH, Gasson JC. Granulocyte-macrophage colony-stimulating factor enhances phagocytosis of bacteria by human neutrophils. *Blood* 1986;68:708-711.

54. Fletcher MP, Gasson JP. Enhancement of neutrophil function by granulocyte-macrophage colony-stimulating factor involves recruitment of a less responsive subpopulation. *Blood* 1988;71:652-658.

55. Baldwin GC, Gasson JC, Quan SG, et al. Granulocyte-macrophage colony-stimulating factor enhances neutrophil function in acquired immunodeficiency syndrome patients. *Proc Natl Acad Sci USA* 1988;85:2763-2766.

56. Church GM, Gilbert W. Genomic sequencing. *Proc Natl Acad Sci USA* 1984;81:1991-1995.

57. Andersen RD, Taplitz SJ, Wong S, Bristol G, Larkin B, Herschman HR. Metal-dependent binding of a factor *in vivo* to the metal responsive elements of the metallothionein-1 (MT-1) gene promoter. *Mol Cell Biol* 1987;7:3574-3581.

58. Holt JT, Gopal TV, Moulton AD, Nienhuis AW. Inducible production of *c-fos* antisense RNA inhibits 3T3 cell proliferation. *Proc Natl Acad Sci USA* 1986;83:4794-4798.

59. Nishikura K, Murray JM. Antisense RNA of proto-oncogene *c-fos* blocks renewed growth of quiescent 3T3 cells. *Mol Cell Biol* 1987;7:639-649.

60. Riabowol KT, Vosatka RJ, Ziff EB, Lamb NJ, Feramisco JR. Microinjection of *fos*-specific antibodies blocks DNA synthesis in fibroblast cells. *Mol Cell Biol* 1988;8:1670-1676.

61. Ryseck R-P, Macdonald-Bravo H, Mattéi M-G, Ruppert S, Bravo R. Structure, mapping and expression of a growth factor inducible gene encoding a putative nuclear hormonal binding receptor. *EMBO J* 1989;8:3327-3335.

62. Christy BA, Lau LF, Nathans D. A gene activated in mouse 3T3 cells by serum growth factors encodes a protein with "zinc finger" sequences. *Proc Natl Acad Sci USA* 1988;85:7857-7861.

63. Janssen-Timmen U, Lemaire P, Mattéi M-G, Revelant O, Charnay P. Structure, chromosome mapping and regulation of the mouse zinc-finger gene *Krox-24*; evidence for a common regulatory pathway for immediate-early serum-response genes. *Gene* 1989;80:325-336.

Advances in Regulation of Cell Growth, Volume 2;
Cell Activation: Genetic Approaches, edited by
James J. Mond, John C. Cambier, and Arthur Weiss.
Raven Press, Ltd., New York © 1991.

6

Identification of G Protein Mutants Encoding Constitutively Active and Dominant Negative α Subunit Polypeptides

Sunil K. Gupta, Joseph M. Lowndes, Shoji Osawa
and Gary L. Johnson

Division of Basic Sciences, National Jewish Center for Immunology and Respiratory Medicine, Denver, Colorado 80206 and Department of Pharmacology, University of Colorado Medical School, Denver, Colorado, 80262

GTP-binding regulatory proteins, referred to as G proteins, provide a signal transduction coupling mechanism for many cell surface receptors. Receptors function as exchange catalysts to mediate the dissociation of GDP bound to the guanine nucleotide binding site allowing GTP to bind to the G protein. Bound GTP induces changes in the G protein such that an activated conformation is assumed which regulates the activity of specific effector enzymes and ion channels. The generally accepted interactions of receptor, G protein and effector enzyme or ion channel during the activation cycle are shown in Fig. 1. G proteins are heterotrimers composed of α, β and γ subunits, and the α-subunit contains the guanine nucleotide binding site which, in the nonactivated state, is occupied by GDP. Agonist or hormone binding to an appropriate receptor leads to exchange of GDP for GTP and the dissociation of the $\beta\gamma$ subunit complex from α-GTP. The α subunit has an intrinsic GTPase activity that hydrolyzes GTP to GDP. The GTPase activity of the α subunit returns the activity of the protein to the inactive α-GDP form. The G protein activation/inactivation cycle will be repeated until the hormone is removed from the system at which time the G proteins accumulate in the inactive state as a result of GTP hydrolysis.

Properties of the known G protein α subunits are listed in Table 1. Available evidence indicates several α subunits are capable of regulating the activity of multiple effector enzymes or ion channels. Four forms of α_s arise from alternative splicing mechanisms (1), while the fifth α_s, referred to as α_{olf}, is the product of a unique gene expressed only in olfactory neurons (2). The $\alpha_{i1, 2 \& 3}$ polypeptides are products of unique genes (3), as are α_o (4), α_z (5) and the retinal rod (α_{t1}) and cone (α_{t2}) subunits, (6,7).

The α_s subunit is clearly the stimulatory G protein which activates adenylyl cyclase and modulates the activity of specific Ca^{2+} (8,9). $\alpha_{i1, 2 \& 3}$ activate specific inward rectifying K^+ channels (10), activate or inhibit different Ca^{2+} channels (11,12), inhibit adenylyl cyclase (10) and are implicated in the regulation of phospholipases C and (13) and A_2 (14). α_o is the most abundant α subunit polypeptide in the brain (15); it is structurally similar to the α_i polypeptides (16), and is involved in the regulation of K^+ and Ca^{2+} channels (17,18) and possibly phospholipases (19). The function of α_z is unclear but was found to be

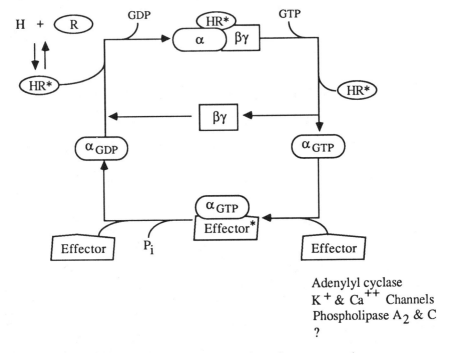

Fig. 1: G protein cycle for activation of enzymes and
ion channels in response to hormone (H) interaction
with its receptor (R).

expressed in many tissues with the highest abundance in brain and liver (20). The α_z polypeptide differs from the other α subunit polypeptides characterized to date in that its intrinsic GTPase activity is several-fold lower, suggesting once it binds GTP it would remain in an activated conformation for a longer duration than that determined for α_s, α_i or α_o (21). Finally, $\alpha_{t1 \& 2}$ are specifically expressed in rods and cones, respectively, and regulate cGMP phosphodiesterase (6,7,22). The α_i and α_o polypeptides, but not α_s and α_t are posttranslationally modified by the addition of myristic acid in the tissues in which they are normally expressed (23). The selective myristylation among the α chain subunits presumably involves the requirement for differential interactions with the plasma membrane and possibly the effector enzyme or ion channel (24). Several additional new α chains have been identified using polymerase chain reaction cloning strategies based on sequence homologies with the known α subunit polypeptides (25). The function of the newly identified α chains is unknown.

More than 80 receptors have been demonstrated to be coupled to G proteins (26). It is clear that multiple responses are elicited by specific receptors coupled to G proteins depending on the repertoire of effector enzymes and ion channels expressed in the responding cell. For example, receptor activation of G_i proteins could potentially inhibit adenylyl cyclase, activate inward rectifying K^+ channels,

Table 1: G Protein α, β and γ subunits

Possibly as many as 20-22 α subunits:

5 α_s: Activates adenylyl cyclase and modulates Ca^{+2} channels.

3 α_i: Inhibit adenylyl cyclase, regulate K^+
& Ca^{+2} channels, activate phospholipase A_2,
   regulate phospholipase C?

2 α_o: Regulate K^+ and Ca^{+2} channels.

1-2 α_z: Function ?

2 α_t: Cone and rod transducin, regulate cGMP PDE.

6-8 α?: Function ?? (phospholipase C, ion channels, etc.)

2-4 β subunits: Required for receptor activation of α subunit, attenuates α activity, may regulate phospholipase A_2 activity.

6-8 γ subunits: Tightly associated with β subunit, probably plays important role in regulating βγ affinity for α subunit.

There potentially could be 4x8= 32 combinations of βγ to interact with a given α subunit. Potential significance of different affinities of βγ for α and receptor selectivity is presently unclear.

inhibit one type of Ca^{2+} channel, activate a second Ca^{2+} channel and stimulate phospholipase A_2 activity. As yet, no cell has been shown to express the entire repertoire of α_i regulated enzymes and ion channels, but the complexity of the signalling cascade which could result from the receptor activation of G proteins is demonstrated with this example. In an attempt to define the role of specific G protein α subunits in the control of cell function, we have employed a genetic strategy to develop constitutively active and dominant negative mutant α chain polypeptides. Characterization of the properties of mutant α_s and α_{i2} polypeptides by expression in fibroblasts and their influence on the regulation of second messenger pathways is discussed in this chapter.

Structure of α_s and α_{i2} polypeptides

The α_s polypeptides of 44.5 and 46 kDa are derived from a single transcript by alternative splicing. In $\alpha_{s44.5}$ a region of mRNA encoding amino acids Gly72-Gly86 within the α_s46 polypeptide is removed. The α_s46 and $\alpha_{s44.5}$ polypeptides are respectively 394 and 379 amino acids in length. Deletion of this 15 amino acid sequence near the amino-terminus was shown to have a small effect on GTPase activity of recombinant $\alpha_{s44.5}$ relative to α_s46 (27), otherwise no significant differences in regulation or function are readily apparent. The α_s46 polypeptide migrates anomalously on SDS-PAGE behaving as a 52 kDa protein, whereas the $\alpha_{s44.5}$ polypeptide more accurately migrates as a 45 kDa protein.

The three α_i polypeptides are respectively 354 (α_{i1}), 355 (α_{i2}) and 354 (α_{i3}) amino acids and migrate as 40-41 kDa polypeptides on SDS-PAGE. The $\alpha_{i1,2\ \&\ 3}$ polypeptides are highly conserved with only 13 non-conserved amino acid substitutions among the three polypeptides. As predicted from their differences in function α_s and α_i polypeptides demonstrate significant diversity within specific regions of their primary sequences. Figure 2 shows the amino acid sequence for α_s46 and α_{i2} with underlined sequences showing nonconserved residues between the two polypeptides. Several interesting differences arise from analysis of the α_s and α_{i2} primary sequences. The α_s polypeptide has three unique inserts at residues 7-13, 72-86 and 324-336. Disregarding the $\alpha_{s46.5}$ sequence Gly72-Gly86, which is the alternative splicing domain that may be removed without noticeable change in α_s function (see above), the two proteins are approximately 65% homologous. The non-conserved residues are fairly evenly distributed between the NH_2- and COOH-terminal moieties of the two α chains. The greatest regions of diversity between the α_s and α_{i2} polypeptides are the extreme NH_2-terminus from residues α_sSer7-His41 (α_{i2}Ala7-Val34), α_sLys88-Arg160 (α_{i2}Cys66-Gly138) and the COOH-terminal region from α_sLeu302-Leu394 (α_{i2}Thr280-Phe355). Within these regions the two polypeptides are approximately 55% non-homologous suggesting they are domains encoding the unique α_s and α_{i2} functions.

Residues involved in GDP/GTP binding have been clearly identified in the refined x-ray crystallographic models for p21ras (28,29) and EF-Tu (30). Homologous residues in G protein α chains are postulated to serve the same functions but no crystal structure has yet been obtained for any α subunit. Present evidence, however, suggests strong similarities in the GDP/GTP binding domains of p21ras, EF-Tu and G protein α chains. X-ray crystallographic analysis of GDP and GMPPNP, a hydrolysis resistant GTP analog, liganded p21ras has localized the activating alterations induced by the binding of GMPPNP around the γ-phosphate group of the bound GMPPNP which adds a negative charge compared to GDP that apparently triggers the conformational changes (29,31). With GMPPNP

αs MGCLGNSKTEDQRNEEKAQREANKKIEKQLQKDKQVYRAT 40
αi2 MGCTVSA-------EDKAAAERSKMIDKNLREDGEKAARE 33

αs HRLLLLGAGESGKSTIVKQMRILHVNGFNGEGGEEDPQAA 80
αi2 VKLLLLGAGESGKSTIVKQMKIIHEDGYSEE--------- 64

αs RSNSDGEKATKVQDIKNNLKEAIETIVAAMSNLVPPVELA 120
αi2 ------ECRQYRAVVYSNTIQSIMAIVKAMGNLQIDFADP 98

αs NPENQFRVDYILSVMNVPNFDFPPEFYEHAKALWEDEGVR 160
αi2 QRADDARQLFALSCAAEEQGMLPEDLSGVIRRLWADHGVQ 138

αs ACYERSNEYQLIDCAQYFLDKIDVIKQADYVPSDQDLLRC 200
αi2 ACFGRSREYQLNDSAAYYLNDLERIAQSDYIPTQQDVLRT 178

αs RVLTSGIFETKFQVDKVNFHMFDVGGQRDERRKWIQCFND 240
αi2 RVKTTGIVETHFTFKDLHFKMFDVGGQRSERKKWIHCFEG 218

αs VTAIIFVVASSSYNMVIREDNQTNRLQEALNLFKSIWNNR 280
αi2 VTAIIFCVALSAYDLVLAEDEEMNRMHESMKLFDSICNNK 258

αs WLRTISVILFLNKQDLLAEKVLAGKSKIEDYFPEFARYTT 320
αi2 WFTDTSIILFLNKKDLFEEKI--TQSPLTICFPEYTGANK 296

αs PEDATPEPGEDPRVTRAKYFIRDEFLRISTASGDGRH-YC 359
αi2 YDEA-----------ASY-IQSKFEDL-NKRKDTKEIYT 322

αs YPHFTCAVDTENIRRVFNDCRDIIQRMHLRQYELL 394
αi2 --HFTCATDTKNVQFVFDAVTDVIIKNNLKDCGLF 355

Fig.2: Amino acid sequence for the αs and αi2 polypeptide chains. Underlined amino acids are nonconserved residues between the two α subunits.

the new contacts that are formed involve Gly10, Ala11, Thr35 and Gly60 (29,31). The interactions induced by GMPPNP binding resulting in activation of p21ras causes conformational changes in the segments Ile21-Gly48 and 57-77. This is consistent with the region between Tyr32 and Tyr40 interacting with GAP (29,31) and the critical region adjacent to the invariant Asp/Gly region (Asp57-Gly60) being critical in the signalling functions of p21ras (29,31).

The common transforming mutations in p21ras are within or near the consensus sequences which comprise the GDP/GTP binding domain and influence GTPase activity (Table 2). For example, Gly12 is part of the PO_4-box involved in forming the binding pocket for the phosphoryls of GDP and is thought to be involved in the hydrolysis of GTP. Mutation of Gly12→Val or most other amino acids disrupts the binding pocket and causes reduced GTPase activity. Two other common transforming mutations in p21ras are Ala59→Thr and Gln61→Leu both of which are involved in the conformational switch which occurs upon binding of GTP. The association of Gly12, Ala59 and Gln61 with the regions undergoing changes in the p21ras tertiary structure following GMPPNP binding indicates that mutation of these amino acid residues must alter the conformation of the protein thus influencing its function including GAP interaction and GTPase activity.

The conserved sequences comprising the GDP/GTP binding domain for G protein α chains and p21ras are compared in Table 2. We (32-34) and others (35) have used site-directed mutagenesis to introduce mutations in the $α_s$ polypeptide which have been shown to alter GDP/GTP interactions and GTPase activity in p21ras in an attempt to probe the GDP/GTP binding domain of G protein α chains. Five point mutations have been introduced in the $α_s$ polypeptide that are transforming when introduced in p21ras due to changes in GDP/GTP binding or GTPase activity. The Gly49→Val mutation ($α_sG49V$) corresponds to residue Gly12 in p21ras that is part of the PO_4-box sequence. Two mutations, Gly225→Thr and Gly226→Ala lie within the invariant Asp-x-x-Gly sequence and Gln227→Leu is directly adjacent. These amino acids correspond to residues 59-61 in the p21ras polypeptide that undergo a significant conformational change upon binding of GTP (31).

Analysis of amino acid mutations on $α_s$ regulation of adenylyl cyclase

One method we have used to examine mutations influencing the sequences involved in forming the GDP/GTP binding domain of G protein α chains is the transient expression of wild-type and mutant $α_s$ polypeptides in COS-1 cells. COS-1 cells express equivalent levels of the 45 and 52 kDa forms of $α_s$, as determined by immunoblotting. When $α_s$ cDNA encoding the 52 kDa form of $α_s$ is used for transfection, a 4-6-fold increase is observed in immunoreactive 52 kDa $α_s$ polypeptide with no detectable change in expression of the 45 kDa $α_s$ or 41 kDa $α_i$ polypeptides. Similar levels of expression were observed for 52 kDa $α_s$ polypeptides encoding specific amino acid substitutions. None of the α-subunit cDNAs altered COS cell expression of β- or γ- subunits during the 65-72 hr incubation following transfection (34). After expression of the $α_s$ polypeptide in transfected COS-1 cells, a 5-6-fold increase in cAMP levels was observed relative to control transfected cells (Table 3). Expression of the $α_{i2}$ polypeptide actually slightly lowers cAMP levels in COS-1 cells. Based on similar levels of expression determined by immunoblotting $α_sG49V$ stimulated cAMP synthesis 2-fold greater than the wild-type $α_s$ cDNA. Similar analysis of the $α_sQ227L$ mutant indicated it

Table 2: Comparison of the amino acid residues comprising the GDP/GTP binding domains of c-H-ras, α_s and α_{i2}.

c-H-ras ^{5}KLVVVGAGGVGKSALT^{20}

α_s ^{42}RLLLLGAGESGKSTIV^{57}

α_{i2} ^{35}KLLLLGAGESGKSTIV^{50}

c-H-ras ^{57}DTAGQ^{61}

α_s ^{223}DVGGQ^{227}

α_{i2} ^{201}DVGGQ^{205}

c-H-ras ^{116}NKCD^{119}

α_s ^{292}NKQD^{295}

α_{i2} ^{269}NKKD^{273}

c-H-ras ^{144}TSAK^{147}

α_s ^{364}TCAV^{367}

α_{i2} ^{325}TCAT^{328}

Mutations in α_s residues G49 → V, G225 → T, G226 → A and Q227 → L resulted in major functional changes in α_s activity as detailed in the text.

Table 3: cAMP levels in COS cells expressing
wild-type and mutant α$_s$ polypeptides

Mutation	pmol cAMP/mg protein
Control (no DNA)	67
pCW1	33
α$_s$	358
G49V	750
G225T	128
G226A	83
Q227L	1767
D295A	296
α$_{i2}$	65

COS-1 cells were transfected with the appropriate
cDNAs encoding the indicated α$_s$ mutants inserted
in the pCW1 expression plasmid. 65 hr after transfection
cells were assayed for cAMP levels in the presence
of the the phosphodiesterase inhibitor isobutyl
methylxanthine. The schematic designates the
location of each point mutation within the α$_s$
polypeptide primary sequence.

was an even stronger activator of cAMP synthesis than α_SG49V, causing an 18-20-fold increase in basal cAMP levels. The relative difference in the ability of α_SQ227L compared to α_SG49V to stimulate cAMP synthesis correlates with their respective GTPase activities, with the G49V and Q227L mutations having approximately 60 and 5% of the wild-type α_S GTPase activity (36). The relative ability of G49V and Q227L mutations to stimulate cAMP synthesis and their respective loss of GTPase activity is similar to the G12V and Q61L mutants of p21ras, where Q61L has a lower GAP-stimulated GTPase activity than G12V and a stronger transformation potential (37). The common influence of these mutations suggests that the GDP/GTP binding and GTPase domains of the α_S and p21ras polypeptides are very similar. However, examination of two additional mutations within conserved regions of the two molecules indicates a significant difference between the two proteins. The mutation Ala59→Thr is a strong transforming mutation due to a loss in the ability of the mutant p21ras polypeptide to hydrolyze GTP (38). The corresponding mutation in α_S is Gly225→Thr. Expression of α_SG225T in COS-1 cells demonstrates that the mutant polypeptide is weakly active and elevates cAMP to less than 25% of the level observed with the wild-type α_S polypeptide. This result suggests the Gly225→Thr mutation is actually inhibitory to the activity of the α_S polypeptide. A potential explanation for this finding comes from the analysis of an additional mutation at α_S residue 226. Mutation of Gly226→Ala was identified as a spontaneous α_S mutant in S49 mouse lymphoma cells which resulted in the inability of the mutant α_S chain to activate adenylyl cyclase in the presence of GTP (39). The basis for the α_SG226A phenotype appears related to the putative "switch" function assigned to the corresponding domain in p21ras that, as discussed above, is involved in the conformational change which occurs with GTP binding. In p21ras Gly60 appears to serve as a pivot point for the downstream amino acid region involved in the altered conformation induced by GTP binding. The Gly226→Ala mutation most likely inhibits the conformational change induced by GTP, thus preventing the α_S activated state to be assumed and the subsequent stimulation of adenylyl cyclase. Consistent with this notion is the finding that expression of α_SG226A in COS-1 cells is unable to stimulate cAMP synthesis. The α_SG225T mutation appears to have a similar structural change, although less dramatic, as the G226A mutation. Amazingly, the α_SQ227L mutation, which is adjacent to the inhibitory G225T and G226A mutations, strongly activates cAMP synthesis. The striking difference between the inhibitory nature of the G225T and G226A mutations and the activating Q227L mutation is the influence of the amino acid changes on the ability of the α_S polypeptide to pivot at residue 226 in response to GTP. These findings demonstrate the critical nature of this region in controlling GTP-induced activation of α_S interaction with adenylyl cyclase and stimulation of GTPase activity. The failure of the G225T mutation to be an activating α_S mutant is in contrast to the properties of the similar mutation in p21ras and indicates that the GDP/GTP binding domains are similar but the regulation of the two polypeptides by GTP-mediated conformational changes involving the "switch box" are not identical.

Mutation of an additional conserved amino acid involved in GDP/GTP binding in p21ras and α_S also defines apparent differences in the guanine nucleotide regulation of the two polypeptides. The consensus sequence Asn-Lys-X-Asp (see Table 2), corresponding to residues 116-119 in p21ras, is involved in determining specificity for guanine nucleotide binding to proteins such as p21ras, EF-Tu and G protein α chains. The Asn within this sequence is critical for stable nucleotide binding because of its hydrogen bonding to the C-6 keto group of the guanine ring (29,31). The Asp residue interacts with the C-2 amino group on the guanine ring to stabilize the binding of guanine nucleotides within the binding pocket. In p21ras, replacement of this Asp residue (Asp119) with an alanine (D119→A) reduced the affinity of both GTP and GDP by a factor of 20 (40). However, the ability of p21rasD119A to induce transformation of NIH3T3 cells was similar to that of the oncogenic p21rasV12, Thr59 mutant. The increased transformation potential observed with the reduced affinity for guanine nucleotides in p21rasD119A was attributed to an increased dissociation rate for bound GDP. Asp119 in p21ras corresponds to Asp295 in the α_S polypeptide. Mutation of Asp295→Ala in the α_S polypeptide had no influence on the ability to stimulate cAMP synthesis relative to the wild-type α_S chain when expressed in COS-1 cells (Table 3). This result indicates that α_SD295A is not an activated α_S mutant in contrast to the results obtained with α_S G49V and α_SQ227L.

The failure of α_SD295A to be an activated α_S mutant may, in fact, result from intrinsic differences in the GTPase regulatory properties of p21ras and G protein α chains. G protein α chains have an intrinsic GTPase activity with a k_{cat} for recombinant wild-type α_S of 3.5 min^{-1} (36). The wild-type p21ras k_{cat} is about 0.02 min^{-1} in the absence of GAP and about 1 min^{-1} in the presence of GAP (41). G proteins do not need interaction with an accessory protein like GAP for the hydrolysis of GTP. Thus, even though the GDP dissociation rate may be enhanced in the α_SD295A mutant, similar to that predicted for p21rasD119A, the combination of a diminished GTP affinity and the intrinsic high GTPase activity of α_S may prevent the Asp295→Ala mutation from activating the α_S polypeptide measured by increased adenylyl cyclase activity. This hypothesis is actually testable by introducing the Asp295→Ala mutation in the α_S cDNA encoding either the α_SG49V or α_SQ227L mutants. The prediction would be that inhibition of the α_S GTPase would allow the altered GDP dissociation induced by the Asp295→Ala mutation to manifest itself by enhancing cAMP synthesis when the double mutant was expressed in COS cells. We are presently making the appropriate constructs to address this question.

The domain responsible for the high intrinsic GTPase activity of G protein α chains has been putatively mapped to a region within α_S that contains the Arg residue ADP-ribosylated by cholera toxin (42). ADP-ribosylation of Arg201 in α_{S46} results in inhibition of α_S GTPase activity. Mutation of Arg201 to almost any other amino acid inhibits α_S GTPase activity (42), even though this region of the α_S polypeptide is not directly involved in the binding of GDP/GTP. Rather, it has been proposed by H. Bourne (43) that this represents a region of the α_S polypeptide controlling a GAP (GTPase activating protein) function intrinsic to G protein α chains. In fact, F. McCormick (44) noted that residues 189-203 surrounding Arg201 in α_{S46} is a region with some homology to the putative GAP binding site in p21ras. Thus, there are substantial differences in the regulation of G protein α subunits and p21ras that are, in part, related to the additional amino acid sequence information encoded in the α chains. The increased amino acid sequence encoded in G protein α chains most certainly results in numerous additional intramolecular contacts in their tertiary structure as well as complex interactions with other proteins including the βγ-subunit complex.

Functional properties of α_s/α_{i2} chimeras

In an attempt to identify regions of α chain primary sequences involved in intra- and intermolecular contacts that are important for regulation and impart the unique properties of the different G proteins we have generated a series of α_s/α_{i2} chimeras and examined their characteristics. The COS cell expression system was initially used to assay three α chain chimeras, where corresponding regions of α_s and α_{i2} were shuttled within cDNAs, in order to define domains in the α_s and α_{i2} polypeptides responsible for regulation of adenylyl cyclase (Fig. 3). One chimera, referred to as $\alpha_{i/s(Bam)}$, encodes the first 212 residues of α_{i2} and the COOH-terminal 160 residues of α_s, yielding an α chain chimera that encodes the first 60% of α_{i2} and the last 40% of α_s. Within the NH$_2$-terminal moiety encoding α_{i2}, approximately 35% of the amino acids are nonconserved relative to the α_s sequence. The second chimera, referred to as $\alpha_{s/i(38)}$, has the last 38 amino acids of α_s substituted with the COOH-terminal 36 residues of α_{i2} (45). Both mutations were also placed within the same cDNA encoding for the polypeptide $\alpha_{i(Bam)/s/i(38)}$. Analysis of each chimeric polypeptide in COS cells demonstrated that each was a functional α_s polypeptide capable of activating adenylyl cyclase. The $\alpha_{i/s(Bam)/s/i(38)}$ polypeptide also appeared to be approximately 2-fold greater in ability to activate adenylyl cyclase at similar levels of expression relative to the $\alpha_{i/s(Bam)}$ and $\alpha_{s/i(38)}$ polypeptides, indicating it was a more active α_s subunit in its ability to activate cAMP synthesis in COS cells. Importantly, pertussis toxin treatment of the cells for 16 hr before assay of cAMP to inhibit α_i activation had little or no influence on the ability of any of the α chain chimeras, α_s, or the point mutants α_sQ227L or α_sG49V to activate adenylyl cyclase. Thus, the chimeras behave as functional α_s polypeptides capable of activating adenylyl cyclase and cumulatively define the α_s activation domain to be encoded within a 122 amino acid core sequence residing within residues Ile235-Arg356 of the α_s polypeptide. Both the NH$_2$-terminal 60% and COOH-terminal 10% of the α_s polypeptide may be substituted with α_{i2} sequences and maintain the ability to activate adenylyl cyclase. Further evidence that the α_s activation domain resides within residues 235-356 was the finding that deletion of a unique 13 residue sequence (G327-Y339) within this 122 amino acid domain resulted in complete loss of adenylyl cyclase activation. This sequence is absent in all α chain polypeptides characterized to date that do not stimulate cAMP synthesis, indicating this domain is critical for stimulation of adenylyl cyclase activity by the α_s core activation domain.

In addition to the $\alpha_{i/s(Bam)}$ chimera, a second NH$_2$-terminal chimera was constructed which had the first 61 amino acids of α_s replaced with the first 54 residues of α_{i2} (Fig. 4). This chimera, referred to as $\alpha_{i(54)/s}$, results in the loss of seven unique α_s amino acids, and 16 of the first 34 α_{i2} residues are nonconserved relative to the α_s sequence. The last 20 α_{i2} amino acids within the chimera are identical or highly conserved when compared to the α_s sequence. We found that the $\alpha_{i(54)/s}$ chimera was a highly robust α_s chain in its ability to activate adenylyl cyclase (33). The activated character of the $\alpha_{i(54)/s}$ polypeptide contrasted with that of the $\alpha_{i/s(Bam)}$ construct, which behaves as a functional wild-type α_s. The kinetics of adenylyl cyclase activation by GTP and GTPγS is markedly altered for the $\alpha_{i(54)/s}$ polypeptide relative to that measured with either the wild-type α_s or $\alpha_{i/s(Bam)}$ polypeptide chains. The $\alpha_{i(54)/s}$ phenotype is manifested in adenylyl cyclase assays as a significant decrease in the time required to achieve maximal cyclase activation by the hydrolysis-resistant GTP analog, GTPγS. The diminished lag time associated with $\alpha_{i(54)/s}$, but not α_s or $\alpha_{i/s(Bam)}$, is explained by an accelerated rate of GDP dissociation, allowing a faster GTPγS binding and adenylyl cyclase activation. Thus, the NH$_2$-terminal moiety of α_{i2} and α_s may be interchanged with normal maintenance of intrinsic α_s regulation, but mutation at the extreme α_s NH$_2$-terminus results in loss of an attenuator function controlling α_s activation of adenylyl cyclase. It was also observed that when the $\alpha_{i(54)/s}$ mutation was placed in the same cDNA as the α_sQ227L point mutation, the resulting

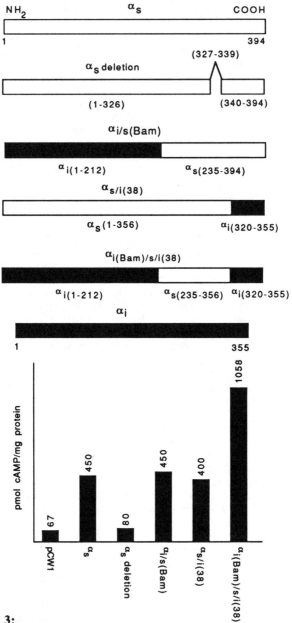

Fig. 3:
Identification of the core αs activation domain using
deletion and αs/αi chimeras. Cyclic AMP was measured
in transfected COS cells as described in the legend for Table 3.
Map of the chimeras and deletion mutant used for the
transfections is shown above.

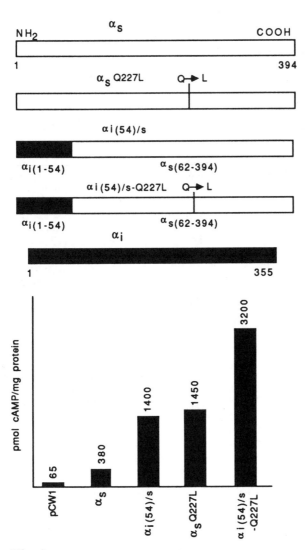

Fig.4:

Activation of cAMP synthesis by expression of the $\alpha_{i(54)/s}$ chimera and its additivity with the α_sQ227L point mutation within the α_s polypeptide. Cyclic AMP was measured 65 hr after transfection of COS cells with each of the constructs as described in the legend to Table 3.

adenylyl cyclase activation and cAMP accumulation was additive relative to each mutation alone (Fig. 4). The enhanced rate of GDP dissociation observed with the $\alpha_{i(54)/s}$ polypeptide and the inhibited GTPase activity resulting from the α_{s}Q227L mutation explains their additivity. By altering the two rate-limiting steps in α chain activation (GDP dissociation and GTPase), a very strong constitutively active α_{s} polypeptide is observed.

Three additional chimeras were constructed with different regions of α_{s} substituted with corresponding NH$_2$-terminal regions of α_{i2} in order to further define the α_{s} region controlling GDP dissociation (Fig. 5). The $\alpha_{i(7)/s}$ cDNA replaces the first 14 residues of α_{s} with the NH$_2$-terminal 7 residues of α_{i2}. Within this sequence α_{s} has 7 residues (Leu4-Glu10) that are absent in α_{i2} (see Fig. 2). The $\alpha_{i(64)/s}$ chimera replaces the first 86 residues of α_{s} with the first 64 residues of α_{i2}. This NH$_2$-terminal region encodes two unique α_{s} sequences (Leu4-Glu10; Gly72-Gly86) accounting for the 22 amino acid difference relative to the α_{i2} sequence. The differences between $\alpha_{i(54)/s}$ and $\alpha_{i(64)/s}$ are the deletion of the Gly72-Gly86 unique α_{s} sequence, which is removed by alternative splicing in some cell types to generate the 45 kd α_{s} polypeptide isoform (27), and substitution of α_{s} residues Ile62-Glu71 with the corresponding region of α_{i2}, in which there are three nonconserved amino acid substitutions. The third construct is $\alpha_{i(122)/s}$, which substitutes the first 144 residues of α_{s} with the corresponding 122 residues of α_{i2} and creates an α_{i2}/α_{s} chimera roughly halfway between $\alpha_{i(54)/s}$ and $\alpha_{i/s(Bam)}$. Within the junctions of the $\alpha_{i(64)/s}$ and $\alpha_{i(122)/s}$ chimeras there are 35 nonconserved residues between α_{i2} and α_{s}, and between the junctions of $\alpha_{i2(122)/s}$ and $\alpha_{i/s(Bam)}$ there are 24 nonconserved amino acid substitutions.

Expression of the three additional chimeras in COS cells elucidated the boundaries within the α_{s} polypeptide sequence that encoded the attenuator function controlling GDP dissociation. The phenotype of the $\alpha_{i(64)/s}$ chimera was similar to $\alpha_{i(54)/s}$ in its enhanced ability to elevate cAMP levels relative to wild-type α_{s} or $\alpha_{i/s(Bam)}$ (Fig. 5). Strikingly, $\alpha_{i(7)/s}$ and $\alpha_{i(122)/s}$ were similar to wild-type α_{s} polypeptides in their ability to stimulate cAMP synthesis. These findings demonstrated that deletion of the unique α_{s} sequences Leu4-Glu10 or Gly72-Gly86 are not responsible for the $\alpha_{i(54)/s}$ and $\alpha_{i(64)/s}$ phenotype. In addition, the $\alpha_{i(122)/s}$ chimera behaved similarly to α_{s} and $\alpha_{i/s(Bam)}$ in its ability to stimulate cAMP synthesis, indicating that the α_{s} sequence Glu145-Trp234 is not involved in the phenotypic differences of $\alpha_{i(54)/s}$ and $\alpha_{i(64)/s}$ relative to the $\alpha_{i/s(Bam)}$ polypeptide. Thus, the domain controlling the rate of GDP dissociation maps to α_{s} residues Gly15-Pro144. Within this region, residues corresponding to α_{s} Arg42-Arg62 and α_{i2} Lys35-Lys55, which contain the GAGESGK phosphate binding sequence, are conserved in the two α chains, indicating that this domain is not involved in the mutant phenotype. The α_{s} domains involved in the activating mutation must be within the sequences Gly15-His41, Ile62-Glu71, Glu87-Pro144. The corresponding sequences in α_{i2} are Glu8-Val34 and Ile55-Glu122.

Pertussis and cholera toxin-catalyzed ADP-ribosylation

Pertussis toxin catalyzes the ADP-ribosylation of α_{i2} but not α_{s}. The ADP-ribosylation reaction requires the presence of the $\beta\gamma$ subunit complex, and in its absence the G protein α subunit is not recognized by pertussis toxin (16). The consensus amino acid in α_{i} and α_{o} ADP-ribosylated by pertussis toxin is a cysteine four residues from the COOH-terminus. The $\alpha_{s/i(38)}$ chimera, therefore, encodes the pertussis toxin ADP-ribosylation site normally found in the α_{i2} polypeptide.

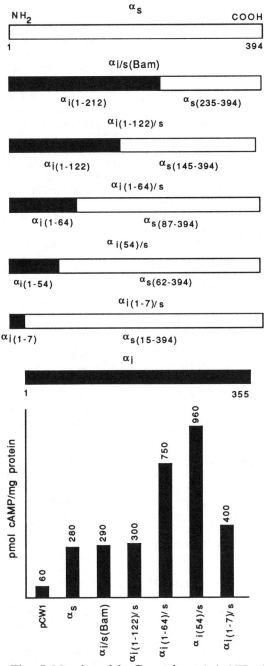

Fig. 5: Mapping of the G protein α chain NH₂ -Terminus attenuation domain using α_i/α_s chimeric polypeptides. Cyclic AMP levels after transfection of COS cells was determined as described in the legend to Table 3.

The $\alpha_{s/i(38)}$ chimeric polypeptide is not ADP-ribosylated by pertussis toxin (45), indicating that the 36 amino acid α_{i2} sequence at the COOH-terminus is not sufficient for pertussis toxin recognition of a G protein α subunit polypeptide. In contrast, the $\alpha_{i(Bam)/s/i(38)}$ and $\alpha_{i(64)/s/i(38)}$ chimeric polypeptides were excellent substrates for pertussis toxin-catalyzed ADP-ribosylation, even though they are functional α_s chains in their ability to activate adenylyl cyclase. Thus, sequences within the NH$_2$-terminal moiety of α_{i2} are required for pertussis toxin-catalyzed ADP-ribosylation of the cysteine four amino acids from the COOH-terminus. Surprisingly, the $\alpha_{i(54)/s/i(38)}$ and $\alpha_{i(122)/s/i(38)}$ polypeptides were found not to be substrates for pertussis toxin. Thus, interaction of the $\alpha_{i(64)/s/i(38)}$ and $\alpha_{i(Bam)/s/i(38)}$ polypeptides with the βγ subunit complex is sufficient to allow pertussis toxin ADP-ribosylation of the G protein. In contrast, $\alpha_{i(54)/s/i(38)}$ and $\alpha_{i(122)/s/i(38)}$, which disrupt the α chain sequence surrounding the NH$_2$-terminal junction for the $\alpha_{i(64)/s/i(38)}$ chimera, are not pertussis toxin substrates. This finding indicates that within the α subunit polypeptide NH$_2$-terminus there must be more than one contact site for the βγ subunit complex, and these sites appear to be disrupted by the $\alpha_{i(54)/s}$ and $\alpha_{i(122)/s}$ chimeric sequences.

Thus, the regulatory properties we assign to the α subunit attenuator domain, which are disrupted in the $\alpha_{i(54)/s}$ and $\alpha_{i(64)/s}$ chimeras but normal in $\alpha_{i(7)/s}$, $\alpha_{i(122)/s}$ and $\alpha_{i/s(Bam)}$, appear to overlap with the functions regulated by the βγ subunit complex. The functions that have been assigned to the βγ subunit complex include attenuation of GDP dissociation from the α subunit, pertussis toxin recognition of α_i and α_o polypeptides, and its requirement for efficient coupling of receptors to α chain activation. The control of GDP dissociation and attenuation of adenylyl cyclase activation by GTP are lost in the $\alpha_{i(54)/s}$ and $\alpha_{i(64)/s}$ chimeras. ADP-ribosylation by pertussis toxin of the $\alpha_{i(64)/s/i(38)}$ polypeptide was similar to that observed with wild-type α_{i2}, but inhibited in the $\alpha_{i(54)/s/i(38)}$ and $\alpha_{i(122)/s/i(38)}$ chimeras. The $\alpha_{i(54)/s}$ polypeptide is, however, efficiently coupled to the β-adrenergic receptor (33), which requires association with the βγ subunit complex (46). The $\alpha_{i(54)/s}$, $\alpha_{i(64)/s}$ and $\alpha_{i(122)/s}$ NH$_2$-terminal mutants, therefore, differentially disrupt two of the three functions assigned to the βγ control of the α subunit polypeptide: attenuation of GDP dissociation and recognition by pertussis toxin.

In contrast to pertussis toxin, cholera toxin recognition of α_s does not require the βγ subunit complex. Cholera toxin ADP-ribosylates α_s at Arg201, a site roughly in the middle of the polypeptide whose flanking amino acid sequence is conserved in α_{i2} (16). Only α_s and α_sQ227L are good substrates for cholera toxin (34). The $\alpha_{i/s(Bam)}$ polypeptide is not recognized by cholera toxin - presumably because α_{i2} normally is not a cholera toxin substrate, and the arginine that would be recognized by cholera toxin is encoded within the α_{i2} moiety of the chimera (Arg179). The $\alpha_{i(54)/s}$ chimera, which has only a small region of α_{i2} at its NH$_2$-terminus, was a poor ADP-ribosylation target for cholera toxin, indicating that disruption of the α subunit polypeptide NH$_2$-terminus affected the toxin recognition of the α_sArg201 residue in the middle of the polypeptide. Introduction of the COOH-terminal α_{i2} sequence in the chimera, $\alpha_{s/i(38)}$, resulted in complete inhibition of cholera toxin ADP-ribosylation of the α_s polypeptide. The COOH-terminal mutation, therefore, abolished the ability of cholera toxin to recognize the chimeric α polypeptide, even though it was a functional α_s polypeptide capable of activating adenylyl cyclase. The multiple nonconserved mutations introduced in the NH$_2$- or COOH-terminus of the $\alpha_{i(54)/s}$ and $\alpha_{s/i(38)}$ polypeptides must therefore introduce intramolecular changes in the structure of the mutant α_s polypeptides so as to diminish or inhibit cholera toxin-catalyzed ADP-ribosylation of Arg201 without significantly influencing the ability of the core activation domain to recognize adenylyl cyclase.

A prediction of G protein α chain polypeptide structure, functional domains and regulation

We can conclude from the properties of the point mutants and α_{i2}/α_s chimeras that the adenylyl cyclase activation domain lies within the α_s residues Ile235-Gly355; sequences in the NH_2-terminus regulate the rate of α_s activation by guanine nucleotides independent of the GTPase shut-off mechanism intrinsic to G protein α subunits; mutation within the NH_2-terminus of α_s enhances the rate of GTPγS activation, defining this domain as a modulator of GDP dissociation and GTP activation; requirements for pertussis toxin-catalyzed ADP-ribosylation of the cysteine that is the fourth amino acid from the COOH-terminus, are encoded within residues Glu8-Trp122 of the α_{i2} polypeptide; and mutation in both the NH_2- and COOH-termini influence cholera toxin-catalyzed ADP-ribosylation of Arg201 in the middle of the α_s chain.

The activated phenotype of the $\alpha_{i(54)/s}$ and $\alpha_{i(64)/s}$ chimeric polypeptides defines the NH_2-terminal region as an attenuator control domain within the α subunit polypeptide. The attenuator control domain is distinct from the GTPase function encoded within the α_s polypeptide. Mutation of both the attenuator control and GTPase domains was shown to generate a very strong constitutively active α subunit polypeptide, indicating that the attenuator function is independent of the GTPase and GAP (GTPase activating protein) domains of the α_s polypeptide. Five regulatory functions which control G protein α chain activation and turn-off and a sixth domain involved in effector activation can be assigned to specific regions of the polypeptide (Fig. 6). These functions are attenuator, GAP, GDP/GTP binding, GTPase, receptor interaction and effector activation domains.

Genetic manipulation of G protein α chain sequences indicates the polypeptide has a structure that may be roughly divided in "halves". The $\alpha_{i/s(Bam)}$ chimera, which functions as a wild-type α_s chain, defines the COOH-terminal moiety of a G protein α chain as encoding the effector enzyme activation and receptor contact sequences (47). The NH_2-terminal moiety of G protein α chains containing the attenuator domain therefore, can be shuttled between the different α subunit polypeptides with the retention of normal regulation of the COOH-terminal effector activation and receptor contact moiety (33). Existing structural data for α chain polypeptides also places the NH_2- and COOH-termini in close proximity to each other and oriented towards the middle of the molecule (33). The close proximity of the α subunit polypeptide NH_2- and COOH-termini places the COOH-terminal receptor recognition sequence near the attenuator and βγ regulatory domains. This is consistent with the demonstration that receptors have contact sites for βγ as well as the α subunit and that the βγ subunit complex is absolutely required for efficient guanine nucleotide exchange (46,48). The changes in both cholera and pertussis toxin-catalyzed ADP-ribosylation patterns observed with the α_{i2}/α_s chimera support the prediction that the NH_2 and COOH-termini are in close proximity and oriented towards the middle of the polypeptide tertiary structure. Mutations at either end of the α_s polypeptide influence cholera toxin recognition of the Arg201 in the middle of the polypeptide primary sequence and appropriate NH_2-terminal sequences are required for pertussis toxin recognition of the cysteine four amino acids from the α subunit COOH-terminus. Since α_s and α_{i2} bind common βγ subunits (16), the changes in toxin ADP-ribosylation patterns defined with the

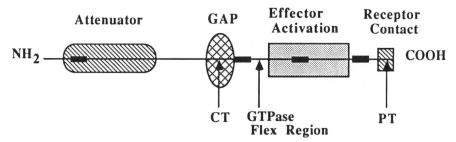

Fig. 6: G protein α chain domains mapped using
amino acid mutation and α_s/α_{i2} chimeras.
The solid bars represent the four regions
forming the GDP/GTP binding domain
within the α subunit primary sequence.

α_{i2}/α_s chimeras must involve changes in intramolecular α chain interactions. Thus, the NH_2- and COOH-termini of the G protein α subunit polypeptide function as modulators of the core attenuator and enzyme activation sequences.

Identification of dominant negative G protein α subunit mutations

In addition to the characterization of constitutively active G protein α subunit mutants we have devoted considerable effort in the identification of dominant negative mutations. Herskowitz (49) has predicted that dominant negative mutants will frequently retain a subset of the intact, functional domains of the wild-type polypeptide, but have some domains either altered or missing. It was also predicted that two mutants having a good chance of creating a dominant negative polypeptide would encode: 1, an extensively mutagenized COOH-terminal segment or 2, an extensively mutagenized NH_2-terminal segment. The $\alpha_{s/i(38)}$ and $\alpha_{i(54)/s}$ chimeras respectively fulfill these properties of a predicted dominant negative G protein α chain. Because the COOH-terminal half of α subunits encodes the effector activation and receptor contact domains it was predicted that mutations at this end of the polypeptide might generate dominant negative mutants.

We had previously shown that $\alpha_{s/i(38)}$ was a functional α_s polypeptide capable of activating adenylyl cyclase because the effector activation domain was intact. It is also known that G_i proteins regulate receptor-coupled phospholipases in specific cell types. In Chinese hamster ovary (CHO) cells G_{i2} regulates the thrombin and type 2 (P_2)-purinergic receptor stimulation of phospholipase A_2. When the $\alpha_{s/i(38)}$ polypeptide was expressed in CHO cells the receptor-stimulated phospholipase A_2 activity was inhibited (14). Expression of the $\alpha_{i(54)/s}$, $\alpha_{i/s(Bam)}$ or overexpression of wild-type α_{i2} in CHO cells did not affect receptor-stimulated phospholipase A_2 activity. Neither the constitutively active $\alpha_s Q227L$ polypeptide nor cyclic AMP analogs, which activate cyclic AMP-dependent protein kinase, had an effect on receptor-stimulated phospholipase A_2 activity. Thus, only the $\alpha_{s/i(38)}$ chimera inhibited receptor-stimulated phospholipase A_2 activity and functioned as a dominant negative α_{i2}-like polypeptide (14). The $\alpha_{s/i(38)}$ polypeptide is unable to activate phospholipase A$_2$ because it lacks the α_{i2} effector activation domain, but the last 36 amino acids of the α_{i2} primary sequence appear to be a critical domain for G protein regulation of phospholipase A$_2$. The $\alpha_{s/i(38)}$ chimera also inhibits the ability of ionomycin, a calcium ionophore, to stimulate phospholipase A$_2$ activity similar to that observed with pertussis toxin treatment of CHO cells. As detailed previously (see above), pertussis toxin catalyzes the ADP-ribosylation of a cysteine four amino acids from the α_{i2} COOH-terminus, consistent with the finding that the α_{i2} COOH-terminus is involved in the α_{i2} control of phospholipase A$_2$ activity. The general utility of this observation is the use of the polymerase chain reaction to construct similar chimeras using COOH-terminal sequences encoded by G protein α chains other than α_s or α_{i2}. These chimeras can be used as dominant negative mutants to define the involvement of a G protein α chain in regulating different enzyme systems such as phospholipase C or a specific ion channel. In combination with the constitutively active G protein α chain mutants equivalent to the α_sQ227L polypeptide dominant negative mutants provide a powerful genetic approach to dissect the G proteins coupled to different effector enzyme systems.

Recently, we have also determined that overexpression of the α_sG225T mutant (see Table 3 and discussion in text) is capable of inhibiting the β-adrenergic receptor stimulation of adenylyl cyclase in COS-cells (50). The α_sG225T mutant was shown to inefficiently activate adenylyl cyclase and its overexpression competitively inhibits the ability of the wild-type α_s polypeptide to activate adenylyl cyclase in response to β-adrenergic receptor stimulation. Since this region of α_s is highly conserved in all G protein α chains, similar mutations to α_sG225T may create dominant negative mutants in other α chains by mutation of the corresponding glycine residue to a threonine. This approach is technically more straightforward than the construction of COOH-terminal chimeras such as the $\alpha_{s/i(38)}$ polypeptide, but appears to require higher expression of the mutant α chain to efficiently inhibit the regulation of effector systems by the endogenous wild-type G protein.

The oncogenic potential of G protein α subunit mutants.

Recently it was demonstrated that mutations in specific G protein α subunit polypeptides occur at very high frequency in certain neuroendocrine tumors (51). The findings strongly implicate specific G protein mutants as oncogenes when expressed in appropriate tissues. Interestingly, the mutations in α_s found in pituitary adenomas and thyroid tumors are α_sR201C and α_sQ227L which map to the GAP and GTPase regulatory domains of the polypeptide (see Fig. 6). Both mutations inhibit GTPase activity resulting in the constitutive activation of the α_s polypeptide. Similar GTPase inhibiting mutations in α_{i2} were found in adrenal cortical and ovarian endocrine tumors.

We have found that activated α_{i2} polypeptides whose GTPase activity is inhibited dramatically alter the growth regulatory properties of cultured fibroblasts (unpublished observations). In fact, in foci formation assays the activated α_{i2}Q205L mutant transforms both NIH3T3 and Rat 1A cells. Expression of α_{i2}Q205L dramatically decreases the serum requirement for growth reminiscent of previously characterized oncogenes such as p21ras and v-src. The growth regulatory properties perturbed by expression of α_{i2}Q205L are just now being defined, but it is becoming apparent that G proteins interact with the tyrosine kinase growth factor receptor family of proteins. Thus, G protein and tyrosine kinase regulated effector signal pathways appear to be integrated to a much greater level than previously appreciated. The characterization of dominant negative and constitutively active G proteins that we have detailed in this chapter allows straightforward genetic strategies to dissect which G proteins interact with specific growth factor receptor signalling systems to control cell growth and differentiated phenotype.

ACKNOWLEDGMENTS

This work was supported by NIH grants GM 30324, DK 37871 and the American Heart Association. We thank Marjorie Greiner for preparation of the manuscript. The first three authors are listed in alphabetical order. S.O. present address is Department of Cell Biology and Anatomy, University of North Carolina, Chapel Hill, NC 27599.

REFERENCES

1. Bray, P., et al., Human cDNA clones for four species of $G\alpha_s$ signal transduction protein. *Proc. Natl. Acad. Sci. U.S.A.* 83, 8893-8897, 1986.
2. Jones, D.T. and Reed, R.R. Golf: An olfactory neuron specific-$G\alpha_s$ protein involved in odorant signal transduction. *Science*, 244, 790-795, 1989.
3. Jones, D.T. and Reed, R.R., Molecular cloning of five GTP-binding protein cDNA species from rat olfactory neuroepithelium, *J. Biol. Chem.*, 262, 14241-14249, 1987.
4. Itoh, et al., Presence of three distinct molecular species of G_i protein α subunit: structure of rat cDNAs and human genomic DNAs. *J. Biol. Chem.*, 263, 6656-6664, 1988.
5. Fong, H.K.W., et al., Identification of a GTP-binding protein α-subunit that lacks an apparent ADP-ribosylation site for pertussis toxin. *Proc. Natl. Acad. Sci. U.S.A.* 85, 3066-3070, 1988.
6. Medynski, D., et al., Amino acid sequence of the α subunit of transducin deduced from the cDNA sequence. *Proc. Natl. Acad. Sci. U.S.A.*, 82, 4311-4315, 1985.
7. Lochrie, M.A., Hurley, J.B. and Simon, M.I., Sequence of the alpha subunit of photoreceptor G-protein: homologies between transducin, *ras* and elongation factors, *Science*, 228, 96-99, 1985.
8. Yatani, et al., A G-protein directly regulates mammalian cardiac calcium channels, *Science*, 238, 1288-1292, 1987.
9. Scott, R.H. and Dolphin, A.C., Activation of a G-protein promotes agonist responses to calcium channel ligands, *Nature*, 330, 760-762, 1987.
10. Yatani, et al., The G protein-gated atrial K^+ channel is stimulated by three distinct $G_i\alpha$-subunits, *Nature* 336, 680-682, 1988.
11. Hescheler, J., et al., Angiotension II-induced stimulation of voltage-dependent Ca^{2+} currents in an adrenal cortical cell line, *EMBO J.*, 7, 619-624, 1988.
12. Ewald, et al., Differential G Protein-mediated Coupling of Neurotransmitter Receptors to Ca^{2+} Channels in Rat Dorsal Root Ganglions in vitro, *Neuron*, Vol. 2, 1185-1193, 1989.

13. Fain J., et al., Evidence for involvement of guanine nucleotide binding regulating protein in the activation of phospholipases, *FASEB J.*, 2, 2569-2574, 1988.

14. Gupta, S.K., et al., A G protein mutant that inhibits thrombin and purinergic receptor activation of phospholipase A2, *Science*, 249, 662-666, 1990.

15. Sternweiss, P.C. and Robishaw, J.D., Isolation of two proteins with high affinity for guanine nucleotides from membranes of bovine brain, *J. Biol., Chem.*, 259, 13806-13813, 1984.

16. Johnson, G.L., and Dhanasekaran, N., The G-Proetin Family and Their Interaction with Receptors. *Endocrine Rev.*, 10, 317-331, 1989.

17. Gilman, A.G., G proteins: transducers of receptor-generated signals, *Annu. Rev. Biochem.*, 56, 615-649, 1987.

18. Birnbaumer, L., et al., Signal transduction by G proteins, *Kindey Int.*, 32, S-24-S-37, 1987.

19. Bloch, D.B., et al., The G Protein α_0 Subunit Alters Morphology, Growth Kinetics, and Phospholipid Metabolism of Somatic Cells, *Mol. Cell. Biol.* 9, 5434-5439, 1989.

20. Spicher, K., et al., Immunochemical detection of the alpha-subunit of the G protein, G_z, in membranes and cytosols of mammalian cells, *Biochem. Biophys. Res. Commun.*, 157, 883-890, 1988.

21. Casey, P.J., et al., G_z, a guanine binding protein with unique biochemical properties, *J. Biol. Chem.*, 265, 2383-2390, 1990.

22. Fung, B.K.-K., Hurley, J.B. and Stryer, L., Flow of information in the light-triggered cyclic nucleotide cascade of vision, *Proc. Natl. Acad. Sci. U.S.A.*, 84, 7493-7497, 1987.

23. Buss, J.B., et al., Myristoylated α subunits of guanine nucleotide-binding regulatory proteins, *Proc. Natl. Acad. Sci. U.S.A.*, 84, 7493-7497, 1987.

24. Buss, J.E., et al., Activation of the cellular proto-oncogene product p21*ras* by addition of a myristylation signal, *Science*, 243, 1600-1603, 1989.

25. Strathmann, M., Wilkie, T.M. and Simon, M.I., Diversity of the G-protein family: sequences from five additional α subunits in the mouse, *Proc. Natl. Acad. Sci. U.S.A.* 86, 7407-7409, 1989.

26. Iyengar, R. and Birnbaumer, L., In:G-Proteins (ed. Iyengar, R. and Birnbaumer, L.), pp. 1-17, Academic Press, New York, 1990.

27. Graziano, M.P., Freissmuth, M., and Gilman, A.G., Expression of $G_s\alpha$ in *E. coli*: purification and properties of two forms of the protein. *J. Biol. Chem.*, 264, 409-418, 1989.

28. De Vos, A.M., et al., Three dimensional structure of an oncogene protein: catalytic domain of human c-H ras p21, *Science*, 239, 888-893, 1988.

29. Pai, et al., Structure of the guanine nucleotide binding domain of the Ha-ras oncogene product p21 in the triphosphate conformation, *Nature*, 341, 209-214, 1989.

30. Jurnak, F., Structure of the GDP domain of EFTμ and location of the amino acids homologous to *ras* oncogene proteins, *Science*, 230, 32-36, 1985.

31. Jurnak, F., Heffron, S. and Bergmann, E., Conformational changes involved in the activation of *ras* p21: Implications for related proteins, *Cell*, 60, 525-528, 1990.

32. Woon, C.W., et al., Mutation of glycine 49 to valine in the α subunit of G_s results in the constitutive elevation of cyclic AMP synthesis, *Biochem.*, 28, 4547-4551, 1989.

33.   Osawa, S., et al., Mutation of the G$_s$ Protein α Subunit NH$_2$ Terminus Relieves an Attenuator Function, Resulting in Constitutive Adenylyl Cyclase Stimulation., *Mol. Cell. Biol.*, 10, 2931-2940, 1990.

34.   Osawa, S., et al., Gα$_i$-Gα$_s$ chimeras define the function of α chain domains in control of G protein activation and βγ subunit complex interactions, *Cell*, 63, 697-706, 1990.

35.   Masters, S.B., et al., Mutations in the GTP-binding site of G$_{s\alpha}$ alter stimulation of adenylyl cyclase, *J., Biol. Chem.*, 264, 15467-15474, 1989.

36.   Graziano, M.P., and Gilman, A.G., Synthesis in *Escherichia coli* of GTPase-deficient mutants of G$_{s\alpha}$. *J. Biol. Chem.*, 264-15475-15482, 1989.

37.   Barbacid, M., *ras* genes, *Annu. Rev. Biochem.*, 56, 779-827, 1987.

38.   Gibbs, J.B., et al., Intrinsic GTPase activity distinguishes normal and oncogenic ras p21 molecules, *Proc. Natl. Acad. Sci. U.S.A.*, 81, 5704-5708, 1984.

39.   Miller, R.T., et al., A mutation that prevents GTP-dependent activation of the α chain of G$_s$, *Nature* (London), 334, 712-715, 1988.

40.   Sigal, I.S., et al., Mutant *ras*-encoded proteins with altered nucleotide binding exert dominant biological effects, *Proc. Natl. Acad. Sci. U.S.A.*, 83, 952-956, 1986.

41.   Trahey, M. and McCormick, F., A cytoplasmic protein stimulates normal N-*ras* p21 GTPase, but does not affect oncogenic mutants, *Science*, 238:542-545, 1987.

42.   Casey, P.J. and Gilman, A.G., G protein involvement in receptor-effector coupling, *J. Biol. Chem.*, 263, 2577-2580, 1988.

43.   Landis, C.A., et al., GTPase inhibiting mutations activate the α chain of G$_s$ and stimulate adenylyl cyclase in human pituitary tumours, *Nature*, 340, 692-696, 1989.

44.   McCormick, F., Gasp: not just another oncogene, *Nature*, 340, 678-679, 1989.

45.   Woon, C.W., et al., Expression of a Gα$_s$/Gα$_i$ chimera that constitutively activates cyclic AMP synthesis, *J. Biol. Chem.*, 264, 5687-5693, 1988.

46.   Kelleher, D.J., and Johnson, G.L., Transducin inhibition of light-dependent rhodopsin phosphorylation: evidence for βγ interaction with rhodopsin, *Mol. Pharmacol.*, 34, 452-460, 1988.

47.   Masters, et al., Carboxyl terminal domain of G$_{s\alpha}$ specifies coupling of receptors to stimulation of adenylyl cyclase, *Science*, 241, 448-451, 1988.

48.   Fung, B.K.-K., Characterization of transducin from bovine retinal rod outer segments: separatation and reconstitution of subunits, *J. Biol. Chem.*, 258, 10495-10502, 1983.

49.   Herskowitz, I., Functional inactivation of genes by dominant negative mutations, *Nature*, 329, 219-222, 1987.

50.   Osawa, S. and Johnson, G.L., A dominant negative Gα$_s$ mutant is rescued by secondary mutation of the α chain amino-terminus, *J. Biol. Chem.*, in press.

51.   Lyons, J., et al., Two G protein oncogenes in human endocrine tumors, *Science*, 249, 655-659, 1990.

Advances in Regulation of Cell Growth, Volume 2;
Cell Activation: Genetic Approaches, edited by
James J. Mond, John C. Cambier, and Arthur Weiss.
Raven Press, Ltd., New York © 1991.

7

The CD28 Signal Transduction Pathway in T Cell Activation

Kelvin Lee M.D.[1], Carl H. June M.D.[2] and Craig B. Thompson M.D.[3]

[1,3]*Howard Hughes Medical Institute, Ann Arbor, Michigan. Department of Internal Medicine, Division of Hematology/Oncology, University of Michigan, 1150 West Medical Center Dr., Ann Arbor, MI 48109.*

During antigen-specific activation of a T cell, signal transduction through the T cell receptor (TCR)/CD3 complex plays a central role in the initiation of a series of biochemical and cellular events that leads to cell proliferation and/or effector function. However, it has become clear that TCR stimulation alone is not sufficient to account for all the events that occur during *in vivo* T cell activation. A number of T cell surface receptors have been shown to contribute to T cell activation (1). Following binding to their own ligand, these surface receptors (or accessory molecules) may function to augment signal transduction through the T cell receptor by facilitation of cell-cell adhesion between the antigen presenting cell (APC) and the T cell, or may initiate a signal transduction pathway that is distinct from those utilized by the TCR/CD3 complex (1).

Recent evidence has demonstrated that the accessory molecule CD28 (previously termed T44 or Tp44) may affect T cell activation through a novel signal transduction pathway. In this chapter, we will review the tissue distribution of CD28 expression and that of the CD28 ligand B7/BB-1, the effect of CD28 activation at a cellular and molecular level, and what is presently known about the components of the CD28 activation pathway.

DISTRIBUTION AND PHENOTYPE

CD28 was first defined as a human T cell-specific surface molecule by the monoclonal antibody 9.3 (2). Since then, a number of other monoclonal antibodies have been raised to CD28 (reviewed in 3), all apparently directed against the same epitope. CD28 homologues have also been identified in mice (4) and macaques (5). CD28 was initially found to be expressed on the surface of 25-56% of thymocytes and 54-86% of peripheral blood T cells (2). Further analysis has defined CD28 expression in thymocyte and mature T cell subpopulations. In the human thymus, virtually all cells which express the TCR/CD3 complex also express CD28 (6,7). On immature thymocytes ($CD3^{-/low}CD4^+CD8^+$ (double positives)), the expression of CD28 is low (6). As thymocytes mature, surface expression of CD28 increases 5 to 10 fold (6). High levels of CD28 expression appears to be restricted to thymocytes which also have high levels of CD3 surface expression, and visa versa (6,7). These $CD3^+CD28^+$ thymocytes are also either $CD4^+$ or $CD8^+$ (single positive) and represent mature thymocytes. This population constitutes approximately one third of the total thymocyte population (6). The majority of thymocytes that express the γ/δ TCR also express CD28, with the highest CD28 expression on $CD3^+$ cells (8).

In peripheral blood, greater than 90% of $CD4^+$ T cells express CD28 (9). Phenotypically, $CD4^+CD28^+$ T cell clones display T helper function (10). The density of CD28 surface expression has been reported to distinguish between two functionally distinct $CD4^+$ subsets (11). One subset displays anti-CD3 mAb-mediated cytotoxicity and can be induced to secrete IL-2, gamma interferon (γIFN) and tumor necrosis factor alpha and beta (TNF α/β) following anti-CD3 mAb and PMA activation. These clones have comparatively low CD28 surface expression. The other subset displays no cytotoxicity, minimal lymphokine production and has comparatively high CD28 expression. Whether one subset is derived from the other following prior stimulation or that these subsets are truly distinct populations remains to be determined. As for the $CD4^+CD28^-$ T cells, one study has found that T cell clones with antigen-specific suppressor function were all CD28 negative, and the majority were $CD4^+$ and $CD8^-$ (10).

Approximately 50% of $CD8^+$ peripheral blood T cells are also positive for CD28 (9). CD28 expression is more common on T cells that express high levels of CD8 (75% of this population) than on T cells that express lower levels (33%) (9). Phenotypically, $CD8^+CD28^+$ T cells have MHC-restricted cellular cytotoxic function (12,13) while $CD8^+CD28^-$ T cells have suppressor function for both immunoglobulin production (12) and alloantigen-induced proliferation (13).

Peripheral blood T cells that are $CD28^-$ express CD11b on their surface (9), and these two antigens define non-overlapping T cell subsets. As previously noted, $CD4^+$ or $CD8^+CD28^-CD11b^+$ T cells have suppressor function. Recently, it has been reported that $CD4^+CD11b^+CD28^-$ lymphocytes have limited TCR diversity and proliferative responses to alloantigens, and may represent a population that does not undergo selection in the thymus (15).

In studies of T cell clones, it was initially reported that CD28 expression could not be detected on CD3+ lymphocytes that did not express the $\alpha\beta$ TCR heterodimer (16), suggesting that lymphocytes bearing the $\gamma\delta$ TCR heterodimer did not express CD28. However, we and others have detected high expression of CD28 on approximately 50% of cells coexpressing the $\gamma\delta$ form of the TCR in bulk cultures of proliferating primary T cells (8 and authors' unpublished observations).

Although CD28 was initially described as a T cell-specific antigen, CD28 expression has been subsequently found on cells representative of mature plasma cells (myeloma cell lines and tissue) but not on B cells of earlier maturational stages (17,28). Resting B cells induced to differentiate in culture by anti-Ig stimulation will express CD28 on their surface (17). However, stimulation of CD28 by mAb 9.3 does not increase proliferation or immunoglobulin synthesis in myeloma cell lines (17) and a functional role for CD28 in normal plasma cells has not yet been determined.

MOLECULAR AND GENETIC CHARACTERISTICS OF CD28

Immunoprecipitation by the anti-CD28 mAb 9.3 yields an 88-90 kDa protein under nonreducing conditions, which resolves to a single 44-45 kDa protein under reducing conditions, from T cell lysates (2,18,19,20), T cell leukemia cell line HBP-ALL (18), the CD3⁻ Jurkat cell line JA3 (21) and plasmacytoma cell line 266BL (17). Thus it appears that CD28 is expressed on the T cell and myeloma cell surface as a disulfide-bonded homodimer. Additional studies indicate that the CD28 antigen is a surface molecule confined to the cell membrane (22). These studies also find that CD28 is associated with the detergent-insoluble cell fraction and may be associated with cytoskeletal elements or other proteins (22). However, pretreatment of cells with cytochalasin B to disrupt cytoskeletal elements prior to isolation has no effect on this insolubility (22). The CD28 antigen has pronounced charge heterogeneity (19,23) which appears to be due in part to variable sialylation (18,24).

The human gene that encodes the CD28 antigen has been identified by cDNA cloning (25) and supports many of the previous findings. The cDNA clone predicts a transmembrane protein of 202 amino acids (mature form). The extracellular domain contains 134 residues (5 cysteine residues) and 5 N-linked glycosylation sites. Comparison to other proteins reveals that CD28 is homologous to immunoglobulin variable domains, making it a member of the immunoglobulin gene superfamily. Transfection of this cDNA into COS cells leads to the surface expression of the homodimeric protein. A cDNA for the murine homologue of human CD28 has also been cloned (4). This cDNA also predicts an integral membrane protein of 218 residues. Nucleotide homology to human is 77%

A.
```
                  110v        120v        130v
     HCD28     TDIYFCKIQVMYPPPYLDNQKSNGTIIH
               TD:YFCKI: MYPPPYLDN::SNGTIIH
     MCD28     TDLYFCKIEFMYPPPYLDNERSNGTIIH
                  110^        120^        130^


                  200v        210v        220v
     HCD28     MNMTPRRPGPTRKHYQPYAPPRDFAAYRSX
               MNMTPRRPG TRK.YQPYAP:RDFAAYR:X
     MCD28     MNMTPRRPGLTRKPYQPYAPARDFAAYRPX
               190^        200^        210^
```

B.
```
                  110v        120v
     HCD28     TdiYfCKiqvMYPPPY
     MCD28     TdLYfCKiEfMYPPPY
     CTLA-4    TgLYlCKvElMYPPPY
                  130^
```

FIG.1. A) Comparison of human and murine CD28 protein sequence
(see text). Amino acid residues are numbered above and
below the respective proteins. B) Comparison of human and
murine CD28 to CTLA-4 protein sequences (see text).
Identical amino acid residues are in bold capitals. Amino
acid residues for human CD28 are numbered above and CTLA-4
below.

within the open reading frame with conservation of the 5 N-linked
glycosylation sites and 6 of 7 cysteine residues. There are two
regions in the 3' untranslated sequence that have >75% nucleotide
homology and are speculated to have some physiologic importance.
On a protein level, overall homology is 68% to human with striking
conservation within two regions of the molecule. One region
within the cytoplasmic domain (murine aa 190-217) is 89%
homologous (figure 1A) and suggests this region may be important
for signal transduction or attachment to cytoskeletal elements.
The other region is in the extracellular domain (murine aa 108-
135) and is 92% homologous (figure 1A). Interestingly, within this
region is a span of 30 residues near the transmembrane domain,
roughly bounded by 2 cysteine residues, that has significant
homology to CTLA-4 (figure 1B), a T cell specific gene that is
structurally similar to CD28 and maps to the same region of
chromosome 2 (q33-34) as CD28 (26,27). A functional role for this
region has not been ascribed, but may be important for intra- or
intermolecular associations.
     Human CD28 is encoded by a single copy gene (25,28) that is
organized into four exons (28) (figure 2). Each exon encodes a
functional domain of the mature peptide: exon 1 encodes the 5'
untranslated region and leader peptide, exon 2 encodes the
majority of the extracellular domain, exon 3 encodes the remainder
of the extracellular domain and the transmembrane region and

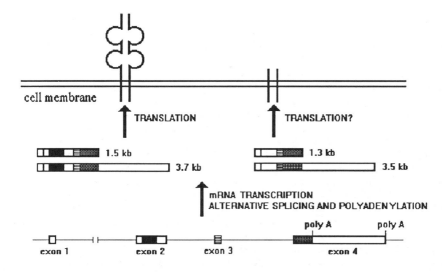

FIG.2.The genomic organization, mRNA and protein expression of CD28.

exon 4 encodes the cytoplasmic domain and 3' untranslated region (28). The four CD28 mRNA species (1.3, 1.5, 3.5 and 3.7 kb) appear to be generated by a combination of two post-transcriptional mechanisms (28). The first involves the use of an alternate, nonconsensus polyadenylation signal that results in the addition of 2167 bp to the 3' untranslated region. This mechanism would account for the size difference between the 1.3/1.5 kb and 3.5/3.7 kb transcripts. The second mechanism involves an internal splicing event that deletes 252 bp within exon 2, resulting in the loss of 84 amino acids, including 4 of 5 extracellular cysteines, and presumably significant changes in secondary structure of the resultant protein. This mechanism would account for the approximately 200 bp size difference between transcripts. This mechanism does not appear to be conserved in the mouse (4). Evidence for the existence and physiological role of this "deleted" form of human CD28 remains to be uncovered.

THE NATURAL LIGAND OF CD28

The immunoglobulin-like structure of CD28, its expression on the surface of the helper cell subset of peripheral blood T lymphocytes and the biological effects of CD28 activation (discussed below) all suggest that CD28 is a receptor for a cell surface molecule or soluble growth factor. Recently, it was found that the transfection and expression of a CD28 cDNA in chinese hamster ovary (CHO) cells could confer the ability upon these cells to adhere to transformed B cell lines, but not to cells of other lineages (29). By utilizing a variety of monoclonal antibodies directed against epitopes on these B cell lines, it was found that this adherence could be blocked by antibodies directed against the B cell-restricted cell surface activation antigen B7/BB-1 (30), as well as by antibodies directed against CD28. B7/BB-1 is expressed on the surface of B cells shortly after they are activated by anti-Ig or EBV (30 and references therein). The B7/BB-1 antigen has also been detected on interlobular thymic tissue (31). Subsequent transfection and expression of a BB-1 cDNA into COS cells conferred the ability to adhere to CHO cells transfected with CD28. Further studies utilizing chimeric molecules (consisting of the extracellular portions of CD28 or B7 fused to IgCγl chains) have confirmed the specificity of the CD28-B7 interaction in both nonlymphoid cells transfected with either CD28 or B7, and in PHA-activated T cells (32). These analysis have also determined that B7 binding to CD28 is mediated through a single class of binding sites with a Kd of 200 nM, an affinity comparable to other lymphoid adhesion molecules (32). More significantly, costimulation with anti-CD3 mAb and immobilized B7 (either adhered to plastic or presented on transfected cell surfaces) resulted in peripheral blood lymphocyte proliferation and IL-2 mRNA accumulation. This indicates that the CD28-B7 interaction results in bonafide biological responses that are consistent with previously reported responses involving CD28 stimulation by monoclonal antibodies (see below). Other authors have shown that IL-2 production by MOLT 14 cells (TCR γ/δ bearing human leukemic T cell line) and MOLT 16 (TCR α/β bearing human leukemic T cell line) can be induced by coculture with BALL-1 cells (human leukemic B cell line), and that this induction requires cell to cell contact and is IL-1 independent (32a). Furthermore, it was shown that this induction could be specifically inhibited by preincubating the T cell lines with soluble, bivalent anti-CD28 mAb, and that this was not due to modulation of CD28. Taken together, these findings argue that the natural ligand for CD28 is B7/BB-1, and that CD28 and B7/BB-1 mediate heterophilic adhesion and biologic effects during T-B interaction.

FUNCTIONAL EFFECTS OF CD28 STIMULATION

CD28 stimulation appears to mediate two distinct effects in T cells. First, in thymocytes and peripheral blood T cells that have received an initial activation signal, CD28 activation by anti-CD28 mAb provides a second signal that allows these cells proliferate. Second, CD28 activation appears to directly enhance the effector function of activated peripheral blood T cells by increasing lymphokine secretion and perhaps cytotoxicity.

Proliferative effects

Thymus

Among the mature thymocyte population (CD2+CD3+CD28+CD1a⁻/CD4+ or CD8+), stimulation with anti-CD2 mAb alone, anti-CD28 mAb alone or a combination of anti-CD2 and anti-CD3 mAbs does not result in proliferation (33). Stimulation by either anti-CD2 or anti-CD3 mAb alone induces expression of IL-2 receptor (but not IL-2) mRNA while anti-CD28 mAb stimulation has no effect (31,33). However, the dual stimulus provided by either anti-CD2 or anti-CD3 and anti-CD28 mAbs results in vigorous proliferation that is IL-2 mediated (7,33). This effect is preceded by a significant increase in expression of IL-2R and IL-2 mRNA and subsequent secretion of IL-2 into the supernatant (33). It therefore appears that the CD28 activation pathway provides an important costimulus for the initiation of autocrine-mediated proliferation in mature thymocytes. Recently, it has been shown that a subset of CD4+CD45RA+ peripheral blood T cells are unresponsive to CD3 stimulation but will proliferate and develop a mature phenotype in response to CD2 and CD28 costimulation (34). These authors have suggested that CD2 and CD28 costimulation may play an important role in peripheral T cells that have not matured fully. Other authors have reported that CD28 is comitogenic with protein kinase C activation by phorbol myristate acetate (PMA) (6), IL-2 (although this may involve CD28 crosslinking) (7), and anti-CD3/CD4 or CD8 heteroconjugate mAb (31) in mature thymocytes. PMA stimulation alone can increase surface expression of CD28 in both mature and immature thymocytes without affecting CD3 expression (6).

CD28 effects in immature thymocytes are less clear. Dual stimulation of the CD2 and CD28 pathways in CD3⁻ thymocytes does not result in proliferation (31,33). Other authors have reported that CD28 is comitogenic with PMA and/or IL-2 in immature, CD3+ thymocytes, although the level of proliferation is 4 to 5 fold less than in mature thymocytes (35). The discovery of B7/BB-1 expression on a subset of cortical thymic tissue suggests that CD28 activation provides a physiologically important progression signal to developing thymocytes.

Peripheral T cells

The stimulation of CD28 by soluble bivalent anti-CD28 mAb alone does not induce peripheral blood T cell proliferation or enhancement of effector function (3). However, it was initially noted that anti-CD28 mAb binding could augment proliferation of T cells stimulated with suboptimal doses of phytohemmagglutin (PHA) (23). Subsequently, it has been shown that CD28 stimulation is a comitogen for peripheral blood leukocytes stimulated by monocytes, PPD or tetanus toxoid (36). In purified resting T cells, CD28 is comitogenic for submaximal amounts of anti-CD2 mAb (37,38), 12-O-tetradecanoyl phorbol-13-acetate (TPA) (20), PMA +/- the calcium ionophore ionomycin (39,40,41,42,43,44,45) and immobilized anti-CD3 mAb (14,19,38,39,41,42,46,47). Proliferation following dual stimulation of CD3 and CD28 is higher and more sustained than that seen following CD3 stimulation alone (19). Dual stimulation of CD3 and CD28 also induces IL-4-mediated proliferation in $CD4^+CD8^-$ $CD29^+CD45R^-$ (memory) T cells (47). In addition, whereas proliferation of resting T cells following anti-CD3 mAb or PMA + ionomycin can be completely abrogated by the addition of the immunosupressant cyclosporin A (CsA), CD28 costimulation confers resistance to the antiproliferative effects of CsA (40,44). Furthermore, costimulation of CD28 confers resistance to the antiproliferative effects of dB-cAMP (39), cholera toxin (43), PGE_2(39), forskolin (39) and glucocorticoids (40) following PMA or CD3 stimulation. These effects serve to distinguish the CD28 receptor from other activational pathways.

CD28 stimulation appears to be dependent on the degree of receptor oligomerization. Activation through crosslinking of CD28 receptors results in additional effects not seen with activation by soluble, bivalent anti-CD28 mAb alone. Stimulation by progressively more highly crosslinked anti-CD28 mAbs alone results in IL-2R expression and increasing sensitivity to the proliferative effects of IL-2 (45,48,49). Highly crosslinked anti-CD28 mAbs can confer upon some resting T cells the ability to proliferate in response to lymphokines (48). Crosslinking of the CD28 receptor results in the proliferation of T cells following exposure to IL-2 (48), IL-4 (47) and IL-6 (45). Finally, it has been reported that the addition of soluble anti-CD28 antibodies suppresses proliferation of MHC class II restricted T_h clones following exposure to alloantigen (36) and in allogeneic and autologous mixed lymphocyte reactions (14). However, additional crosslinking of anti-CD28 mAb results in a reversal of this inhibition (14). These findings suggest that, in addition to a physiologic signal delivered by the crosslinking of CD28 receptors, another function mediated by CD28 involves adhesion to antigen presenting cells that can be blocked by soluble, non-crosslinked mAb.

As noted below, CD28-mediated proliferation following submaximal stimulation by anti-CD2, anti-CD3 mAbs, mitogenic lectins and phorbol esters may be primarily the consequence of its modulation of the effector functions of the target T cell, particularily IL-2 and other lymphokine production.

Effector function

When bulk cultures of resting T cells are stimulated to proliferate with optimal doses of PMA + ionomycin or anti-CD3 mAb, the addition of anti-CD28 mAb does not augment proliferation, based on ^{3}H thymidine incorporation. However, even at maximal proliferation, CD28 stimulation augments ^{3}H uridine incorporation (42). In addition, cell volume (i.e. blastogenesis) increases significantly on a per cell basis following costimulation of CD28 compared to CD3 (46) or PMA (44) stimulation alone. Taken together, these findings suggest that even at maximal proliferation, CD28 costimulation can increase the metabolic activity of T cells. Furthermore, these studies (in conjunction with unpublished data) indicate that CD28 can regulate cellular growth independently from cellular proliferation.

A number of these activities have now been identified. As previously noted, stimulation of resting T cells by soluble bivalent anti-CD28 mAb alone has no effect. Whereas CD3 stimulation alone results in IL-2R expression and modest IL-2 production, costimulation by CD28 results in a marked (10 to 50 fold) increase in IL-2 secretion (40,42,44,46). Similar results have been reported following costimulation by anti-CD28 mAb and anti-CD2 mAb (37), concanavalin A (46), TPA (20), PHA (46), PMA (40,42,44,46) and PMA + ionomycin (40). This last report also notes that while maximal proliferation can be obtained by PMA + ionomycin or PMA + anti-CD28 mAb, maximal IL-2 production requires all three signals.

Regardless of which surface activation pathways are used to induce T cell proliferation, it has been found that the mRNA expression and secretion of IL-2 and other lymphokines, including TNF-α, γIFN, lymphotoxin, IL-3 and GM-CSF, can be significantly augmented by CD28 costimulation (42,43). This augmentation has specificity for lymphokine genes, as mRNA expression of other activation genes, including c-myc, c-fos, and the 4F2 heavy chain, is not affected (50). In mice, expression of this set of lymphokines, designated T_H1 lymphokines, appears to be limited to a particular subset of helper T cells (51). Although the human equivalent of T_H1 lymphocytes has not been identified, it is intriguing that CD28 costimulation recapitulates the phenotype of murine T_H1 clones. Recently, the expression of five additional inducible, T cell-specific genes was found to be enhanced by PMA and CD28 costimulation (52). The significance of this finding awaits further characterization of these genes.

The above findings suggest that the CD28 activation pathway plays an important role in regulating lymphokine production during T cell activation (figure 3). Primary stimulation through the TCR/CD3 pathway results in, at most, only a low level of lymphokine secretion (40,42) that is sufficient to stimulate autocrine-mediated proliferation but not to result in accumulation of lymphokine in the surrounding environment. This hypothesis predicts that costimulation of the CD28 pathway results in marked augmentation of lymphokine mRNA expression and a subsequent shift from autocrine to paracrine production of lymphokines, thus conferring effector function to antigen-stimulated T cells.

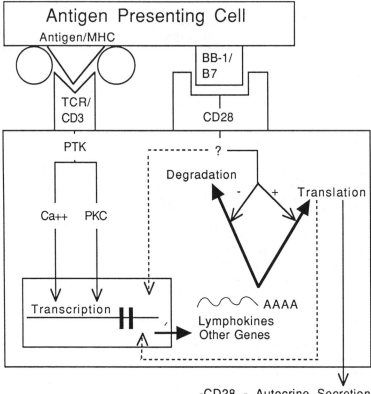

-CD28 - Autocrine Secretion
+CD28 - Paracrine Secretion

FIG 3. Proposed role of the CD28 receptor. Engagement of
antigen/MHC by TCR/CD3 receptor by results in signal
transduction, mediated by protein-tyrosine kinase (PTK),
intracellular calcium (Ca^{2+}) and protein kinase C (PKC), to
the nucleus. This in turn results in transcription of a
number of activation genes, including lymphokines. The fate
of these transcripts is primarily to be degraded, with only
minimal translation and only autocrine effects.
Simultaneous engagement of BB-1/B7 by CD28 results in a yet
undefined second signal, which leads to decreased
degradation and enhanced translation of these transcripts,
resulting in increase protein secretion and paracrine
effects. In addition, CD28 activation may directly affect
transcription, or may indirectly affect transcription
through its effects on mRNA degradation/translation of
transcription factors (dotted lines).

A major mechanism by which CD28 costimulation augments lymphokine production in mature T cells is by inhibiting degradation of lymphokine mRNA (50). This in turn results in higher steady state levels mRNA levels, which enhances translation and augments protein secretion. Mechanisms other than mRNA stabilization may also be involved in the CD28-mediated augmentation of lymphokine production (40). It has been recently reported that stimulation by anti-CD28 mAb + PMA + ionomycin results in a five-fold increase in IL-2 enhancer activity (53). This appears to be mediated by CD28-induced proteins that bind to a unique region within the IL-2 enhancer. The DNA motif to which this complex binds, AAAGAAATTCC, is similar to other motifs found in the 5' flanking region of several other lymphokines (GM-CSF, IL-3, G-CSF, γIFN) (53) and may represent another common element through which multiple lymphokines are affected by CD28 stimulation. Whether this DNA binding activity is the result of protein activation, *de novo* transcription or mRNA stabilization remains to be determined.

Lymphokine expression following stimulation by anti-CD3 mAb, anti-CD3 mAb + PMA, or PMA + ionomycin can be completely suppressed by cyclosporin A (40,42,43), consistent with its immunosuppressive effects. Costimulation with soluble, bivalent anti-CD28 mAb results in lymphokine expression that is resistant to this suppression (40,42,43). This finding is predicted by the similar effect of CD28 costimulation on CsA-mediated suppression of proliferation (40,44). Lymphokine mRNA expression following anti-CD28 mAb + PMA stimulation is completely unaffected by CsA (42). Interestingly, CsA partially suppresses IL-2 mRNA expression stimulated by ionomycin + PMA + anti-CD28 mAb down to levels comparable to those induced by PMA + anti-CD28 mAb (40). This is consistent with the hypothesis that CsA-mediated suppression involves a component of the calcium-dependent intracellular activation pathway (3). This is further supported by the finding that cyclosporin A suppresses, but does not abrogate, lymphokine mRNA expression following CD3 and CD28 costimulation.

Stimulation of the TCR/CD3 complex results in protein-tyrosine kinase activation and subsequent protein kinase C activation and increases in intracellular calcium (54). It is presumably the synergy between protein kinase C and CD28 activation that confers resistance to suppression by CsA. Since the induction of IL-2 gene expression during peripheral blood T cell activation requires transcriptional activation, these results suggest that PMA + CD28 might involve a novel, CsA-resistant, mechanism of IL-2 transcriptional activation. However, to date this mechanism remains undefined. The recently described CD28-responsive IL-2 enhancer binding elements are not induced by PMA + anti-CD28 mAb alone (53). It remains unresolved whether the fundamental mechanism by which CD28 costimulation leads to altered transcription is primarily a direct effect on the target genes or

is a consequence of its effects on mRNA stability and/or
translation of transcriptional regulatory genes.
       The effects of CD28 stimulation on cytotoxic effector
function have been less well characterized. The addition of anti-
CD3/anti-melanoma and anti-CD28/anti-melanoma mAb heteroconjugates
to purified T cells and melanoma cells results in T cell
proliferation and potent tumor cell killing (55). Stimulation of
purified T cells with immobilized anti-CD3 mAb and soluble anti-
CD28 mAb appears to result in potent lytic activity against tumor
targets that does not require heteroconjugate mAb targeting (3).

CD28 SIGNAL TRANSDUCTION PATHWAY

       Experiments that have examined the effects of CD28
activation suggest that CD28 regulates a signal transduction
pathway that is distinct from those induced by the TCR/CD3
complex. To define an accessory pathway involved in T cell
activation requires the resolution of the question: Does
signalling through the accessory molecule simply enhance or
sustain signals provided by the TCR/CD3 complex, or does it
represent a distinct pathway? One would anticipate that pathways
that simply enhance TCR/CD3 signalling would neither function in
the absence of the TCR/CD3 complex or signalling nor would they
mediate novel effects distinct from those resulting from TCR/CD3
signalling. Conversely, pathways that are distinct from the
TCR/CD3 pathway could be expected to either function in the
absence of TCR/CD3 signalling and/or mediate effects that are not
seen with TCR/CD3 stimulation alone.
       There are several lines of evidence that suggest that the
CD28 pathway is distinct from other known pathways. First,
modulation of CD3 or CD2 off the surface of T cells does not
affect surface expression of CD28, and visa versa (20,21,56),
suggesting the lack of close physical association. In addition,
modulation of the TCR/CD3 complex does not affect the ability of T
cell to proliferate in response to crosslinked anti-CD28 mAb and
IL-2 (48) or to soluble, bivalent anti-CD28 mAb and TPA (20),
indicating that the actual TCR/CD3 complex is not required for
CD28-mediated effects. T cell lines that do not have surface
expression of CD3 nonetheless express CD28 and can be stimulated
to produce IL-2 by anti-CD28 mAb and protein kinase C activation
by PMA (41), suggesting no functional dependence on cell surface
structures associated with the TCR/CD3 complex. Second, it is
clear that CD28 stimulation synergizes with maximal TCR/CD3 or
PMA/ionomycin stimulation to augment IL-2 production in purified T
cells (40,42,44,46). Finally, CD28 stimulation results in the
development of resistance to the suppressive effects of
cyclosporin A, glucocorticoids and PGE_2 (3). In contrast, the
proliferative responses of T cells stimulated through the TCR/CD3
complex alone are suppressed by either CsA, glucocorticoid or
PGE_2. Also, the ability of highly crosslinked anti-CD28 mAb to

mediate distinct effects, such as IL-2 and IL-6 responsiveness, without additional signals is further evidence that CD28 stimulation involves a distinct pathway. Altogether, these lines of evidence argue strongly that the CD28 receptor mediates a signal transduction pathway that is distinct from the pathway activated by the TCR/CD3 complex.

What are the components of the CD28 pathway? One effect of CD28 stimulation is the stabilization and enhanced translation of lymphokine mRNA, which results in the paracrine secretion of T_H1 lymphokines (5). In addition, CD28 stimulation appears to affect transcriptional regulation. CD28 stimulation in the Jurkat cell line costimulated with PMA + ionomycin appears to result in the synthesis of novel proteins which bind to the IL-2 enhancer and increase its activity (53).

CD28-induced lymphokine mRNA stabilization may involve the synthesis of novel proteins that bind to elements within lymphokine mRNA, such as AU-rich motifs (50), and mediate stability. Alternatively, mRNA stabilization may be mediated by elements of the protein translation machinery itself, such as initiation factors. Enhanced translation might secondarily protect lymphokine mRNA's from degradation. Consistant with this is the recent finding that messages containing AU-rich motifs are susceptible to inducible alterations in translation efficiency (57). Enhanced translational efficiency may also explain the augmented metabolic activity and blastogenesis that result from CD28 costimulation.

As for the initial components of the CD28 pathway, the biochemical nature of the signals generated following binding to the CD28 receptor remains unclear. The stimulus caused by the binding of soluble, bivalent anti-CD28 mAb or F(ab)'$_2$ fragments, which is sufficient to serve as a comitogen and mediate lymphokine expression, does not result in an increase in intracellular calcium (48,49) or production of inositol phosphates (InsP) in resting T cells (48). Soluble anti-CD28 mAb binding also has no effect on cAMP levels (39). In fact, proliferation induced by CD28 and CD3 costimulation is completely resistant to the antiproliferative effects of the adenylate cyclase agonist PGE$_2$, unlike that induced by CD3 stimulation alone (39). Furthermore, proliferation and lymphokine expression (TNF-α, lymphotoxin) induced by CD28 and PMA costimulation is resistant cholera toxin, which causes increases in cAMP through G protein activation (43). Soluble anti-CD28 mAb binding causes a small increase in cGMP levels in the Jurkat T cell line (39). However, this is not thought to mediate the effect of CD28 stimulation on lymphokines since treatment of T cells with agents that increase cellular cGMP does not augment lymphokine expression (3). It has been recently observed that the protein-tyrosine kinase inhibitor genistein (4,5,7 trihydroxyisoflavone), in combination with cyclosporin A, could completely inhibit IL-2 production stimulated by PMA + ionomycin + anti-CD28 mAb (58). The CD28 receptor does not have significant homology to any known protein-tyrosine kinase and attempts to coprecipitate protein-tyrosine kinase activity with CD28 have been unsuccessful to date (58).

Unlike soluble anti-CD28 mAb, stimulation by crosslinked anti-CD28 mAb results in an increase in intracellular calcium that is insensitive to extracellular calcium depletion (48,49). This effect can be abrogated by pertussis toxin (49). Crosslinking CD28 receptors also results in sustained inositol phosphate production (48). However, the inositol phosphate production stimulated by crosslinking CD28 appears to be different from that induced by TCR/CD3 stimulation. Inositol phosphate production following CD28 stimulation is enhanced by pretreatment with phorbol esters, whereas phorbol ester pretreatment results in suppression of TCR/CD3-induced InsP production (39).

It is presently unclear whether signalling under physiologic conditions through the CD28 pathway is mediated primarily through receptor occupancy alone, or primarily through receptor crosslinking alone or whether both mechanisms play important roles. How might the observed differences between the effect of soluble versus crosslinked CD28 stimulation be reconciled? One possibility is that CD28 regulates two signal transduction pathways - one which requires a high degree of receptor crosslinking, results in IL-2 and IL-6 sensitivity and is mediated by inositol phosphate production and calcium mobilization, and one which requires minimal receptor oligomerization, is independent of the TCR/CD3 complex, results in specific increases in lymphokine production and is mediated by a yet uncharacterized second messenger. Alternatively, signal transduction through the CD28 receptor may be mediated by a single mechanism that is sensitive to the degree of receptor oligomerization and is capable delivering a graded signal.

In conclusion, it seems clear that the CD28 activation pathway mediates important, novel biologic events involved in T cell activation and development of effector function. Many components of this pathway have been defined. Many other components, as well as the normal and pathophysiologic role of CD28 activation, await further study. In particular, the cellular interactions between cells expressing CD28, the CD28 homologue CTLA-4 and the CD28 ligand BB-1/B7 remain to be defined. It is likely that these cell surface molecules will play important roles in lymphocyte development and in regulation of cell-cell interactions during an in vivo immune response.

REFERENCES

1. Weiss A. T lymphocyte activation. In:Paul WE, ed. Fundamental Immunology, second edition. New York: Raven Press, 1989.

2. Hansen JA, Martin PJ, Nowinski RC. Monoclonal antibodies identify a novel T-cell antigen and Ia antigens of human lymphocytes. Immunogenetics 1980;10:247-260.

3. June CH, Ledbetter JA, Linsley PS, Thompson CB. Role of the CD28 receptor in T-cell activation. Immunol. Today 1990;11:211-216.

4. Gross JA, St. John T, Allison JP. The murine homologue of the T lymphocyte antigen CD28. J. Immunol. 1990;144:3201-3210.

5. Clark EA, Draves KE. Activation of macaque T cells and B cells with agonistic monoclonal antibodies. Eur. J. Immunol. 1987;17:1799-1805.

6. Turka LA, Ledbetter JA, Lee K, June CH, Thompson CB. CD28 is an inducible T cell surface antigen that transduces a proliferative signal in CD3+ mature thymocytes. J. Immunol. 1990;144:1646-1653.

7. Pierres A, Cerdan C, Lopez M, Mawas C, Olive D. "CD3low" human thymocyte populations can readily be triggered via the CD2 and/or CD28 activation pathways whereas the CD3 pathway remains nonfunctional. J. Immunol. 1990;144:1202-1207.

8. Testi R, Lanier LL. Functional expression of CD28 on T cell antigen receptor γ/δ-bearing T lymphocytes. Eur. J. Immunol. 1989;19:185-188.

9. Yamada H, Martin PJ, Bean MA, Braun MP, Beatty PG, Sadamoto K, Hansen JA. Monoclonal antibody 9.3 and anti-CD11 antibodies define reciprocal subsets of lymphocytes. Eur. J. Immunol. 1985;15:1164-1168.

10. Li SG, Ottenhoff THM, Van den Elsen P, Koning F, Zhang L, Mak T, De Vries RRP. Human suppressor T cell clones lack CD28. Eur. J. Immunol. 1990;20:1281-1288.

11. Rotteveel FTM, Kokkelink I, Van Lier RAW, Kuenen B, Meager A, Miedema F, Lucas CJ. Clonal analysis of functionally distinct human CD4+ T cell subsets. J. Exp. Med. 1988;168:1659-1673.

12. Lum LG, Orcutt-Thordarson N, Seigneuret MC, Hansen JA. *In vitro* regulation of immunoglobulin synthesis by T-cell

subpopulations defined by a new human T-cell antigen (9.3). Cell. Immunol. 1982;72:122-129.

13. Damle NK, Mohagheghpour N, Hansen JA, Engleman EG. Alloantigen-specific cytotoxic and suppressor T lymphocytes are derived from phenotypically distinct precursors. J. Immunol. 1983;131:2296-2299.

14. Damle NK, Doyle LV, Grosmaire LS, Ledbetter JA. Differential regulatory signals delivered by antibody binding to the CD28 (Tp44) molecule during the activation of human T lymphocytes. J. Immunol. 1988;140:1753-1761.

15. Morishita Y, Sao H, Hansen JA, Martin PJ. A distinct subset of human CD4+ cells with a limited alloreactive T cell receptor repertoire. J. Immunol. 1989;143:2783-2789.

16. Poggi A, Bottino C, Zocchi MR, Pantaleo G, Ciccone E, Mingari C, Moretta L, Moretta A. CD3+WT31- peripheral T lymphocytes lack T44 (CD28), a surface molecule involved in activation of T cells bearing the α/β heterodimer. Eur. J. Immunol. 1987;17:1065-1068.

17. Kozbor D, Moretta A, Messner HA, Moretta L, Croce CM. Tp44 molecules involved in antigen-independent T cell activation are expressed on human plasma cells. J. Immunol. 1987;138:4128-4132.

18. Lesslauer W, Gmunder H, Bohlen P. Purification and N-terminal amino acid sequence of the human T90/44 (CD28) antigen. Immunogenetics 1988;27:388-391.

19. Martin PJ, Ledbetter JA, Morishita Y, June CH, Beatty PG, Hansen JA. A 44 kilodalton cell surface homodimer regulates interleukin 2 production by activated human T lymphocytes. J. Immunol. 1986;136:3282-3287.

20. Hara T, Fu SM, Hansen JA. Human T cell activation II: A new activation pathway used by a major T cell population via a disulfide-bonded dimer of a 44 kilodalton polypeptide (9.3 antigen). J. Exp. Med. 1985;161:1513-1524.

21. Moretta A, Pantaleo G, Lopez-Botet M, Moretta L. Involvement of T44 molecules in an antigen-independent pathway of T cell activation. J. Exp. Med. 1985;162:823-838.

22. Lesslauer W, Lerch P, Gmunder H. Two detergent-insoluble proteins of the human lymphocyte membrane are enriched in an isolated membrane fraction. Biochem. Biophys. Acta 1982;693:351-358.

23. Gmunder H, Lesslauer W. A 45-kDa human T-cell membrane glycoprotein functions in the regulation of cell proliferative responses. Eur. J. Immunol. 1984;142:153-160.

24. Gmunder H, Lerch P, Lesslauer W. Human lymphocyte membrane proteins treated with neuraminidase. Biochem. Biophys. Acta 1982;693:359-363.

25. Aruffo A, Seed B. Molecular cloning of a CD28 cDNA by a high-efficiency COS cell expression system. Proc. Natl. Acad. Sci. USA 1987;84:8573-8577.

26. Lafage-Pochitaloff M, Costello R, Couez D, Simonetti J, Mannoni P, Mawas C, Olive D. Human CD28 and CTLA-4 Ig superfamily genes are located on chromosome 2 on bands q33-34. Immunogenetics 1990;31:198-201.

27. Dariavach P, Mattei M, Golstein P, Lefranc M. Human Ig superfamily CTLA-4 gene: chromosomal localization and identity of protein sequence between murine and human CTLA-4 cytoplasmic domains. Eur. J. Immunol. 1988;18:1901-1905.

28. Lee KP, Taylor C, Petryniak B, Turka LA, June CH, Thompson CB. The genomic organization of the CD28 gene: Implications for the regulation of CD28 mRNA expression and heterogeneity. J. Immunol. 1990;145:344-352.

29. Linsley PS, Clark EA, Ledbetter JA. T-cell antigen CD28 mediates adhesion with B cells by interacting with activation antigen B7/BB-1. Proc. Natl. Acad Sci. USA 1990;87:5031-5035.

30. Freeman GJ, Freedman AS, Segil JM, Lee G, Whitman JF, Nadler LM. B7, a new member of the Ig superfamily with unique expression on activated and neoplastic B cells. J. Immunol. 1989;143:2714-2722.

31. Turka LA, Linsley PS, Paine R, Schieven GL, Thompson CB, Ledbetter JA. Signal transduction via CD4, CD8, and CD28 in mature and immature thymocytes: Implications for thymic selection. J. Immunol. in press.

32. Linsley PS, Brady W, Grosmaire L, Aruffo A, Damle NK, Ledbetter JA. Binding of the B cell activation antigen B7 to CD28 costimulates cytokine mRNA accumulation and T cell proliferation. J. Exp. Med. in press.

32a. Kohno K, Shibata Y, Matsuo Y, Minowada J. CD28 molecule as a receptor-like function for accessory signals in cell-mediated augmentation of IL-2 production. Cell. Immunol. 1990;131:1-10.

33. Yang SY, Denning SM, Mizuno S, DuPont B, Haynes BF. A novel activation pathway for mature thymocytes: Costimulation of CD2 (T,p50) and CD28 (T,p44) induces autocrine interleukin 2/interleukin 2 receptor-mediated cell proliferation. J. Exp. Med. 1988;168:1457-1468.

34. Koulova L, Yang SY, Dupont B. Identification of the anti-CD3-unresponsive subpopulation of CD4+, CD45RA+ peripheral T lymphocytes. J. Immunol. 1990;145:2035-2043.

35. Zocchi MR, Marelli F, Poggi A. CD1+ thymocytes proliferate and give rise to functional cells after stimulation with monoclonal antibodies recognizing CD3, CD2 or CD28 surface molecules. Cell. Immunol. 1990;129:394-403.

36. Lesslauer W, Koning F, Ottenhoff T, Giphart M, Goulmy E, van Rood JJ. T90/44 (9.3 antigen). A cell surface molecule with a function in human T cell activation. Eur. J. Immunol. 1986;16:1289-1296.

37. Van Lier RAW, Brouwer M, Aarden LA. Signals involved in T cell activation. T cell proliferation induced through the synergistic action of anti-CD28 and anti-CD2 monoclonal antibodies. Eur. J Immunol. 1988;18:167-172.

38. Ledbetter JA, Martin PJ, Spooner CE, Wofsy D, Tsu TT, Beatty PG, Gladstone P. Antibodies to Tp67 and Tp44 augment and sustain proliferative responses of activated T cells. J. Immunol. 1985;135:2331\2336.

39. Ledbetter JA, Parsons M, Martin PJ, Hansen JA, Rabinovitch PS, June CH. Antibody binding to CD5 (Tp67) and Tp44 T cell surface molecules: Effects on cyclic nucleotides, cytoplasmic free calcium, and cAMP-mediated suppression. J. Immunol. 1986;137:3299-3305.

40. June CH, Ledbetter JA, Lindsten T, Thompson CB. Evidence for the involvement of three distinct signals in the induction of IL-2 gene expression in human T lymphocytes. J. Immunol. 1989;143:153-161.

41. Weiss A, Manger B, Imboden J. Synergy between the T3/antigen receptor complex and Tp44 in the activation of human T cells. J. Immunol. 1986;137:819-825.

42. Thompson CB, Lindsten T, Ledbetter JA, Kunkel SL, Young HA, Emerson SG, Leiden JM, June CH. CD28 activation pathway regulates the production of multiple T-cell-derived lymphokines/cytokines. Proc. Natl. Acad. Sci. USA 1989;86:1333-1337.

43. Bjorndahl JM, Sung SJ, Hansen JA, Fu SM. Human T cell activation: differential response to anti-CD28 as compared to anti-CD3 monoclonal antibodies. Eur. J. Immunol. 1989;19:881-887.

44. June CH, Ledbetter JA, Gillespie MM, Lindsten T, Thompson CB. T-cell proliferation involving the CD28 pathway is associated with cyclosporin-resistant interleukin 2 gene expression. Mol. Cell. Bio. 1987;7:4472-4481.

45. Baroja ML, Ceuppens JL, Van Damme J, Billiau A. Cooperation between an anti-T cell (anti-CD28) monoclonal antibody and monocyte-produced IL-6 in the induction of T cell responsiveness to IL-2. J. Immunol. 1988;141:1502-1507.

46. Baroja ML, Lorre K, Van Vaeck F, Ceuppens JL. The anti-T cell monoclonal antibody 9.3 (anti-CD28) provides a helper signal and bypasses the need for accessory cells in T cell activation with immobilized anti-CD3 and mitogens. Cell. Immunol. 1989;120:205-217.

47. Damle NK, Doyle LV. Stimulation via the CD3 and CD28 molecules induces responsiveness to IL-4 in CD4+CD29+CD45R− memory T lymphocytes. J. Immunol. 1989;143:1761-1767.

48. Ledbetter JA, Imboden JB, Schieven GL, Grosmaire LS, Rabinovitch PS, Lindsten T, Thompson CB, June CH. CD28 ligation in T-cell activation: Evidence for two signal transduction pathways. Blood 1990;75:1531-1539.

49. Ledbetter JA, June CH, Grosmaire LS, Rabinovitch PS. Crosslinking of surface antigens causes mobilization of intracellular ionized calcium in T lymphocytes. Proc. Natl. Acad. Sci. USA 1987;84:1384-1388.

50. Lindsten T, June CH, Ledbetter JA, Stella G, Thompson CB. Regulation of lymphokine messenger RNA stability by a surface-mediated T cell activation pathway. Science 1989;244:339-343.

51. Mosmann TR, Coffman RL. Heterogeneity of cytokine secretion patterns and function of helper T cells. Annu. Rev. Immunol. 1989;7:145-173.

52. Irving SG, June CH, Zipfel PF, Siebenlist U, Kelly K. Mitogen-induced genes are subject to multiple pathways of regulation in the initial stages of T-cell activation. Mol. Cell Biol. 1989;9:1034-1040.

53. Fraser JD, Irving BA, Crabtree GR, Weiss A. Regulation of interleukin-2 gene enhancer activity by the T cell accessory molecule CD28. Science, in press.

54.   Altman A, Coggeshall KM, Mustelin T.   Molecular events
mediating T cell activation. In:Dixon FJ, ed.   Advances in
Immunology, vol 48. San Diego: Academic Press, 1990.

55.   Jung G, Ledbetter JA, Muller-Eberhard HJ.   Induction of
cytotoxicity in resting human T lymphocytes bound to tumor cells
by antibody heteroconjugates.   Proc. Natl. Acad. Sci. USA
1987;84:4611-4615.

56.   Moretta A, Olive D, Poggi A, Pantaleo G, Mawas C, Moretta L.
Modulation of surface T11 molecules induced by monoclonal
antibodies: analysis of the functional relationship between
antigen-dependent and antigen-independent pathways of human T cell
activation.   Eur. J. Immunol. 1986;16:1427-1432.

57.   Han J, Brown T, Beutler B.   Endotoxin-responsive sequences
control cachectin/tumor necrosis factor biosynthesis at the
translational level.   J. Exp. Med. 1990;171:465-475.

58.   Trevillyan J, Lu Y, Bjorndahl J, Phillips C, Atluru R.   T-
cell activation via the CD28 pathway is blocked by a selective
protein-tyrosine kinase inhibitor.   FASEB J. 1990;4:A2200.

Advances in Regulation of Cell Growth, Volume 2;
Cell Activation: Genetic Approaches, edited by
James J. Mond, John C. Cambier, and Arthur Weiss.
Raven Press, Ltd., New York © 1991.

8

Molecular Approaches to the Study of Adrenergic Receptors

Brian K. Kobilka

Departments of Molecular and Cellular Physiology, and Cardiology, and the Howard Hughes Medical Institute, Stanford University Medical Center, Stanford, California 94305

Adrenergic receptors are plasma membrane proteins which respond to catecholamine hormones and neurotransmitters. Studying the structure and function of these receptors has been challenging because they are not naturally abundant proteins, and they require a lipid environment to be fully active. Furthermore, pharmacologic data have suggested that there may be more subtypes of adrenergic receptors than previously reported. Yet, the synthetic ligands which are capable of detecting subtle differences between receptor subtypes are not selective enough to pharmacologically isolate specific receptor subtypes for study in vivo. Applying the techniques of molecular biology to the study of adrenergic receptors has helped overcome many of the technical difficulties that had complicated previous efforts at biochemical characterization of these proteins. This chapter will begin with a brief overview of what is currently known about adrenergic receptors, followed by a review of the various ways that the techniques of molecular biology have been applied to the study of adrenergic receptors. More comprehensive reviews of adrenergic receptor structure, function and regulation have recently been published (1,2).

GENERAL PROPERTIES OF ADRENERGIC RECEPTORS

Adrenergic receptors are plasma membrane proteins which are activated by the catecholamines epinephrine and norepinephrine. These receptors form the interface between the sympathetic nervous system and the viscera. The activity of the sympathetic nervous system is regulated in part by adrenergic receptors in the vasomotor centers of the brainstem. Sympathetic nerve terminals in peripheral tissues release catecholamines which activate postsynaptic adrenergic receptors. These receptors play an important role in sympathetic control of systemic blood pressure, cardiac rate and contractility, fat and carbohydrate metabolism, as well as fluid and electrolyte balance.

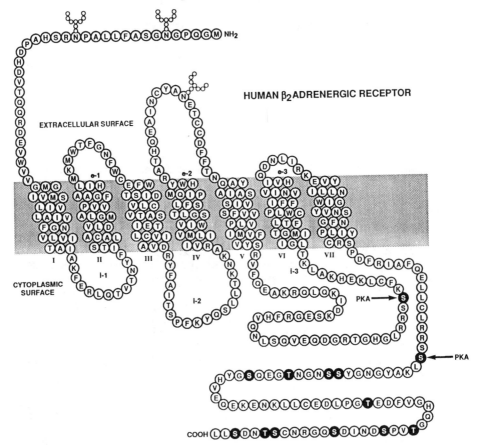

FIG. 1. Diagram of the proposed membrane topology for the human ß-2 adrenergic receptor. Circles with letters represent amino acids, identified by the single letter amino acid code. Proposed membrane-spanning domains are numbered with roman numerals. Extracellular domains are designated by a prefix **e-**. Intracellular domains are designated by a prefix **i-**. Black circles with white letters indicate potential phosphorylation sites. Potential sites for phosphorylation by protein kinase are indicated by **PKA**.

Stimulation of adrenergic receptors by catecholamines leads to a functional interaction between these receptors and GTP binding proteins or G proteins. Receptor activated G proteins modulate the activity of a variety of cellular enzymes such as adenylyl cyclase, phospholipase A, and phospholipase C, as well as K^+ and Ca^{++} channels. Several comprehensive reviews of G proteins have recently been published (3,4).

To date, the genes and/or cDNAs for nine types of adrenergic receptors have been cloned. These include the avian ß receptor (5); mammalian ß-1(6), ß-2(7,8) and ß-3 (9) receptors; two subtypes of mammalian α-1 adrenergic receptors (10,11), and three subtypes of mammalian α-2 adrenergic receptors (12,13,14,15). The classification into α-1, α-2, ß-1,-2, and -3 is based on the pharmacologic properties of these receptors. Figure 1 is a diagram of the proposed membrane topology for the human ß-2 adrenergic receptor. All of the adrenergic receptors have similar structural features. There are seven clusters of hydrophobic amino acids which are thought to be membrane spanning domains. The amino terminus is extracellular and the carboxyl terminus lies in the cytoplasm. Receptors with similar pharmacologic properties share the greatest homology within the hydrophobic domains. Evidence from a variety of studies (some to be discussed below) indicate that the hydrophobic domains form the ligand binding pocket. Key residues involved in catechol binding include an aspartate residue in the third hydrophobic domain (16), and two serine residues in the fifth hydrophobic domain (17). The aspartate has been proposed to act as the counter ion for the amine nitrogen on the catecholamine (16). The serine residues may form hydrogen bond with hydroxyl groups on the catechol ring (17). Portions of the second and third intracellular loop and the carboxyl terminus are involved in interaction with G proteins (18,19,20,21). These domains may also interact with kinases (22,23) and cytoskeletal proteins (24) involved in regulation of receptor function.

CLONING OF ADRENERGIC RECEPTORS

Cloning has had a tremendous impact on the field of adrenergic receptors. As indicated above, 9 types of adrenergic receptor genes and/or cDNA's have been cloned. Of these, only four were cloned as a result of using protein sequence data to design oligonucleotide probes for library screening. These receptors include the turkey ß receptor (5), the hamster ß-2 receptor (7), the hamster α-1 receptor (10), and the human α-2 receptor (12). Obtaining the protein sequence for these receptors was particularly challenging because they are not naturally abundant proteins; and because of the difficulties encountered in purifying the protein and in purifying the extremely hydrophobic peptides for sequence analysis. For example, the hamster α-1 receptor was purified from DDT$_1$MF-2 cells. To obtain sufficient protein for sequence analysis, 1600 liters of cultured cells were used (10). Purification of sufficient quantities of the human α-2 receptor to obtain protein sequence required 1400 units of outdated human platelet (12).

Clones for those adrenergic receptors for which protein sequence was not available were obtained by using probes based on the sequence of the cloned adrenergic receptors. These probes were used to screen libraries at reduced stringency (6,9,11,13), or were used as primers to amplify sequences from cDNA or genomic libraries using the polymerase chain reaction (14). The cloning of the different adrenergic receptor genes has confirmed the existence of receptor subtypes which had been proposed to exist on the basis of pharmacologic data (25,26). These pharmacologic studies were complicated by the fact that animal tissues often contained very low levels of binding activity or contained more than one adrenergic receptor subtype. Furthermore, available agonists and antagonists could not pharmacologically isolate specific subtypes in whole animals or in tissues where more than one subtype was expressed. The cloning data prove that these subtype differences are due to different gene products and not to differences in posttranslational modification of receptors. The clones provide extremely valuable tools for understanding the function of these closely related subtypes. Individual subtypes can be expressed separately in tissue culture cell lines which do not normally express adrenergic receptors. Thus the cellular physiology and pharmacology of individual subtypes can be studied without the need for highly selective subtype specific ligands.

ANALYSIS OF RECEPTOR STRUCTURE

The Use Of Sequence Information From Cloned Receptors To Interpret Biochemical Studies

The amino acid sequence derived from the nucleotide sequence of the cDNA and genomic clones provided the first clue to the structure of adrenergic receptors. The observation that the ß-2 receptor shared similar structural features with bacteriorhodopsin led to the proposal that ß-2 receptors consisted of seven membrane spanning domains (7). The primary amino acid sequence deduced from the nucleotide sequence of cloned receptors has also been used to interpret biochemical studies investigating receptor structure and function. Three groups have used photo-activatable crosslinking antagonists to locate receptor domains involved in ligand binding to the turkey ß receptor (27), the hamster ß-2 receptor (28), and the human platelet α-2 receptor (29). The goal of these studies was to identify the receptor domain which contained the covalently bound radioactively labeled ligand. In two cases it was possible to identify the labeled peptide by obtaining amino acid sequence (27,28). In the case of the turkey ß receptor label was associated with the seventh hydrophobic domain. In the hamster ß-2 receptor, the antagonist was bound to a peptide in the second hydrophobic domain. In the case of the human α-2 receptor, it was not possible to identify the labeled peptide by sequence analysis. However, the amino acid sequence of this receptor was used to predict proteolytic cleavage sites and identify the labeled peptide by its physical characteristics (29). The studies identified the labeled α-2 receptor peptide as being from the third hydrophobic domain. Each of the studies suggests a different site for labeling of the receptor with a photo-activatable ligand. This may be due to differences in the binding site for each receptor. More likely, it suggests that several hydrophobic domains participate in the formation of the ligand binding pocket. The latter interpretation is consistent with data from mutagenesis studies to be discussed below.

The amino acid sequence has also been used to interpret studies using limited proteolysis to probe receptor structure. This approach has been used to locate the site of glycosylation of the hamster ß-2 receptor in the amino terminus and sites of regulatory phosphorylation by ß adrenergic receptor kinase in the carboxyl terminus (30). It has also been used to identify essential structural elements required for function of the turkey ß receptor (31) and the porcine α-2 receptor (32). In these studies proteolysis was carried out under conditions which removed functionally nonessential portions of the receptor protein leaving a core structure which retained functional properties such as ligand binding and G protein activation. The components of this functional core were determined by peptide mapping, and by the use of antibodies capable of recognizing specific peptide sequences from the receptor. Rubenstein and colleagues demonstrated that most of the carboxyl terminus, the third intracellular loop and the amino terminus could be removed from the turkey erythrocyte ß receptor without affecting the ability of the receptor to bind ligands or activate G_s. Wilson and colleagues carried out similar studies on the porcine α-2 receptor and obtained data suggesting that removal of the last two hydrophobic domains by proteolysis did not affect the ability of the first five membrane spanning domains of this receptor to bind antagonists (32). This result is somewhat surprising in light of the photoaffinity labeling studies on the turkey ß receptor described above and studies on chimeric receptors to be discussed below. These approaches to the study of receptor structure complement the use of mutagenesis. Removal of such large domains by deletion mutagenesis may result in nonfunctional protein because the deleted domains may be necessary for proper protein folding. However, removal of the domains by protease treatment after folding is complete may be tolerated.

Site Directed Antibodies

Wang and colleagues have used the sequence for the cloned hamster ß-2 adrenergic receptor to obtain domain-specific antibodies which were used to study the membrane topology of this receptor (33). Synthetic peptides representing portions of the hydrophilic domains of the receptor were used to immunize rabbits to obtain polyclonal antiserum. These antisera were then used to determine if the receptor domains recognized by these antibodies were located on the extracellular surface or on the cytoplasmic side of the plasma membrane. Indirect immunofluorescence of fixed cultured cells expressing up to 2 million receptors per cell was carried out with each of the antisera. The results from this study confirmed the membrane topology shown in figure 1. Antibodies to proposed extracellular domains were capable of detecting receptor on fixed, nonpermeablized cells. Antiserum to putative intracellular domains produced fluorescence only on fixed cells that were permeablized with detergent permitting the antibody to pass through the lipid bilayer.

EXPRESSION TECHNOLOGY

Cloned adrenergic receptors have been expressed in a wide variety of cells. Expression of cloned receptor genes is an extremely valuable tool for studying the structure and function of these proteins. Different types of expression systems have different applications. Some systems provide a physiologic environment in which studies of signal transduction and receptor regulation can be carried out. Other systems are capable of producing large quantities of receptor protein which may be used to study receptor structure. In the sections which follow, the different expression systems that have been used to study adrenergic receptors will be discussed.

Cultured Mammalian Cells

Transient Expression

A variety of eukaryotic cell lines have been used to express adrenergic receptors. Cos-7 cells (SV40 transformed african green monkey kidney cells) are frequently used to express wild type and mutant receptors for the purpose of studying their pharmacologic properties. Plasmid DNA can be easily and efficiently inserted into these cells by a variety of techniques (34). Using immunofluorescence to detect expressed ß-2 adrenergic receptor, we have observed incorporation of DNA into approximately 5-10% of cells following DEAE-DEXTRAN mediated transfection (M. von Zastrow, B. Kobilka, unpublished observation). Once inside these cells, plasmids carrying the SV40 origin of replication are replicated many times as episomes. The high efficiency of transfection combined with the high copy number of plasmids in the transfected cells results in high levels of expression by the third day following transfection. The amplification of the transfected plasmid eventually leads to cell death after 5-7 days; therefore, stable cell lines cannot be obtained. Maximal expression results in approximately 10 - 30 pmole receptor per mg crude membrane protein. The level of expression can be modified by adjusting the amount of DNA used in the transfection (35). The amount of receptor produced by these cells is more than adequate for pharmacologic studies on cloned receptors. However, because less than 10 % of the cells contain recombinant receptors, studying the activation of signal transduction pathways by receptors may be complicated by a relatively high level of basal activity contributed by nontransfected cells. For example, we have not observed significant ß-2 adrenergic stimulation of adenylyl cyclase in membranes from transfected Cos-7 cells. However, receptor mediated changes in cytosolic cAMP, as measured by radioimmunoassay, have been reported (35). These cells have also been used to study receptor mediated changes in polyphosphoinositide metabolism (36).

Stable expression

Plasmids containing recombinant adrenergic receptor DNA can be stably integrated into the chromosomal DNA of cultured cells along with a selectable marker, such as the gene for neomycin resistance (34). A variety of cell lines have been used including CHO, CHW, HELA, mouse L cells. These cell lines were chosen both for the ease with which they can be transfected, and because they do not normally express adrenergic receptors. The advantage of stable expression is that all cells are genetically identical and express the same amount of recombinant receptor. These cell lines are particularly useful for studies of receptor mediated signal transduction and receptor regulation since they will have reproducible properties over several passages. The principal drawback to this type of expression system is the time and effort required to generate these cell lines (3-6 weeks). This may be impractical and expensive when large numbers of mutants need to be evaluated.

Xenopus laevis Oocytes

Xenopus laevis oocytes have been used for expressing ß-2 receptors and receptor mutants for binding and G protein coupling studies (18,19,37). Synthetic RNA is transcribed from a linearized plasmid template and injected into stage V - stage VI oocytes. It is possible to inject as many as 1000 oocytes in an hour with an inexpensive micromanipulator and a dissecting microscope. Maximal expression is observed within 24-48 hours. The level of expression of ß-2 receptor ranges from 250-750 fmole/mg and depends on the quality of the oocytes and the amount of RNA injected. ß-2 adrenergic receptors expressed in oocytes couple efficiently to endogenous Gs and stimulate adenylyl cyclase following the addition of agonist (37). The disadvantage of this expression system is the initial cost of obtaining the equipment for injection, and the seasonal variability in the quality of oocytes. However, this system is useful for analyzing the binding and coupling properties of a large number of mutants in a relatively short time. Functional data for a mutant can be obtained within three days of having the purified plasmid DNA encoding the recombinant receptor. In contrast, stable expression of mutants in mammalian cells requires 3-6 weeks. Transient expression in Cos-7 cells can also be achieved in three days; however, as noted above, this system may not be as useful for studying receptor - G protein interactions.

Insect Cells

High levels of expression of the ß-2 adrenergic receptor has been achieved in insect cells (sf9) infected with recombinant baculovirus (38). The recombinant virus is obtained by homologous recombination in vivo. Viral DNA is cotransfected with a plasmid containing a portion of the viral genome in which the viral sequence encoding a polyhedron protein (which is not essential for virus function) is replaced with the receptor coding sequence under control of the polyhedron protein promoter. Homologous recombination between the viral DNA and the plasmid results in recombinant virus (39). While the frequency of recombination is low (approximately 1:1000), recombinant virus can be detected by differences in the morphology of infected cells in a plaque assay. Pure recombinant virus stocks can be obtained 3-4 weeks after the initial transfection. The sf9 cells can be grown to high density (up to 10 million cells per ml) in a completely defined medium prior to infection with the recombinant virus. The ß-2 receptor expressed in these insect cells differs from that produced in mammalian cells in the extent of processing of the asparagine-linked sugars, however this does not adversely affect receptor function. This expression system is potentially most useful for the production of large quantities of protein for biochemical studies. Up to 40 nmole of ß-2 receptor could be produced in a liter of insect cell culture (38). This represents approximately 2 mg of receptor protein.

Yeast

Recently, expression of the human ß-2 receptor has been achieved in yeast (40). A chimeric receptor molecule was made by fusing a short segment of the amino terminus of the yeast mating factor receptor (a yeast G protein coupled receptor) with the amino terminus of the human ß-2 adrenergic receptor. The chimeric receptor produced in this system exhibits pharmacologic characteristics identical to those of ß-2 receptors expressed in mammalian cells. Furthermore, the ß-2 receptor could be shown to couple to rat $G_{s\alpha}$ which was coexpressed with the ß-2 receptor in a GPA1 mutant yeast strain. This strain is deficient in the α subunit for the G protein which couples to the yeast mating factor receptor. Stimulating the ß receptor in these cells with the agonist isoproterenol induced shmoo formation. Shmoo formation is a morphologic change seen in wild type yeast cells following stimulation of the yeast mating factor receptor. This demonstrates compatibility between components of a mammalian signal transduction system with components of a yeast signal transduction system. This system offers potential for using well established techniques of yeast genetics for carrying out mutagenesis studies on adrenergic receptors and mammalian G protein α subunits. This system also has potential for large scale production of recombinant receptor protein, since yeast cells grow to high density in relatively inexpensive medium.

Prokaryotic Expression

Expression of functional adrenergic receptors has been achieved in E. coli (41-44). Initial expression studies made use of a fusion of the ß-2 receptor with ß-galactosidase (41) or with lamB (42); however, later studies achieved functional expression of a non-fused receptor protein in an expression vector under the control of a T7 promoter (43). Strassberg and colleagues have demonstrated that both ß-1 and ß-2 receptors expressed in E. coli exhibit the pharmacologic properties of receptors expressed in eukaryotic cells (44). To date, prokaryotic expression has not been widely used in structure/function studies of adrenergic receptors because these cells lack G proteins and second messenger systems capable of interacting with adrenergic receptors. Furthermore, the level of expression in this system is relatively low (approximately 250 functional receptors/cell) (43), as such it is not useful for large scale production of receptor protein for biochemical studies. However, the low cost of growing bacterial cells makes this system potentially useful for producing receptor for screening drugs (41). Perhaps the most exciting application of this expression system is for screening mutant receptors following saturation mutagenesis (43). This technique will be discussed below.

Cell Free Expression

Expression of functional ß receptors in a cell-free expression system has been demonstrated (45). This expression system consists of synthetic RNA, rabbit reticulocyte lysate and a source of microsomal membranes prepared from Xenopus laevis oocytes. Attempts to obtain expression using canine pancreatic microsomal membranes were unsuccessful. A level of expression of up to 500 fm/mg membrane protein or 1 picomole per ml of translation can be achieved. This cell free expression system was used for studying the biosynthesis of the ß-2 adrenergic receptor (45). Receptors are efficiently translocated into the endoplasmic reticulum and glycosylated within 20 minutes following the addition of RNA to the translation system; however, the receptor is unable to bind ligands at this point. Additional processing is needed to produce functional receptor. This posttranslational processing requires intact microsomal membranes, ATP and one or more high molecular weight cytosolic factors, and takes approximately 30 minutes to complete. In addition to studying receptor biosynthesis, this system may also be useful for producing receptor labeled to high specific activity with radioactive amino acids for biochemical studies.

MUTAGENESIS

Mutagenesis studies on adrenergic receptors have yielded valuable information about the functional role of various structural domains. Changes in the primary amino acid sequence of receptors are accomplished by modifying the DNA sequence of the cloned gene or cDNA. The modified genes can then be expressed in one of the systems described above, and the functional properties of the mutant receptor can be determined. This section will focus on different approaches to designing and studying mutant receptors.

Mutation studies are generally carried out to locate structural domains which are directly involved in a specific function such as ligand binding or G protein activation. Mutations may produce several possible outcomes. [1] The mutation may have no effect on receptor function, suggesting that the modified domain is not essential for receptor function. [2] The mutation may result in a completely nonfunctional receptor. This result may be interpreted as indicating that the modified domain was a common component of all functional domains, or that the modification led to improper folding or processing of the receptor. Improperly folded proteins may be degraded or have incompletely processed asparagine-linked glycosylation. These structural changes may be detected by western blot analysis. [3] The mutation may modify one functional trait while preserving at least one of the other functional characteristics of the wild type receptor. This result is most useful as it indicates that the mutation affected a specific functional domain and not the overall structure of the receptor.

Deletion Mutagenesis

Deletion mutagenesis is a useful approach for identifying the functional significance of relatively large structural domains and is therefore often the first approach used in studying a protein. In deletion mutagenesis, the coding sequence of cloned receptor genes are modified to encode receptors lacking one or more amino acids. Large deletion mutations which do not modify receptor function are often more informative than those which lead to loss of function. Deletions may cause significant changes in the conformation of a protein. Therefore, the loss of a particular function in a deletion mutation is likely to be a consequence of improper processing or allosteric effects. However, if function is preserved, the deleted domain is likely to have little functional or structural significance.

Richard Dixon and his colleagues carried out studies on a comprehensive series of deletion mutations of the hamster ß-2 receptor (21,46). These mutant receptors were expressed in tissue culture cells and their functional properties were determined. Mutants which were not functional were analyzed by western blot analysis to determine if they were processed normally. These deletion studies demonstrated the importance of the hydrophobic domains for proper processing of the receptor (46). Most of the extracellular domains could be deleted without loss of function, suggesting that the hydrophobic domains were involved in forming the ligand binding pocket. These studies also identified regions of the hydrophilic domains involved in G protein coupling (21).

Chimeric Receptors

Another approach to obtaining information about the functional importance of relatively large domains is to construct chimeric receptors from two closely related receptors with distinct functional properties. By correlating the functional properties of the chimeric receptors with their structural components, it is possible to assign functions to specific structural domains. The first such experiments carried out on this class of receptors studied the properties of chimeric receptors constructed from the human ß-2 and human α-2 adrenergic receptors (18). The functional properties of the chimeric receptors were correlated with their content of α-2 and ß-2 receptor structure. The results demonstrated that the seventh hydrophobic domain contained important determinants of binding specificity and that the third cytoplasmic loop was involved in G protein coupling specificity. The specific region in the third cytoplasmic domain was further narrowed in a series of experiments in which portions of the human M1-muscarinic cholinergic receptor third cytoplasmic loop were replaced by sequences from the turkey ß receptor (20). It was determined that the exchange of a sequence of 12 amino acids in the amino terminal region of this domain was capable of altering G-protein coupling specificity.

Chimeric studies have also been done to determine the domains which are important for determining differences in binding specificity between the ß-1 and ß-2 receptors (44,47). These two receptors are functionally and structurally more closely related than are the ß-2 and α-2 receptors. The results from these studies indicate that factors that determine more subtle differences in ligand binding specificity are contributed by multiple structural domains.

Single Amino Acid Substitutions

More specific information regarding the functional role of a specific domain can be obtained by making single amino acid substitutions. The choice of target amino acids to be modified may be based on a comparison of sequences of functionally related receptors, as well as on data from deletion or chimeric studies. When a specific amino acid is chosen for study, consideration must be given to the amino acid which will replace it. An hypothesis is made regarding the importance of the chemical properties of the wild type amino acid in the function of the receptor. The amino acid chosen to substitute for the wild type amino acid should be similar in size, but differ in the essential chemical characteristic necessary for the hypothetical function of the wild type amino acid. For example, if the wild type amino acid is negatively charged, the substitute amino acid might be of similar size, but neutral or positively charged.

A recent study by Strader and colleagues (17) illustrates the power of this approach to study receptor structure and function. These investigators hypothesized that two serine residues in the fifth membrane spanning domain of the hamster ß-2 adrenergic receptor (serine 204 and 207) may be involved in forming the agonist binding pocket by forming hydrogen bonds with the meta- and para-hydroxyl groups on the catechol ring. The effect of mutations of these serine residues on the activation of the receptor by isoproterenol and by derivatives of isoproterenol lacking either the meta- or para-hydroxyl was determined. These studies support the hypothesis that the meta-hydroxyl on the catechol ring forms a hydrogen bond with the hydroxyl on serine 204 while the para-hydroxyl on the catechol ring forms a hydrogen bond with the hydroxyl on serine 207. The wild type receptor could be activated by isoproterenol and by isoproterenol lacking either the meta- or para-hydroxyl. The lower binding affinity observed for the hydroxyl-minus derivatives of isoproterenol relative to isoproterenol is consistent with the loss of energy from one hydrogen bond (17). ß-2 receptor in which serine 204 was replaced by alanine could be activated by isoproterenol and the isoproterenol derivative lacking the meta-hydroxyl, but not by the isoproterenol derivative lacking the para-hydroxyl. Conversely, the receptor in which serine 207 was replaced by alanine could be activated by isoproterenol and the isoproterenol derivative lacking the para-hydroxyl, but not by the isoproterenol derivative lacking the meta-hydroxyl. Studying interactions between modified receptors and modified ligands is a very powerful approach to probing the boundaries of the ligand binding pocket.

Saturation Mutagenesis

Saturation mytageneses is the process of studying the functional consequences of a large number of amino acid substitutions in a small structural domain of a protein (48,49). This approach has been particularly useful in understanding the importance of the chemical nature of specific amino acids on the functional properties of a protein (48). It may also provide information about the likely orientation of an amino acid side chain with respect to the interior of the protein. Mutations of amino acids in structural domains which have amino acid side chains facing the exterior of the protein often do not adversely affect the structure or function of the protein.

Saturation mutagenesis can be accomplished by engineering restriction sites around the coding sequence of the structural domain of interest, and using these sites to insert oligonucleotide cassettes containing degenerate codons at one or more positions (50). This strategy permits generation of mutants having all possible amino acid substitutions at specific sites within the structural domain. The large number of mutations that result have limited this approach to the study of bacterial proteins for which a selection strategy can be devised based on nutritional requirements or antibiotic resistance. Bacterial cells expressing mutations having the desired functional characteristics will survive. The mutant DNA can then be isolated from individual bacterial colonies and the nature of the mutation determined by DNA sequence analysis.

Recent success in expressing adrenergic receptors in E. coli has made it possible to apply saturation mutagenesis to the study of ß adrenergic receptor structure (43). Strosberg and colleagues developed an approach for screening a large number of ß-2 receptor mutations by expressing them in bacteria, transferring the colonies to nitrocellulose filters, and identifying colonies expressing functional receptor by incubating the filters with the ß receptor antagonist [125I]cyanopindolol. Colonies expressing functional receptor bind to the iodinated antagonist and can be identified by autoradiography. This permits the screening of hundreds of mutants for the ability to bind to a ß receptor antagonist. The sequence extending from amino acid 76 through 83 in the second hydrophobic domain of the human ß-2 receptor was studied using cassette mutagenesis. The results indicate that within this sequence, positions 76, 78, 80 and 82 were conserved, that is, only a few mutations in these positions were observed in functional receptors. This suggests that amino acids in these positions are either required for ligand binding or maintaining the structure of the receptor (43). Amino acid substitutions were more frequent at positions 77, 79 and 81, suggesting that these residues are not critical for the proper folding of the receptor or for antagonist binding. The use of E. coli as an expression system limits the scope of these studies to analyzing the effect of mutations on ligand binding. As noted above, both ß-2 receptor and rat $G_{s\alpha}$ have recently been expressed in yeast, and a functional interaction between these proteins led to growth arrest. This raises the possibility of performing saturation mutagenesis on receptor domains involved in receptor-G protein interaction and screening mutants in yeast.

CONCLUSION

The techniques of molecular biology have been used to overcome some of the technical difficulties encountered in studying adrenergic receptors. Tremendous progress has been made over the past five years towards understanding the structure, function and regulation of these receptors. While much has been learned, many questions have yet to be answered, and some new questions have arisen as a result of recent progress in this field. Continued success towards understanding the molecular basis of adrenergic receptor function in health and disease will require creative use of techniques of molecular biology, molecular genetics, pharmacology, biochemistry and cell biology.

ACKNOWLEDGMENTS

The author wishes to thank Tong Sun Kobilka for helpful comments.

REFERENCES

1.   Strader CD, Irving SS, Dixon RAF. Structural basis of ß-adrenergic receptor function. FASEB J 1989;3:1825-1832.

2.   Hausdorff WP, Caron MG, Lefkowitz RJ. Turning off the signal: desensitization of ß-adrenergic receptor function. FASEB J 1990;4:2881-2889.

3.   Birnbaumer L, Abramowitz J, Brown AM. Receptor-effector coupling by G proteins. Biochimica et Biophysica Acta 1990;1031:163-224.

4.   Freissmuth M, Casey PJ, Gilman AG. G proteins control diverse pathways of transmembrane signaling. FASEB J 1989;3:2125-2131.

5.   Yarden Y, Rodriguez H, Wong SK, et al. The avian ß-adrenergic receptor: primary structure and membrane topology. Proc Natl Acad Sci USA 1986;83:6795-6799.

6.   Frielle T, Collins S, Daniel KW, Caron MG, Lefkowitz, RJ, Kobilka, BK. Cloning of the cDNA for the human β_1-adrenergic receptor. Proc Natl Acad Sci USA 1987;84:7920-7924.

7.   Dixon RAF, Kobilka BK, Strader DJ, Benovic JL, et al. Cloning of the gene and cDNA for mammalian β_2-adrenergic receptor and homology with rhodopsin. Nature 1986;321:75-79.

8.   Kobilka BK, Dixon RAF, Frielle T, et al. cDNA for the human β_2-adrenergic receptor: a protein with multiple membrane-spanning domains and encoded by a gene whose chromosomal location is shared with that of the receptor for platelet-derived growth factor. Proc Natl Acad Sci USA 1987;84:46-50.

9.   Emorine LJ, Marullo S, Briend-Sutren MM, et al. Molecular characterization of the human β_3-adrenergic receptor. Science 1989;245:1118-21.

10.   Cotecchia S, Schwinn DA, Randall RR, Lefkowitz JL, Caron MC, Kobilka BK. Molecular cloning and expression of the cDNA for the hamster α_1-adrenergic receptor. Proc Natl Acad Sci USA 1988;85:7159-7163.

11.   Lomasney JW, Leeb-Lundberg F, Cotecchia S, et al. Mammalian α_1-adrenergic receptor. J Biol Chem 1986;261:7710-7716.

12. Kobilka BK, Matsui H, Kobilka TS, et al. Cloning, sequencing, and expression of the gene coding for the human platelet α_2-adrenergic receptor. Science 1987;238:650-656.

13. Regan JW, Kobilka TS, Yang-Feng TL, Caron MG, Lefkowitz RJ, Kobilka BK. Cloning and expression of a human kidney cDNA for an α_2-adrenergic receptor subtype. Proc Natl Acad Sci USA 1988;85:6301-6305.

14. Lomasney JW, Lorenz W, Allen LF, et al. Expansion of the α_2-adrenergic receptor family: cloning and characterization of a human α_2-adrenergic receptor subtype, the gene for which is located on chromosome 2. Proc Natl Acad Sci USA 1990;87:5094-5098.

15. Weinshank RL, Zgombick JM, Macchi M, et al. Cloning, expression, and pharmacological characterization of a human α_{2B}-adrenergic receptor. Molecular Pharmacology 1990;38:681-688.

16. Strader CD, Sigal IS, Candelore MR, Rands E, Hill WS, Dixon RAF. Conserved asparic acid residues 79 and 113 of the ß-adrenergic receptor have different roles in receptor function. J Biol Chem 1988;263:10267-10271.

17. Strader CD, Candelore MR, Hill WS, et al. Identification of two serine residues involved in agonist activation of the ß-adrenergic receptor. J Biol Chem 1989;264:13572-13578.

18. Kobilka BK, Kobilka TS, Daniel K, Regan JW, Caron MG, Lefkowitz RJ. Chimeric α_2-,β_2-adrenergic receptors:delineation of domains involved in effector coupling and ligand binding specificity. Science 1988;240:1310-1316.

19. O'Dowd BF, Hnatowich M, Regan JW, Leader WM, Caron MG, Lefkowitz RJ. Site-directed mutagenesis of the cytoplasmic domains of the human ß$_2$-adrenergic receptor. J Biol Chem 1988;263:15985-15992.

20. Wong SK, Parker EM, Ross Em. Chimeric muscarinic cholinergic:ß-adrenergic receptors that activate Gs in response to muscarinic agonists. J Biol Chem 1990;265:6219-6224.

21. Strader DS, Dixon RAF, Cheung AH, Candelore MR, Blake AD, Sigal IS. Mutations that uncouple the ß-adrenergic receptor from G$_s$ and increase agonist affinity. J Biol Chem 1987;262:16439-16443.

22.   Bouvier M, Collins S, O'Dowd BF, et al. Two distinct pathways for cAMP-mediated down-regulation of the β_2-adrenergic receptor. J Biol Chem 1989;264:16786-16792.

23.   Bouvier M, Hausdorrf WP, Di Blasi A, et al. Mutations of the β_2-adrenergic receptor which remove phosphorylation sites delay the onset of agonist promoted desensitization. Nature 1988;333:370-372.

24.   Valiquette M, Bonin H, Hnatowich M, Caron MG, Lefkowitz RJ, Bouvier M. Involvement of tyrosine residues located in the carboxy tail of the human β_2-adenergic receptor in agonist-induced down-regulation of the receptor. Proc Natl Acad Sci USA 1990;87:5089-5093.

25.   Bylund DB. Subtypes of α_2-adrenoceptors: pharmacological and molecular biological evidence converge. TIPS 1988;9:356-361.

26.   Han C, Abael PW, Minniman KP. α_1-adrenoceptor subtypes linked to different mechanisms for increasing intracellular Ca^{2+} in smooth muscle. Nature 1987;329:333-335.

27.   Wong SKF, Slaughter C, Ruoho AE, Ross, EM. The catecholamine binding site of the β-adrenergic receptor is formed by juxtaposed membrane-spanning domains. J Biol Chem 1988;263:7925-7928.

28.   Dohlman, HG, Caron MG, Strader CD, Amlaiky N, Lefkowitz RJ. Identification and sequence of a binding site peptide of the β_2-adrenergic receptor. Biochemistry 1988;27:1813-1817.

29.   Matsui H, Lefkowitz RJ, Caron MG, Regan JW. Localization of the fourth membrane spanning domain as a ligand binding site in the human platelet α_2-adrenergic receptor. Biochemistry 1989;28:4125-30.

30.   Dohlman HG, Bouvier M, Benovic JL, Caron MG, Lefkowitz RJ. The multiple membrane spanning topography of the β_2-adrenergic receptor. J Biol Chem 1987;262:14282-14288.

31.   Rubenstein RC, Wong SKF, Ross EM. The hydrophobic tryptic core of the β-adrenergic receptor retains G_s regulatory activity in response to agonists and thiols. J Biol Chem 1987;262:16655-16662.

32. Wilson AL, Guyer CA, Cragoe Jr EJ, Limbird LE. The hydrophobic tryptic core of the porcine α_2-adrenergic receptor retains allosteric modulation of binding by Na$^+$, H$^+$, and 5-amino-substituted amiloride analogs. J Biol Chem 1990;265:17318-17322.

33. Wang H, Lipfert L, Malbon CC, Bahouth S. Site-directed anti-peptide antibodies define the topography of the ß-adrenergic receptor. J Biol Chem 1989;264: 14424-14431.

34. Cullen BR. The use of eukaryotic expression technology in the functional analysis of cloned genes. Methods Enxymol 1987;152:684-705.

35. Cotecchia S, Kobilka BK, Daniel KW, et al. Multiple second messenger pathways of α-adrenergic receptor subtypes expressed in eukaryotic cells. J Biol Chem 1990;265:63-69.

36. Fargin A, Raymond JR, Regan JW, Cotecchia S, Lefkowitz RJ, Caron MG. Effector coupling mechanisms of the cloned 5-HT1A receptor. J Biol Chem 1989;264:14848-14852.

37. Kobilka BK, MacGregor CH, Daniel KW, Kobilka TS, Caron MG, Lefkowitz RJ. Functional activity and regulation of human ß$_2$-adrenergic receptors expresseed in xenopus oocytes. J Biol Chem 1987;262:15796-15802.

38. George ST, Arbabian MA, Ruoho AE, Kiely J, Malbon CC. High-efficiency expression of mammalian ß-adrenergic receptors in baculovirus-infected insect cells. Biochemical Biophysical Research Communications 1989;163:1265-1269.

39. Luckow VA, Summers MD. Trends in the development of baculovirus expression vectors. Bio/Technology 1988;6:47-55.

40. King K, Dohlman HG, Thorner J, Caron MG, Lefkowitz RJ. Control of yeast mating signal transduction by a mammalian ß$_2$-adrenergic receptor and Gs α subunit. Science 1990;250:121-123.

41. Marullo S, Delavier-Klutchko C, Eshdat Y, Strosberg AD, Emorine L. Human ß$_2$-adrenergic receptors expressed in Escherichia coli membranes retain their pharmacological properties. Proc Natl Acad Sci USA 1988;85:7551-7555.

42.   Chapot MP, Eshdat Y, Marullo S, et al. Localization and characterization of three different ß-adrenergic receptors expressed in <u>Escherichia coli</u>. Eur J Biochem 1990;187:137-144.

43.   Breyer RM, Strosberg AD, Guillet JG. Mutational analysis of ligand binding activity of ß$_2$ adrenergic receptor expressed in <u>Escherichia coli.</u> EMBO J 1990;9:2679-2684.

44.   Marullo S, Emorine LJ, Strosberg AD, Delavier-Klutcho C. Selective binding of ligands to ß1, ß2 or chimeric ß1/ß2-adrenergic receptors involves multiple subsites. <u>EMBO J</u> 1990;9:1471-1476.

45.   Kobilka, B. The role of cytosolic and membrane factors in processing of the human ß$_2$ adrenergic receptor following translocation anad glycosylation in a cell-free system. <u>J Biol Chem</u> 1990;265:7610-7618.

46.   Dixon RAF, Sigal IS, Candelore MR, et al. Structural features required for ligand binding to the ß-adrenergic receptor. <u>EMBO J</u> 1987;6:3269-3275.

47.   Frielle T, Daniel KW, Caron MG, Lefkowitz RJ. Structural basis of ß-adrenergic receptor subtype specificity studied with chimeric ß$_1$/ß$_2$-adrenergic receptor chimers. <u>Proc Natl Acad Sci USA</u> 1988;85:9494-8.

48.   Reidhaar-Olson JF, Sauer RT. Combinatorial cassette mutagenesis as a probe of the informational content of protein sequences. <u>Science</u> 1988;241:53-57.

49.   Shortle, D. Probing the determinants of protein folding and stability with amino acid substitutions. <u>J Biol Chem</u> 1989;264:5315-5318.

50.   Hill DE, Oliphant AR, Struhl D. Mutagenesis with degenerate oligonucleotides: an efficient method for saturating a defined DNA region with base pair substitutions. <u>Methods Enzymol</u> 1987;155:558-68.

Advances in Regulation of Cell Growth, Volume 2;
Cell Activation: Genetic Approaches, edited by
James J. Mond, John C. Cambier, and Arthur Weiss.
Raven Press, Ltd., New York © 1991.

9

Function of CD4 and CD8 in T-cell Activation and Differentiation

Julia M. Turner[1] and Dan R. Littman[2]

[1]Department of Pathology, University of Cambridge, Tennis Court Road,
Cambridge CB2 1QP, UK [2]Howard Hughes Medical Institute and Departments of
Microbiology and Immunology and of Biochemistry and Biophysics, University of
California, San Francisco, California 94143-0414

The complex process of T lymphocyte development gives rise to cells that are tolerant to self antigens yet respond to foreign antigens complexed to host major histocompatibility (MHC) molecules (1,2). This ability of T cells to discriminate between self and nonself requires them to respond differentially to diverse environmental cues. For example, distinct signals must exist for clonal deletion of self-reactive thymocytes (3-6), for positive selection of thymocytes that can interact with self-MHC and foreign antigen (7-10), for activation of mature T lymphocytes by foreign antigen (11), and for induction of paralysis or anergy if the same mature T cells encounter self antigen (12-14). These varied responses are initiated by specific cell surface receptors that interact with molecules on the surface of other cells or with soluble lymphokines, but are also thought to be influenced by spatial and temporal constraints. Little is currently known about the nature of signals involved in thymic differentiation and induction of anergy. However, some of the key molecules involved in transducing signals in responses of T cells to specific antigen have recently been identified. Although most studies have utilized transformed T cells and hybridomas, the results are likely to be relevant to understanding normal T cell activation and decision-making processes in the thymus. This chapter focuses on the roles of several T cell surface molecules in antigen specific signal transduction in T cells and thymocytes. In particular, approaches aimed at defining the precise roles of the transmembrane glycoproteins CD4 and CD8 in T-cell activation are described and we speculate as to how these molecules may be involved in specifying developmental pathways in the thymus.

THE T CELL RECEPTOR COMPLEX AND ITS CORECEPTORS

The events involved in T cell maturation include: (i) rearrangement and expression of the T cell antigen receptor (TCR) genes; (ii) interaction of the TCR's with MHC molecules (of which there are two structurally distinct classes - I and II) expressed by specialized cells within the thymus; (iii) clonal deletion of thymocytes whose TCR's react with self antigens complexed to host MHC molecules; (iv) expansion of clones with TCR's that react with host MHC molecules on thymic epithelial cells and (v) induction of anergy in clones that are released from the thymus but bear autoreactive TCR's (reviewed in ref. 1). During maturation, precursor thymocytes that express both the CD4 and CD8 cell surface glycoproteins (double positive cells) give rise to mature cells that express either one or the other molecule: the CD4+ T cells, primarily helper cells whose TCR's are specific for class II MHC molecules; and the CD8+ cells, almost exclusively cytotoxic cells with receptors recognizing class I MHC molecules (15,16). While specific recognition of the antigen/MHC complex is achieved by the TCR's clonally distinct αβ heterodimer, transduction of signals across the plasma membrane is believed to occur through a heterodimer-associated complex of at least five different subunits, collectively referred to as CD3 (Fig. 1) (10,11). The αβ TCR heterodimer interacts with polymorphic residues on the surface of the MHC molecule and peptide antigens bound adjacent to these residues within a cleft

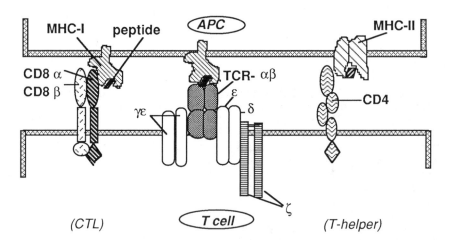

FIG.1. T-cell recognition. The TCR-αβ heterodimer, shown in association with the CD3 complex (γ,δ,ε) and ζ homodimer, is coexpressed with either CD8 (this would be a cytotoxic T cell, CTL) or CD4 (a T-helper cell). The TCR binds to a complex of a peptide and either a class I or II MHC molecule on the surface of an antigen presenting cell (APC); CD8 and CD4 bind to monomorphic regions of class I and class II MHC molecules, respectively.

in the MHC protein. In addition to the TCR heterodimer, CD4 and CD8 also bind to MHC molecules (17-19) and are likely to have key roles in signal transduction in developing thymocytes and mature T cells. Each TCR recognizes either a class I or a class II MHC molecule, a phenomenon known as MHC restriction. Studies of mice bearing transgenic TCRs restricted to class I or II MHC imply that the specificity (MHC restriction) of the TCR determines whether the mature T cell expresses CD4 or CD8 (reviewed by Schwartz, ref. 2). The involvement of CD4 and CD8 in this developmental process is suggested by experiments in which antibodies to CD4 or CD8 block the appearance of single positive (CD4+CD8- or CD4-CD8+) populations (20).

How, then, do CD4 and CD8 participate in the establishment of single positive, MHC restricted T cells? As discussed below, CD4 and CD8 molecules have been shown to interact directly with monomorphic domains of the class II and class I MHC molecules, respectively (17-19). In addition, both of these molecules are associated with the cytoplasmic protein tyrosine kinase, $p56^{lck}$ (21-23). These observations have given rise to speculation that antigen-specific signal transduction in T cells requires association of the CD4 or CD8 "co-receptor" molecules with the TCR complex. There is now extensive evidence to support this hypothesis as well as mounting evidence that the protein tyrosine kinase associated with the co-receptors plays a key role in this process.

Effective ligation of the TCR and co-receptors at the T cell surface results in the generation of transmembrane signals, which in turn generate cytoplasmic second messengers via at least two main pathways: i) activation of a phospholipase C (PLC) that cleaves phosphatidylinositol 4,5-biphosphate into 1,2-sn-diacylglycerol (DAG) and 1,4,5-trisphosphate (IP_3) which, in turn, activate the serine/threonine kinase protein kinase C and increase intracellular free calcium, respectively (11); and ii) activation of tyrosine kinases (11,24,25). The relative contribution of these two pathways to initiation of transcriptional changes required for T cell proliferation and differentiation is not yet clear. Recent studies with cultured T cells suggest that PLC activation is dependent on prior triggering of the tyrosine kinase pathway (26), but it is not known whether parallel phosphorylation of substrates not involved in the PI pathway is also required.

CD4 AND CD8 INTERACT WITH MHC MOLECULES

The T-cell specific glycoproteins CD4 and CD8 are both members of the immunoglobulin (Ig) superfamily, yet are structurally distinct and share only very limited sequence homology (15,16). The CD4 glycoprotein appears to be a monomer composed of four Ig-family domains, a transmembrane domain, and a cytoplasmic tail of 38 amino acids. In contrast, CD8 can exist as disulfide-linked aa homodimers or ab heterodimers containing α and β chains encoded by closely linked genes; the α and β chains each consists of a single Ig-family domain followed by a spacer region of 40-50 residues, a transmembrane domain, and a short cytoplasmic tail (28 residues in CD8α and19 residues in CD8β). The strict correlation between expression of CD4 or CD8 and the class of MHC molecule recognized by the TCR suggested that these molecules, like the TCR, also bind to MHC proteins (27). Cell adhesion molecules usually bind their ligands with low affinity; hence such interactions are difficult to define. However, evidence for interaction of CD4 and CD8 glycoproteins with class II and class I MHC molecules has been obtained recently using cells engineered to express high levels of the various components at the cell surface. CV-1 cells infected with an SV40-CD4 recombinant virus and expressing high levels of human CD4 were shown to bind specifically only B cells bearing surface class II MHC molecules (17). Similarly, radiolabelled class I MHC-expressing B cells bound to CHO cells cotransfected with human $CD8\alpha$ and an amplifiable DHFR minigene (18). Binding of class I-expressing B cells was detected only when amplification of $CD8\alpha$ expression passed a threshold, beyond which the extent of cell-cell binding correlated with the relative level of $CD8\alpha$ expression on the CHO cells. This $CD8\alpha$-class I MHC interaction was specifically and completely inhibited by antibodies to class I MHC and to $CD8\alpha$. These results demonstrated clearly that CD4 and $CD8\alpha$ can directly bind to class II and class I MHC molecules, respectively, but that this binding is of low avidity in the absence of the TCR and any other T cell surface molecules which may participate in cell-cell interactions.

To define more precisely the interaction of $CD8\alpha$ with MHC molecules, B cells transfected with naturally occurring or engineered point mutants of a human class I MHC molecule, were tested for binding in the $CD8\alpha$-expressing CHO cell system described above (28,29). The binding site for $CD8\alpha$ mapped to the membrane-proximal monomorphic $\alpha3$ domain of this class I molecule. Cells expressing mutants of class I which did not bind $CD8\alpha$ were also found to be poor targets for cytotoxic T cells, presumably because $CD8\alpha$ could not participate in activation of the CTLs. Similarly, cells expressing a murine class I molecule with a mutation in the $\alpha3$ domain were found not to be recognized by CD8-dependent CTLs (30). These studies provide strong evidence for a direct interaction of CD8 with the $\alpha3$ domain of class I MHC molecules. As the TCR binds to peptides within the groove formed by the $\alpha1$ and $\alpha2$ domains of MHC class I molecules, CD8 and the TCR could theoretically bind to a single class I molecule simultaneously. This has significance for the mechanism of CD4/CD8 cooperation with the TCR, as discussed below.

CD4 AND CD8: ACCESSORY MOLECULES AND CORECEPTORS

Antibodies against CD4 or CD8 have been shown to block the antigen-specific responses of some, but not all, T cells. This variation was proposed to be a function of the affinity of individual TCRs, CD4 and CD8 serving as "accessory molecules" by stabilizing and increasing the affinity between the TCR and the antigen/MHC complex (31). Demonstration of the interaction of CD4 and $CD8\alpha$ with MHC molecules, described above, is consistent with this view of CD4 and CD8 as adhesion molecules. Yet certain observations indicated that the role of CD4 and CD8 may be more complex. Certainly, the requirement for CD4 or CD8 in antigen-specific responses could also been shown by genetic means in murine T cell hybridomas. For example, transfection of genes for a class I-reactive TCR into a class II-reactive T cell confered class I-reactivity upon the recipient cell only if the $CD8\alpha$ gene was also transfected into the host cell (32). Similarly, a hybridoma with a TCR specific for pork insulin showed antigen-mediated activation only when transfected with CD4 (33). However, these studies in reconstituted systems did not address whether CD4 and CD8 were solely contributing to cell-cell adhesion or were also involved in signal transduction through the TCR. Infact, it was difficult to reconcile the correlation of CD4/CD8 expression and class of MHC restriction simply by adhesion. For example, when both class I and II MHC molecules were expressed on antigen presenting cells

(APC's), only the accessory molecule appropriate for the MHC specificity of the TCR (CD4 for class II-restricted TCRs and CD8 for class I-restricted TCRs) could fully reconstitute the antigen-specific response, and antibodies to the appropriate coreceptor were shown to be much more inhibitory than those to the inappropriate coreceptor (34,35). These results suggested that CD4 and CD8 do more than simply contribute to the avidity of the cell-cell interaction.

An additional dimension in the function of CD4 and CD8 has been emerging as a consequence of evidence that these molecules physically interact with the TCR. For example, antibody-mediated crosslinking of CD4 or CD8 to the TCR complex results in an augmented activation signal, as compared to crosslinking of either surface molecule alone (36). Some TCR-specific antibodies that are potent T cell activators induce cell surface co-capping and co-modulation of CD4 with the TCR, perhaps by effecting a conformational change in the TCR and/or CD4 which allows their association (37). In addition, some anti-CD4 antibodies inhibit TCR-mediated T-cell activation, probably via steric hindrance of the CD4/TCR interaction (38,39). Recent studies using energy transfer analysis also suggest that CD4 and the TCR complex are associated (40,41). These data lend support to a model in which CD4 and CD8 act as "coreceptors" for the TCR (reviewed in ref.42). In this model, CD4 or CD8 binds to the same MHC molecule as the TCR (as in Fig. 3) and the resulting association of CD4/CD8 with the TCR generates a signal more potent than ligation of the TCR alone and hence sufficient for T-cell activation. The recent definition of the binding site for CD8a on a class I molecule, described above (29) and the observed deleterious effect of mutation of this site for T-cell activation (43), also strongly support the possibility and significance of corecognition of MHC molecules by CD8 and the TCR.

The model of CD4 and CD8 as coreceptors for the TCR, involving their binding to a common MHC molecule and consequent physical association, begs the question of whether CD4 and CD8 participate in transmembrane signalling via the TCR and subsequent activation of the T cell. The genetic approaches that we and others have taken to begin to answer this question are discussed below.

CD4 AND CD8 ARE ASSOCIATED WITH A TYROSINE KINASE

Once the ligands of CD4 and CD8 had been defined, we proceeded to investigate the potential role of these transmembrane molecules in signal transduction. Cross-species comparison of the amino acid sequences of CD4 and CD8α reveals that the cytoplasmic domains are significantly more conserved than the external and transmembrane domains, suggesting a stringent requirement for specific sequences in the cytoplasmic tail (15). In contrast, there is no obvious homology between the tails of CD4 and CD8 from any species for which the sequences are available. Insight into how CD4 and CD8 may participate in signal transduction came with the discovery by Rudd et al. (21) that the cytoplasmic protein tyrosine kinase p56^{lck} was coimmunoprecipitated with CD4 in resting human T cells. The same was subsequently shown to be true of CD4 and CD8 in human and mouse cells (22,23). p56^{lck} is a member of the *src* family of cytoplasmic protein-tyrosine kinases, having the characteristic *src* homology regions SH2 and SH3 (44), which may be involved in regulation of the enzyme, and a highly conserved C-terminal kinase domain. Like the other members of the *src* family, p56^{lck} is distinguished by a unique N-terminal sequence (reviewed in ref. 45). p56^{lck} is myristylated and membrane-associated and is expressed specifically in lymphocytes, particularly T cells (45-49).

p56^{lck} was initially identified as a putative oncoprotein overexpressed due to retroviral insertion in a murine lymphoma cell line and was subsequently shown to be transcriptionally regulated during T-cell activation. It was therefore proposed that p56^{lck} and perhaps other *src*-like kinases are involved in T-cell activation (47,48,50). Physical association of p56^{lck} with CD4/CD8 supported this idea, particularly since antibody-mediated crosslinking of CD4 was reported to up-regulate the activity of p56^{lck} (51). To investigate the role of p56^{lck} in T-cell activation we began with two basic questions: (i) How do CD4 and CD8, which have minimal sequence homology, interact directly or indirectly with a common molecule, and (ii) what is the functional significance of the association of these potential coreceptor molecules with a tyrosine kinase? Initially, we used a molecular genetic approach to define the interaction; for ease and simplicity, these experiments were carried out in a non-lymphoid system. Murine CD4 or CD8α and p56^{lck} were transiently expressed at high levels in COS-7 cells using an SV40-based expression system (52). The ability of the proteins to associate was assayed by immunoprecipitation of CD4 or CD8α and detection of coprecipitated p56^{lck} by immunoblotting with an antiserum specific for p56^{lck} and by protein kinase assays (autophosphorylation of *src* family members can be readily detected using ^{32}P-γ-ATP). CD4 and CD8α were found to associate non-covalently with p56^{lck} in the absence of lymphoid factors, notably the TCR complex, as has also been shown with CD4 in 3T3 cells (53). Moreover, the second chain in the CD8 complex was found to be dispensable for interaction with p56^{lck}; in fact, heterodimers that lacked the cytoplasmic tail of CD8α but had the complete CD8β chain did not associate with the kinase, indicating that the CD8β molecule does not bind p56^{lck} (54).

The ability of CD4 and CD8α to associate with p56^{lck} when expressed in COS-7 cells provided us with a system to map their interactions in molecular terms. Using chimeras of p56^{lck} with p60^{c-src} (p60^{c-src} did not associate with CD4 or CD8α) and CD4 or CD8α molecules bearing mutations in their cytoplasmic tails (those of CD4 are shown in Fig. 2), we determined that the unique N-terminal region of p56^{lck} binds to the membrane-proximal 10 and 28 cytoplasmic residues of CD8α and CD4, respectively (52) A separate study by Shaw et al. (55) further localized the binding site for CD4 to the N-terminal 32 residues of p56^{lck}. Two cysteine residues in each of the critical sequences in CD4, CD8α and p56^{lck} were found by site-directed mutagenesis to be required for association (52). These two cysteines in CD4 and CD8α lie within the region most highly conserved between species and containing the only similarities between the tails of CD4 and CD8α: a motif ++XcysXcys, where "X" denotes a nonconserved residue and "+" a basic residue. Further studies by Shaw et al. (56) indicated that mutations of both basic residues or of a proline, which often occurs immediately C-terminal to the cysteines, have no effect on the association of CD4 with p56^{lck}. Remarkably, the six residues of the cysteine motif of CD8α (but not CD4) were shown to be sufficient to confer upon a heterologous protein the ability to associate with p56^{lck}; sequences more membrane proximal are likely required in addition to the cysteine residues in CD4. Another molecule with the ++cysXcys motif, CD1a, does not associate with p56^{lck}, indicating that the association is specific to CD4 and CD8α (J.M.T. and D.R.L., unpublished results). It is possible that the cysteine residues of p56^{lck} and CD4 or CD8α collaborate in the coordination of a metal ion or in the

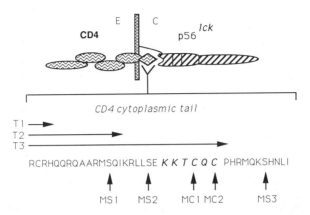

FIG. 2 **Mapping the CD4-p56**^{lck} **association.** Physical interaction of the tail of CD4 and the amino-terminus of p56^{lck} on the cytoplasmic (C) side of the plasma membrane is shown schematically. The amino acid sequence of the tail of murine CD4 is expanded; above it are shown the truncation mutants T1 and T2, which have the first 4 and 15 residues of the tail, respectively, but do not associate with p56^{lck}; and below are the serine and cysteine mutants discussed in the text. The cysteine motif is shown in bold italics.

maintenance of particular conformations permissive for stable protein-protein interactions; the biochemistry of the association has yet to be determined.

REQUIREMENT FOR A CD4-p56^{lck} COMPLEX IN T CELL ACTIVATION

That CD4 and CD8 may themselves transmit a signal as a part of their coreceptor function was suggested by the observation that CD8α was functionally impaired when it lacked the cytoplasmic domain (54). Having defined individual residues within the cytoplasmic domain of CD4 that are essential for its association with p56^{lck}, we began to study the requirement for formation of this complex in T-cell activation. The various mutant CD4 molecules were expressed in two murine T cell lymphomas that were dependent on the presence of CD4 for their response to antigen (57). These lymphomas were CD4-negative but were transfected with genes encoding TCR α and β chains known to be specific for characterized peptides of hen egg lysozyme (HEL). The response of the hybridomas to the HEL peptides, presented by the appropriate class II MHC molecules on transfected L cells, was assayed by measuring production of IL-2. In the absence of CD4, or

CD4 molecule	association with p56^{lck}	IL-2 response to peptide Ag (20 µM HEL) as units /ml
wild-type	+	105
T1	-	4
T2	-	ND
T3	(+)	22
MCA1	-	4
MCA2	-	4
MCA1+2 (double mutant)	-	4
MS1	+	77
MS2	+	90
MS3	+	80
MS1/2/3 (triple mutant)	(+)	8
no CD4	-	<1

Table 1. The CD4-p56^{lck} association in T-cell activation. Mutants of CD4 (see Fig.2) expressed in a peptide antigen (Ag) - HEL-specific hybridoma 171.3, are shown as positive (+) or negative (-) for association with p56^{lck} (parentheses indicate a degree of association). Activation of the T-cell populations expressing these mutants is given as IL-2 production. (ND - not determined).

when expressing CD8α, the hybridomas did not respond to presented peptide; when expressing wild-type CD4, however, they produced high levels of IL-2 (Table 1). Expression of the cytoplasmic mutants of CD4, in place of wild-type CD4, indicated not only that these mutants behaved in T cells just as in COS-7 cells in terms of their ability to associate with p56^{lck} (assayed as before by coprecipitation), but that only those CD4 molecules fully able to associate with p56^{lck} could allow antigen-specific activation of the hybridomas equivalent to wild-type CD4 (Table 1; mutants MS1-3 and MS1/2/3). Very low levels of IL-2 production were observed in hybridomas expressing forms of CD4 that could not bind the kinase (T1, T2, MCA1, MCA2 and MCA1/2); this minor response was presumably mediated by the adhesion function of the external domain on ligation of class II MHC molecules. However, this putative adhesion-potentiated response was always much lower than that of CD4 molecules able to associate with p56^{lck} and was also dependent on the concentration of MHC molecules on the APC's.

We inferred from these genetic studies that optimal antigen-mediated T-cell activation requires formation of a complex of CD4 with the kinase p56^{lck}. The fact that CD8α could not substitute for CD4 in these antigen-specific hybridomas, despite its association with p56^{lck} and the presence of class I MHC molecules on the APC's, supports the corecognition model in which CD4 and the TCR must both bind to the same antigen/MHC complex. Corecognition appears to result in a partial response by the T cell, but the activation signal is markedly enhanced by formation of p56^{lck}/CD4 complexes which, presumably, are involved in the triggering of a biochemical signal transduction pathway. The relative effects of the adhesion and coreceptor functions of CD4 and CD8 most likely vary with the TCR and antigen/MHC complex in question. For example, Zamoyska et al. (54) found that expression of chimeric CD8α/CD4 could substitute for CD8α in one hybridoma bearing a class I MHC-restricted TCR. Prior specific interactions of the T cell may also be significant; a recent study detected an increase in the affinity of the CD8-class I MHC interaction after ligation of the TCR. A major caveat of these functional studies is that the immortalized T cell hybridomas may not be representative of T cell clones *in vivo*. It will therefore be necessary to study the function of CD4 in mice expressing transgenic wild-type or mutant forms of the glycoprotein; analysis of T cells from these animals will permit dissection of the relative significance of adhesion and coreceptor functions of CD4 in the development and antigen reactivity of T cells.

MODELS OF CD4 AND CD8 FUNCTION IN T CELLS

a)*Activation of Mature T cells*

The studies described above strongly suggest that activation of peripheral T cells involves crosslinking of the coreceptor/p56^{lck} complex to the TCR complex. In this scenario, p56^{lck}, which is associated with the coreceptor molecules in resting T cells, is brought into contact with the polypeptides of the TCR complex upon co-recognition of an appropriate MHC/antigen combination by the TCR heterodimer and CD4 or CD8 (Fig.3). This may result in activation of the kinase and/or in its exposure to specific regulatory proteins or substrates. Antibody-mediated crosslinking of CD4 on the surface of peripheral T cells has been reported to increase the activity of the associated kinase (51). However, this result may not reflect an event in physiological T cell activation, where aggregation of CD4 seems less likely to occur than its association with the TCR. It has not yet been determined if crosslinking of CD4 to the TCR results in changes in the activity of p56^{lck}.

There is much current interest in identifying enzymes or cofactors that can modulate the activity of p56^{lck} as well as in isolating physiologically relevant substrates of the kinase. Studies of p56^{lck} activity have been limited to comparisons with the prototype cytoplasmic tyrosine kinase, p60^{c-src}. Like other members of this family of kinases, p56^{lck} has a tyrosine (position 394) that is autophosphorylated (or transphosphorylated) *in vitro* and a C-terminal tyrosine (position 505) that, upon phosphorylation, is thought to down-regulate catalytic activity (58). It is common amongst receptor tyrosine kinases for their activity to be regulated by transphosphorylation (59) It is possible that the phosphorylation state of Y394 is critical for positive regulation of p56^{lck}. Two or more molecules of p56^{lck} could be brought together as CD4 enters the TCR complex; more than one molecule of CD4 appears to associate with a single TCR (41). Subsequent transphosphorylation at Y394 may thus lead to transactivation of p56^{lck}. p60^{c-src} is activated by dephosphorylation of a C-terminal tyrosine residue (Y527) and a consequent conformational change in the protein (60). The kinase activity of p56^{lck} is enhanced and its oncogenic potential revealed by mutation of the equivalent tyrosine (Y505) to phenylalanine (61,62); dephosphorylation of Y505 is thus viewed as a likely positive regulation of p56^{lck} activity. However, the relative state of phosphorylation of Y394 and Y505 in p56^{lck} during T cell activation has not yet been determined.

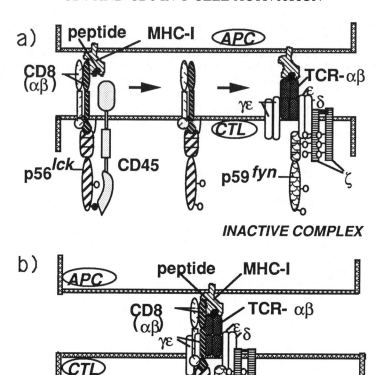

INACTIVE COMPLEX

ACTIVE COMPLEX

FIG. 3. Co-recognition model of T-cell activation.
a) CD8 and the TCR of a cytotoxic T cell (CTL) are shown recognizing different MHC molecules on an antigen presenting cell (APC). CD45 may dephosphorylate p56^{lck} at Y505 (filled circles - phosphate groups; empty circles - dephosphorylated sites); this change in the phosphorylation state may prepare the kinase for activation and entry into the TCR complex. Before co-recognition occurs, however, there is no signal transmission through the TCR.
b) A complex active in signal transduction is created via CD8 and the TCR binding the same MHC-I/peptide molecule; their resultant juxtaposition allows complete activation of p56^{lck}, perhaps involving phosphorylation at Y394 via transphosphorylation or p59^{fyn}; a change in the phosphorylation of p59^{fyn}; and/or relocation of the two tyrosine kinases to pertinent substrates; plus the phosphorylation of the TCR-ζ chain.

A strong candidate for an enzyme that may regulate the state of phosphorylation, and hence the activity, of $p56^{lck}$ is the protein phosphotyrosine phosphatase CD45. This transmembrane protein has several isoforms that differ in their ectodomains and are expressed at high levels in T cells (63). Cells mutated and selected for loss of CD45 expression fail to respond to antigen or antibody stimulation; this defect, which results in loss of PI turnover, IL-2 production, and proliferation, can be corrected by subsequent expression of exogenous CD45 (64,65). In addition, the level of CD45 expression has been shown to correlate inversely with the level of Y505 phosphorylation of $p56^{lck}$ (66); and purified CD45 can dephosphorylate $p56^{lck}$ in vitro and upregulate its activity to some extent (67). The sequence of events leading to activation of $p56^{lck}$ is not yet clear but we may hypothesize that, in the right environment (such as the TCR complex or ligated CD4), CD45 may dephosphorylate Y505. This may create a state of partial activation for $p56^{lck}$, such that phosphorylation of Y394 (discussed below) could then complete the activation of the kinase.(Fig.3).

Another protein that may influence the state of phosphorylation of $p56^{lck}$ is $p59^{fyn}$, also a member of the *src* family of cytoplasmic tyrosine kinases. There is abundant circumstantial evidence that $p59^{fyn}$ has a specialized role in signal transduction in T cells: a T cell-specific splicing event results in a unique isoform of this kinase (68); a strain of mice exists with a genetic defect that results in non-responsive T cells which have elevated levels of $p59^{fyn}$ (69); and components of the TCR complex have been reported to be co-immunoprecipitated with $p59^{fyn}$ (70). The interaction of $p56^{lck}$ with the TCR complex may therefore allow $p59^{fyn}$ to activate $p56^{lck}$ by phosphorylation of Y394 (Fig. 4). Alternatively, $p56^{lck}$ may cause its own down-regulation by activating $p59^{fyn}$; this tyrosine kinase could phosphorylate $p56^{lck}$ at the proposed site of negative regulation, Y505.

The phosphorylation state of tyrosines in $p56^{lck}$ may affect not only kinase activity, but also interaction with other molecules involved in signal transduction; these may be substrates of regulators of the activity and specificity of $p56^{lck}$. For example, the protein p62, which interacts with the SH2 region of $p60^{c-src}$ and *c-fps*, has been implicated in regulation of these cytoplasmic kinases (71). Such interactions of $p56^{lck}$ with regulatory proteins may also be influenced by the cytoplasmic tail sequences of CD4 and CD8. This was suggested by our observation that mere association of the coreceptors with $p56^{lck}$ is not sufficient for maximal T-cell activation; hybridomas expressing CD4 molecules truncated just downstream of the cysteine motif or bearing substitutions of leucine for all three cytoplasmic serine residues gave a suboptimal IL-2 response to antigen/MHC, despite the ability of these CD4 mutants to associate with $p56^{lck}$ (57). Wild-type CD4 and CD8 may thus serve to present $p56^{lck}$ in particular conformations conducive to its productive interaction with regulatory or substrate proteins.

Little is currently known of the physiological substrates of $p56^{lck}$ and hence of the intracellular signaling pathways in which it is involved. Tyrosine phosphorylation is an early event in T-cell activation and appears to precede phospholipase C-induced polyphosphoinositide hydrolysis (25). Treatment of T cells with herbimycin A, an inhibitor of tyrosine kinase activity, blocks subsequent T cell activation and cleavage of phosphatidylinositol bisphosphate (26), suggesting that tyrosine phosphorylation is required to activate PLC. As yet, we can only speculate on the role of $p56^{lck}$ in this process, bearing in mind that T cells express other tyrosine kinases, one of which, $p59^{fyn}$, may be associated with the CD3 complex. Either $p56^{lck}$ or $p59^{fyn}$, or both, may couple the TCR

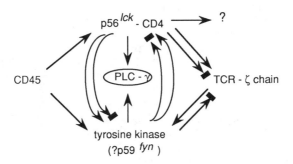

FIG. 4. **Signal transduction in T-cell activation.** Just 5 proteins known
    to be significant in T-cell activation are shown with their many
    possible inter-relationships. Arrows represent activation; arrows
    ending in a block represent inhibition.

complex with a "signalling complex" such as has been defined for the
transmembrane tyrosine kinases, EGF and PDGF receptors (72). This complex
includes the serine/threonine kinase *raf*, phosphatidylinositol-3-kinase (PI-3
kinase), GTPase activating protein (GAP) and phospholipase C γ (PLC-γ). In the
absence of any defined G protein coupling the TCR to PI hydrolysis via activation
of PLC (73), it is tempting to speculate that one or both of the T cell kinases
functions like the EGFR and PDGFR, phosphorylating and activating PLC-γ and
hence resulting in PI hydrolysis.

There is evidence that some T cell hybridomas can be effectively activated in the
absence of CD4 or CD8; and our data shows that, even in the absence of CD4
association with p56^{lck}, there is a baseline level of response to antigen. It is
therefore possible that p59^{fyn} is sufficient for transmitting an activation signal, but
that its activity often needs to be amplified; this amplification may come via
phosphorylation by p56^{lck} subsequent to approximation of the two kinases on co-
recognition of antigen by CD4 and the TCR (Fig. 4). Another possibility is that
p56^{lck} phosphorylates the TCR-ζ chain, which is known to be rapidly
phosphorylated on tyrosines upon T-cell activation (74). The finding that the ζ
chain is also phosphorylated on tyrosines upon crosslinking of CD4 has prompted
speculation that it is a substrate of the p56^{lck} kinase (51). Recent results
demonstrate that the ζ chain is the key component in coupling the TCR heterodimer
to the downstream events, including tyrosine kinase activation and PI turnover
(75). It is as yet a mystery, however, what the activity of the ζ chain is; it does not
show any homolgy to known kinases or phosphatases. Similarly, the significance
of tyrosine phosphorylation on the signal transducing activity of ζ has not yet been
determined. It is possible that the phosphorylation reflects a feedback via an
activated tyrosine kinase, resulting in down-regulation of the activation signal.

CD4 and CD8 are phosphorylated on serine residues and CD4 internalized on T-
cell activation (76,77). Our studies of mutant CD4 molecules in antigen-specific T
cell hybridomas indicated that mutation of any one of the three cytoplasmic serine

residues had no effect on association with p56^{lck} or on antigen-induced IL-2 production and that internalization of CD4 is not required for normal IL-2 production (57). It can therefore be inferred that neither phosphorylation or internalization of CD4 is required for participation of the CD4/p56^{lck} association in T cell activation. That internalization of the CD4/p56^{lck} complex is not required implies that substrates of p56^{lck} reside at the membrane and, as such, may include any of the molecules mentioned above (the ζ chain, *raf*, PLC-γ, and p59^{fyn}). However, PKC-induced internalization of CD4, followed by dissociation of p56^{lck} may be involved in subsequent down-regulation of the activation signal (81).

b) Thymic ontogeny

Models of thymic ontogeny must seek to describe how a double positive (CD4+,CD8+) thymocyte switches to a single positive phenotype that corresponds to the MHC specificity of the TCR (reviewed in ref. 82). Corecognition of MHC molecules by CD4 or CD8 and the TCR is likely to be instrumental in this developmental decision. As diagramed in Fig. 5, two general mechanisms can be envisioned for this process (78): (i) single positive T cells are derived via a stochastic mechanism, in which one of the coreceptors is switched off; if the expressed coreceptor is appropriate to the restriction of the TCR, then the thymocyte is positively selected; or (ii) corecognition of an MHC molecule by CD4 or CD8 and the TCR results in a signal via the engaged coreceptor which actively switches off expression of the unligated coreceptor. The first "selection" model does not discriminate between the signals mediated by CD4 and CD8; the second "instruction" model, however, requires that the two coreceptors mediate distinct signals. In either case, it is anticipated that the association of CD4/CD8 with p56^{lck} and physical association of the complex with the TCR are involved in the signal transduction.

Studies on the interaction of CD4 and CD8α with p56^{lck} indicate that only a few shared residues in the cytoplasmic domains are required for association with the kinase. It is therefore possible that non-shared residues in the tails of CD4 and CD8α direct the phosphorylation of different substrates by p56^{lck}, resulting in the activation of distinct intracellular signalling pathways. Another potential factor in discrimination between CD4- and CD8-mediated signals is the presence of the CD8β chain in the CD8 heterodimer (Fig. 1). This polypeptide does not bind p56^{lck}, but it may regulate its activity and substrate specificity when bound to CD8α; alternatively, the cytoplasmic domain of CD8β may interact with distinct cytoplasmic components that, upon translocation into the TCR complex, activate a signalling pathway distinct from that activated by p56^{lck}. These signals may effect the switch from the double to single positive phenotype, but could also be involved in the functional maturation of thymocytes by inducing the increased surface TCR expression coincident with the switch to the single positive phenotype (79,80) and by initiating the functional developmental programs of thymocytes, resulting in helper functions in most CD4+ cells and cytotoxic functions in most CD8+ cells.

Of course, models of differential signalling via CD4 and CD8 remain to be substantiated, but there are suggestions of subtle differences in the association of p56^{lck} with CD4 versus CD8α. Firstly, in both COS-7 and lymphoid cells, we consistently observed lower levels of p56^{lck} co-immunoprecipitating with CD8 as

SELECTION MODEL (SINGLE POSITIVE CELLS)

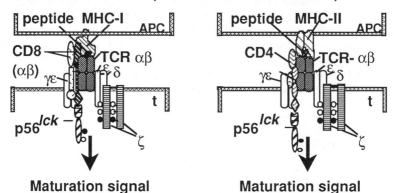

INSTRUCTION MODEL (DOUBLE POSITIVE CELLS)

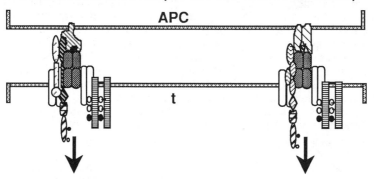

Signal 1:Shut off CD4 gene
Turn on Tc functions

Signal 2:Shut off CD8 gene
Turn on Th functions

FIG. 5. **Development of single positive T cells.** In both models, matching of the coreceptor to the MHC-restriction of the TCR is achieved by co-recognition. Pathways of development for CD8+ and CD4+ cells are shown on the left and right, respectively For the selection model, two thymocytes (t), already single positive due to random loss of expression of one coreceptor, are selected because the remaining coreceptor is appropriate to the specificity of the TCR. In the instruction model, a double positive thymocyte retains expression of whichever coreceptor can bind to the same MHC molecule as the TCR - both eventualities are shown, although in reality the thymocyte would bear TCRs of only one specificity. Signals 1 and 2 may involve phosphorylation of distinct substrates by p56^{lck}, as directed by its association with CD8 versus CD4, possibly leading to transcriptional changes and the determination of the T-cell phenotype as cytotoxic (Tc) or helper (Th).

compared to CD4, and estimates of the proportion of p56^{lck} associated with each coreceptor in normal T cells suggest the same (22,53,57); however, some of these studies do not consider the possibility that the kinase and CD4 or CD8 may become associated after solubilization of the cell, and may therefore not reflect the true state of association in the intact cell. Nevertheless, the affinity of p56^{lck} for CD4 does appear to be greater than for CD8. Secondly, we noted a low level of associated p56^{lck} when only one of the cysteine residues of CD8α was mutated; mutation of both resulted in the complete abolition of association as was observed on mutation of either cysteine residue of CD4. Thirdly, CD4, but not CD8, appeared to require membrane proximal amino acid residues N-terminal to the cysteine motif for association with p56^{lck} (52). These observations suggest that there may be differences in the forces governing association of p56^{lck} with CD4 and CD8α. Fourthly, activation of protein kinase C in T cells causes dissociation of p56^{lck} from CD4 but not from CD8, suggesting differential mechanisms of regulation of the two coreceptor complexes (81). Clearly, however, there is much to be learned about the regulation and substrate specificity of p56^{lck} before any detailed model of differential signalling via CD4 and CD8 can be drawn up.

The interaction of coreceptor with p56^{lck} is also likely to be required in signals that initiate clonal deletion of autoreactive thymocytes. Antibodies against CD4 block clonal deletion of self-reactive thymocytes (83,84); thymocytes expressing an autoreactive MHC-I specific TCR fail to be deleted if the class I MHC molecule has a mutated CD8 binding site (Killeen, N., Moriarty, A. and Littman, D.R., unpublished results); and overexpression of a CD4 transgene in thymocytes abrogates the ability of cells with autoreactive MHC-I specific TCR's to be deleted (Teh et al., unpublished results). This latter result is not observed with tailless CD4 molecules, suggesting that excess CD4 competes for p56^{lck} which is required for the CD8-mediated signal for clonal deletion.

An understanding of the precise means through which CD4 and CD8 transmit signals during positive and negative selection in the thymus awaits future studies using mice in which the various genes involved in these processes have been ablated or mutated. With the rapid pace of progress in the field of mouse developmental biology, solutions are likely to be around the corner.

REFERENCES

1.   Blackman M, Kappler J, Marrack P. The role of the T cell receptor in positive and negative selection of developing T cells. *Science* 1990;248:1335-1341.
2.   Schwartz RH.   Acquisition of immunologic self-tolerance.   *Cell* 1989;57:1073-1081.
3.   Kappler JW, Roehm N, Marrack PC. T cell tolerance by clonal elimination in the thymus. *Cell* 1987;49:273-280.
4.   Kisielow P, Bluthmann H, Staerz U, Steinmetz M, von Boehmer H. Tolerance in T-cell-receptor transgenic mice involves deletion of nonmature CD4+8+ thymocytes. *Nature* 1988;333:742-746.
5.   Kappler JW, Staerz U, White J, Marrack, PC. Self-tolerance elminates T cells specific for Mls-modified products of the major histocompatibility complex. *Nature* 1988;332:35-40.

6.   MacDonald HR, Schneider R, Lees RK, Howe RC, Acha-Orbea H, Festenstein H, Zinkernagel RM, Hengartner H. T-cell receptor Vb use predicts reactivity and tolerance to Mls a-encoded antigens. *Nature* 1988;332:40-45.
7.   Teh HS, Kisielow P, Scott B, Kishi H, Uematsu Y, Bluthmann H, von Boehmer H. Thymic major histocompatibility complex antigens and the alpha/beta T-cell receptor determine the CD4/CD8 phenotype of T cells. *Nature* 1988;335:229-233.
8.   Sha WC, Nelson CA, Newberry RD, Kranz DM, Russell JH, Loh DY. Positive and negative selection of an antigen receptor on T cells in transgenic mice. *Nature* 1988;336:73-76.
9.   Berg LJ, Pullen AM, Fazekas de St. Groth B, Mathis D, Benoist C, Davis MM. Antigen/MHC-specific T cells are preferentially exported from the thymus in the presence of their MHC ligand. *Cell* 1989;58:1035-1046.
10.   Kaye J, Hsu M-L, Sauron M-E, Jameson SC, Gascoigne, NRJ, Hedrick SM. Selective development of CD4+ T cells in transgenic mice expressing a class II MHC-restricted antigen receptor. *Nature* 1989;341:746-749.
11.   Weiss A. Structure and function of the T cell antigen receptor. *J. Clin. Invest.* 1990;86:1015-1022.
12.   Schwartz RH. A cell culture model for T lympyhocyte clonal anergy. *Nature* 1990;348:1349-1356.
13.   Rammensee H-G, Kroschewski R, Frangoulis B. Clonal anergy induced in mature Vβ6+ T lymphocytes on immunizing Mls-1b mice with Mls-1a expressing cells. *Nature* 1989;339:541-544.
14.   Ramsdell F, Lantz T, Fowlkes BJ. A nondeletional mechanism of thymic self tolerance. *Science* 1989;246:1038-1041.
15.   Littman DR. The structue of the CD4 and CD8 genes. *Annu. Rev. Immunol.* 1987;5:561-584.
16.   Parnes JR. Molecular biology and function of CD4 and CD8. *Adv. Immunol.* 1989;44:265-311.
17.   Doyle C, Strominger JL. Interaction between CD4 and class II MHC molecules mediates cell adhesion. *Nature* 1987;330:256-259.
18.   Norment AM, Salter RD, Parham P, Engelhard V, Littman DR. Cell-cell adhesion mediated by CD8 and MHC class I molecules. *Nature* 1988;336:79-81.
19.   Rosenstein Y, Ratnofsky S, Burakoff SJ, Herrmann SH. Direct evidence for binding of CD8 to HLA class I antigens. *J. Exp. Med.* 1989;169:149-160.
20.   Ramsdell F, Fowlkes BJ. Engagement of CD4 and CD8 accessory molecules is required for T cell maturation. *J. Immunol.* 1989;143:1467-1471.
21.   Rudd CE, Trevilyan JM, Dasgupta JD, Wong LL, Schlossman SF. The CD4 receptor is complexed in detergent lysates to a protein-tyrosine kinase (pp58) from human T lymphocytes. *Proc. Natl. Acad. Sci. USA* 1988;85:5190-5194.
22.   Veillette A, Bookman MA, Horak EM, Bolen JB. The CD4 and CD8 T cell surface antigens are associated with the internal membrane tyrosine-protein kinase p56^{lck}. *Cell* 1990;55:301-308.
23.   Barber EK, Dasgupta JD, Schlossman SF, Trevillyan JM, Rudd CE. The CD4 and CD8 antigens are coupled to a protein-tyrosine kinase (p56lck) that phosphorylates the CD3 complex. *Proc. Natl. Acad. Sci. USA* 1989;86:3277-3281.
24.   Samelson LE, Patel MD, Weissman AM, Harford JB, Klausner RD. Antigen activation of murine T cells induces tyrosine phosphorylation of a polypeptide associated with the T cell antigen receptor. *Cell* 1986;46:1083-1090.

25. June CH, Fletcher MC, Ledbetter JA, Samelson LE. Increases in tyrosine phosphorylation are detectable before phospholipase C activation after T cell receptor stimulation. *J. Immunol.* 1990;144:1591-1599.

26. June CH, Fletcher MC, Ledbetter JA, et al. Inhibition of tyrosine phosphorylation prevents T-cell receptor-mediated signal transduction. *Proc. Natl. Acad. Sci. USA* 1990;87:7722-7726.

27. Swain SL. T cell subsets and the recognition of MHC class. *Immunol. Rev.* 1983;74:129-142.

28. Salter RD, Norment AM, Chen BP, et al. Polymorphism in the alpha 3 domain of HLA-A molecules affects binding to CD8. *Nature* 1989;338:345-347.

29. Salter RD, Benjamin RJ, Wesley PK, et al. A binding site for the T-cell co-receptor CD8 on the a3 domain of HLA-A2. *Nature* 1990;345:41-46.

30. Potter TA, Bluestone JA, Rajan TV. A single amino acid substitution in the alpha 3 domain of an H-2 class I molecule abrogates reactivity with CTL. *J. Exp. Med.* 1987;166:956-966.

31. Marrack P, Endres R, Schimonkevitz R, et al. The major histocompatibility complex-restricted antigen receptor on T cells. II. Role of the L3T4 product. *J. Exp. Med.* 1983;158:1077-1091.

32. Gabert J, Langlet C, Zamoyska R, Parnes JR, Schmitt-Verhulst A-M, Malissen B. Reconstitution of MHC class I specificity by transfer of the T cell receptor and Lyt-2 genes. *Cell* 1987;50:545-554.

33. Ballhausen WG, Reske-Kunz AB, Tourvieille B, Ohashi PS, Parnes JR, Mak TW. Acquisition of an additional antigen specificity after mouse CD4 gene transfer into a T helper hybridoma. *J. Exp. Med.* 1988;167:1493-1498.

34. Fazekas de St. Groth, Gallagher PF, Miller JF. Involvement of Lyt-2 and L3T4 in activation of hapten-specific Lyt2+ L3T4+ T-cell clones. *Proc. Natl. Acad. Sci. USA* 1986;83:2594-2598.

35. Jones b, Khavari PA, Conrad PJ, Janeway, CA Jr. Differential effects of antibodies to Lyt-2 and L3T4 on cytolysis by cloned, Ia-restricted T cells expressing both proteins. *J. Immunol.* 1987;139:380-384.

36. Emmerich F, Strittmatter U, Eichmann K. Synergism in the activation of human CD8 T cells by cross-linking the T-cell receptor complex with the CD8 differentiation antigen. *Proc. Natl. Acad. Sci. USA* 1986;83:8298-8302.

37. Saizawa K, Rojo J, Janeway CA Jr. Evidence for a physical association of CD4 and the CD3:α:β T-cell receptor. *Nature* 1987;328:260-263.

38. Bank I, Chess L. Perturbation of the T4 molecule transmits a negative signal to T cells. *J. Exp. Med.* 1985;162:1294-1303.

39. Haque S, Saizawa K, Rojo J, Janeway, C Jr. The influence of valence on the functional activities of monoclonal anti-L3T4 antibodies. *J. Immunol.* 1987;139:3207-3212.

40. Mittler RS, Goldman SJ, Spitalny GL, Burakoff SJ. T-cell receptor-CD4 physical association in a murine T-cell hybridoma: induction by antigen receptor ligation. *Proc. Natl. Acad. Sci. USA* 1989;86:8531-8535.

41. Chuck RS, Cantor CR, Tse DB. CD-4-T-cell antigen receptor complexes on human leukemia T cells. *Proc. Natl. Acad. Sci. USA* 1990;87:5021-5025.

42. Janeway CA, Jr, Rojo J, Saizawa K, et al. The co-receptor function of murine CD4. *Immunological Reviews* 1989;109:77-92.

43. Connolly JM, Hansen TH, Ingold AL, Potter TA. Recognition by CD8 on cytotoxic T lymphocytes is ablated by several substitutions in the class I α3 domain: CD8 and the T-cell receptor recognize the same class I molecule. *Proc. Natl. Acad. Sci. USA* 1990;87:2137-2141.

44.   Sadowski I, Stone JC, Pawson T. A noncatalytic domain conserved among cytoplasmic protein-tyrosine kinases modifies the kinase function and transforming activity of Fujinami sarcoma virus P130gag-fps. *Mol. Cell. Biol.* 1986;6:4396-4408.
45.   Perlmutter RM, Marth JD, Ziegler SF, et al. Specialized protein tyrosine kinase proto-oncogenes in hematopoietic cells. *Biochim, Biophys. Acta* 1988b;948:245-262.
46.   Marchildon GA, Casnelie JE, Walsh KA, Krebs EG. Covalently bound myristate in a lymphoma tyrosine protein kinase. *Proc. Natl. Acad. Sci. USA* 1984;81:7679-7682.
47.   Marth JD, Peet R, Krebs EG, Perlmutter RM. A lymphocyte-specific protein-tyrosine kinase gene is rearranged and overexpressed in the murine T cell lymphoma LSTRA. *Cell* 1990;43:393-404.
48.   Voronova AF, Sefton BM. Expression of a new tyrosine protein kinase is stimulated by retrovirus promoter insertion. *Nature* 1986;319:682-685.
49.   Koga Y, Caccia N, Toyonaga B, et al. A human T-cell specific cDNA clone (YT16) encodes a protein with extensive homology to a family of protein tyrosine kinases. *Eur. J. Immunol.* 1986;16:1643-1646.
50.   Marth JD, Lewis DB, Wilson CB, Gearn ME, Krebs EG, Perlmutter RM. Regulation of pp56^{lck} during T-cell activation: functional implications for the src-like protein tyrosine kinases. *EMBO J* 1987;6:2727-2734.
51.   Veillette A, Bookman MA, Horak EM, Samelson LE, Bolen JB. Signal transduction through the CD4 receptor involves the activation of the internal membrane tyrosine-protein kinase p56^{lck}. *Nature* 1990;338:257-259.
52.   Turner JM, Brodsky MH, Irving BA, Levin SD, Perlmutter RM, Littman DR. Interaction of the unique N-terminal region of tyrosine kinase p56^{lck} with cytoplasmic domains of CD4 and CD8 is mediated by cysteone motifs. *Cell* 1990;60:755-765.
53.   Simpson SC, Bolen JB, Veillette A. CD4 and p56^{lck} can stably associate when co-expressed in NIH3T3 cells. *Oncogene* 1990;4:1141-1143.
54.   Zamoyska R, Derham P, Gorman SD, et al. Inability of CD8α' polypeptides to associate with p56^{lck} correlates with impaired function in vitro and lack of expression in vivo. *Nature* 1989;342:278-281.
55.   Shaw AS, Amrein KE, Hammond C, Stern DF, Sefton BM, Rose JK. The lck tyrosine protein kinase interacts with the cytoplasmic tail of the CD4 glycoprotein through its unique amino-terminal domain. *Cell* 1989;59:627-636.
56.   Shaw AS, Chalupny J, Whitney JA, et al. Short related sequences in the cytoplasmic domains of CD4 and CD8 mediate binding to the amino-terminal domain of the p56^{lck} tyrosine protein kinase. *Mol. Cell. Biol.* 1990;10:1853-1862.
57.   Glaichenhaus N, Shastri N, Littman DR, Turner JM. Antigen-specific signal transduction via the T-cell receptor requires association of p56^{lck} with CD4. *Cell* 1991; (in press).
58.   Cooper JA. The src-family of protein tyrosine kinases. In: Kemp B, Alewood PF, eds. *Peptides and Protein Phosphorylation,* Boca Raton, Florida:CRC Press, Inc., 1990.
59.   Williams LT. Signal transduction by the platelet derived growth factor receptor. *Science* 1989;243:1564-1570.
60.   MacAuley A, Cooper JA. The carboxy-terminal sequence of p56^{lck} can regulate p60src. *Mol. Cell. Biol.* 1988;8:3560-3564.

61.   Marth JD, Cooper JA, King CS, et al. Neoplastic transformation induced by an activated lymphocyte-specific protein tyrosine kinase (pp56^{lck}). *Mol. Cell. Biol.* 1988;8:540-550.

62.   Amrein KE, Sefton BM. Mutation of a site of tyrosine phosphorylation in the lymphocyte-specific tyrosine protein kinase, p56^{lck}, reveals its oncogenic potential in fibroblasts. *Proc. Natl. Acad. Sci.* 1988;85:4247-4251.

63.   Thomas M. The leukocyte common antigen family. *Ann. Rev. Immunol.* 1989;7:339-369.

64.   Pingel JT, Thomas ML. Evidence that the leukocyte-common antigen is required for antigen-induced T lymphocyte proliferation. *Cell* 1989;58:1055-1065.

65.   Koretzky GA, Picus J, Thomas ML, Weiss A. Tyrosine phosphatase CD45 is essential for coupling T-cell antigen receptor to the phosphatidyl inositol pathway. *Nature* 1990;346:66-68.

66.   Ostergaard HL, Trowbridge IS. Coclustering CD45 with CD4 or CD8 alters the phosphorylation and kinase activity of p56^{lck}. *J. Exp. Med.* 1990;172:347-350.

67.   Mustelin T, Coggeshall KM, Altman A. Rapid activation of the T-cell tyrosine protein kinase p56^{lck} by the CD45 phorphotyrosine phosphatase. *Proc. Natl. Acad. Sci. USA* 1989;86:6302-6306.

68.   Cooke MP, Perlmutter RM. Expression of a novel form of the fyn proto-oncogene in hematopoietic cells. *New Biol.* 1989;1:66-74.

69.   Katagiri T, Urakawa K, Yamanashi Y, et al. Over-expression of src family gene for tyrosine-kinase p59^{fyn} in CD4- CD8- T cells of mice with a lymphoproliferative disorder. *Proc. Natl. Acad. Sci. USA* 1989;86:10064-10068.

70.   Samelson LE, Phillips AF, Luong ET, Klausner RD. Association of the fyn protein-tyrosine kinase with the T-cell antigen receptor. *Proc. Natl. Acad. Sci. USA* 1990;87:4358-4362.

71.   Pawson T. Non-catalytic domains of cytoplasmic protein-tyrosine kinases: regulatory elements in signal transduction. *Oncogene* 1988;3:491-495.

72.   Anderson D, Koch CA, Grey L, Ellis C, Moran MF, Pawson T. Binding of SH2 domains of phospholipase C gamma 1, GAP, and Src to activated growth factor receptors. *Science* 1990;250:979-982.

73.   Bourne HR, Sanders DA, McCormick F. The GTPase superfamily: a conserved switch for diverse cell functions. *Nature* 1990;348:125-132.

74   Baniyash M, Garcia-Morales P, Samelson LE, Klausner RD. The T cell antigen receptor zeta chain is tyrosine phosphorylated upon activation. *J. Biol. Chem.* 1988;263:18225-18230.

75.   Irving BA, Weiss A. The cytoplasmic domain of the T cell receptor zeta chain is sufficient to couple to receptor-associated signal transduction pathways. *Cell* 1991; (in press).

76.   Acres RB, Conlon PJ, Mochizuk DY, Gallis B. Rapid phosphorylation and modulation of the T4 antigen on cloned helper T cells induced by phorbol myristate acetate or antigen. *J. Biol. Chem.* 1986;261:16210-16214.

77.   Acres RB, Conlon PJ, Mochizuk DY, Gallis B. Phosphorylation of the CD8 antigen on cytotosic human T cells in response to phorbol myristate acetate or antigen presenting B cells. *J. Immunol.* 1987;139:2268-2274.

78.   Robey E, Axel R. CD4: collaborator in immune recognition and HIV infection. *Cell* 1990;60:697-700.

79.   Bonifacino JS, McCarthy SA, Maguire JE et al. Novel post-translational regulation of TCR expression in CD4+CD8+ thymocytes influenced by CD4. *Nature* 1990;344:247-251.

80.   McCarthy SA, Kruisbeek AM, Uppenkamp IK, Sharrow SO, Singer A. Engagement of the CD4 molecule influences cell surface expression of the T-cell receptor on thymocytes. Nature 1988;336:76-79.

81.   Hurley TR, Luo K, Sefton BM. Activators of protein kinase C induce dissociation of CD4, but not CD8, from p56^{lck}. *Science* 1989;245:407-409.

82.   von Boehmer H. Developmental biology of T cells in T cell-receptor transgenic mice. *Annu. Rev. Immunol.* 1990;8:532-536.

83.   Fowlkes BJ, Schwartz RH, Pardoll DM. Deletion of self-reactive thymocytes occurs at a CD4+8+ precursor stage. *Nature* 1988;334:620-623.

84.   MacDonald HR, Hengartner H, Pedrazzini, T. Intrathymic deletion of self-reactive cells prevented by neonatal anti-CD4 antibody treatment. *Nature* 1988;335:174-176.

Advances in Regulation of Cell Growth, Volume 2;
Cell Activation: Genetic Approaches, edited by
James J. Mond, John C. Cambier, and Arthur Weiss.
Raven Press, Ltd., New York © 1991.

10

Developmental Regulation of the *egr-1* Gene as a Mechanism for Ensuring Unresponsiveness to Antigen in the Immature B Lymphocyte

Vicki L. Seyfert and John G. Monroe

*Department of Pathology and Laboratory Medicine, University of Pennsylvania School
of Medicine, Philadelphia, PA 19104*

The developmental progression of a pluripotential hematopoetic stem cell into a fully mature murine B cell involves a series of sequential events. During this progression, the B cell progenitor acquires a number of lymphocyte or B cell specific markers as well as the expression of the B cell antigen receptor, surface immunoglobulin (sIg). The functions of the mature murine B cell, antigen recognition, processing and presentation of antigen, and/ or antibody production, necessitate that the developing B cell traverse through a stage where those cells expressing sIg with specificity for self-antigens can be either deleted or functionally turned-off (1-4). Hence, the maturational program of the murine B cell not only involves the expression of B cell antigen receptor, for development into a mature immunocompetent cell, but is further complicated by the selective process that these cells must undergo to prevent self-reactive B cells from becoming functional (4). The process of clonal elimination (tolerance) through either functional silencing (5) or deletion (6,7) is believed to occur primarily at the immature stage of B cell development. At this stage, the B cell expresses only the sIgM antigen receptor in contrast to mature B cells which co-express sIgM and sIgD.

A fundamental question in B cell biology is how signaling through the antigen receptor, sIgM, can elicit different cellular responses depending on the developmental stage of the B cell. While immature B cells are rendered tolerant following stimulation through sIgM (1-4), mature B cells are activated to enter cell cycle (8-13) and become competent for proliferation (9,11,14). We have addressed the issue of what defines sIgM induced positive and negative growth responses (activation or unresponsiveness, respectively) by discerning whether there are differences in the signal transduction mechanisms used by immature versus mature B cells. We have primarily focused on the role of growth associated immediate/early genes in determining B cell growth responses.

Translation of sIgM-generated signals into activation versus tolerogenic responses necessitates the selected expression of genes which typify the nature of the B cell response. Coordinate regulation of these sets of genes likely occurs through expression of specific immediate/early type genes (15), encoding transcriptional regulatory factors (16-17) and whose expression is coupled to receptor-generated signals. One such gene, early growth response gene-1 (*egr-1*) (18-21) encodes a sequence-specific DNA binding protein (22) with transcriptional regulatory activity (23). It is expressed following growth factor or mitogen stimulation in a number of cell types (18-21), including B cells (24). A role for *egr-1* expression in determining the nature of the B cell response to sIgM signaling is suggested by experiments showing that tolerance-susceptible immature B cells from adult mouse bone marrow do not express *egr-1* following sIgM crosslinking (25). In contrast, signaling through sIgM on mature B cells leads to activation of these cells as determined by entry into cell cycle (G_0 to G_1 phase transition) (8-13), preceded by induction of *egr-1* transcription (24).

The above findings suggest that differential expression of *egr-1* is involved in the translation of sIgM signals into qualitatively different long-term B cell responses. Furthermore, these results suggest developmental stage-specific regulation for inducible expression of *egr-1* in B cells. We have addressed the differential regulation of *egr-1* expression in mature and immature B cells and have explored the association between the expression of this gene and the ensuing B cell activation response. Our findings indicate that lack of inducible *egr-1* gene expression in immature B cells is regulated in *cis* by a mechanism involving specific DNA methylation (25). Furthermore, our studies suggest that there is a requirement for *egr-1* expression in a positive growth response to sIgM-generated signals (24-26).

In this review, we present our experimental results which have led to these conclusions. We will also discuss the developmental regulation of tolerance susceptibility in B cells, drawing from our interpretations of our studies of immediate/early gene expression and cellular studies by ourselves and others.

RESULTS

Early growth-factor gene-1.

The early growth-factor response gene-1 (*egr-1*) is an immediate early gene originally cloned by Sukhatme et al., (18) from a differential screening of a serum-stimulated fibroblast cDNA library. Using the same strategy, *egr-1* was also cloned by Lau and Nathans, (19) who refer to this gene as *zif268*, and Chavrier et al. (21), who call this gene *Krox-20*. Milbrandt (20) cloned the same gene while differentially screening a cDNA library from nerve growth factor stimulated PC12 cells, and named this gene *NGF1-A*. Thus, inducible expression of the *egr-1* gene is ubiquitous, occuring when cells of multiple lineages are stimulated to grow and divide. Like other immediate early genes, *egr-1* mRNA is rapidly and transiently induced in response to receptor-generated signals (18-19) and is "superinduced" in the presence of the protein synthesis inhibitor cycloheximide (18-21). Expression of this gene in the presence of cycloheximide demonstrates that *egr-1* is a primary response gene not requiring *de novo* protein synthesis for its induction.

Structural information about the *egr-1* gene was determined from a genomic clone (a 6.6 XbaI fragment) isolated from a fibroblast genomic DNA library (27). The *egr-1* gene consists of two exons with a single 700 bp intron between nucleotide position 556 and 557. Analysis of approximately 1 kb of 5' flanking region from this clone revealed that the *egr-1* promoter contains several sequences which are homologous to known transcriptional regulatory elements (enhancers). These enhancer elements include: 6 serum response elements (SRE), elements previously shown to be important in the serum induction of *c-fos* in fibroblasts (28); 2 AP-1 sites, binding sites for the transcription factor AP-1/*c-jun* and *c-fos* complex (29-32); 4 Sp-1 binding sites (33); and two cAMP response elements (34). By deletion analysis of the *egr-1* promoter Seyfert *et al.* (24), have identified the region between -388 and -359 to be necessary for anti-μ or PMA induced transcription of *egr-1* in B cells. This region contains one of the SREs although its role in regulating transcription in this system has not been directly addressed. Interestingly, this region does not contain any of the *egr-1* gene's two AP-1 elements, which have previously been shown to be the PMA response element.

An interesting feature of the *egr-1* promoter is its extremely high GC content, 60.5%, organized into areas of GC rich sequences called CpG or HTF islands (35). CpG islands have been found in a number of genes, including the immediate/early genes *c-myc* and *c-fos*, and are believed to be indicative of ubiquitously expressed genes (35). It is currently accepted that CpG islands are important in the regulation of these genes, particularly by DNA methylation. Transcription of genes with CpG islands is inhibited when the islands are methylated. Also, transfection experiments using *in vitro* methylated CpG island containing genes, have shown that methylation of the coding exons does not interfere with transcription, while methylation of the 5' flanking region containing CpG islands does block transcription (36). An inhibitory effect of methylation at islands has been demonstrated by studies of genes such as the HPRT gene (37). HPRT is methylated and transcriptionally silent on the inactive X chromosome, and analysis of this gene showed that methylation of non-island regions does not correlate with gene inactivity . In several studies it was shown that this gene could be rendered transcriptionally active by treatment with 5-azacytidine which inhibits DNA methylation (38). Similarly, we have found that DNA methylation can play a role in the regulation of *egr-1* (25).

The ubiquitous expression of *egr-1* following growth inducing stimuli has lead to the hypothesis that this factor may play a key role in the control of cell growth. The first evidence supporting this idea was the finding that the predicted EGR-1 protein contains three zinc fingers of the C_2H_2 motif, suggesting that *egr-1* encodes a DNA binding protein (18,21). Since then, it has been established that the EGR-1 protein is a transcriptional activator (23), that binds the consensus DNA sequence GCG(G/T)GGGCG with high affinity (22). Interestingly, this sequence has been found in the promoter regions of many genes, including proto-oncogenes, class II genes, and the transferrin receptor gene (23). EGR-1 does not appear to possess a transcriptional activation domain of a type characterized so far, acidic regions or glutamine- or proline-rich stretches (17). Further analyses have revealed that two proteins are encoded by *egr-1* mRNA. The first is 82 kDa, resulting from translation from the first AUG codon of the open reading frame, and the second is 88 kDa, resulting from translation from a non-AUG codon located upstream (23). These proteins have been found to be phosphorylated on serine residues, a modification that may be important for EGR-1 activity (23). Taken together, these characteristics of the EGR-1 protein suggest that it may be involved in growth-factor or mitogen induced cellular activation functioning as an intracellular "third messenger" which regulates the expression of secondary response genes that define the nature of the cellular response.

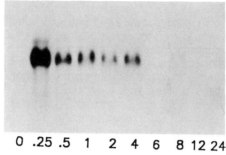

0 .25 .5 1 2 4 6 8 12 24

Anti−μ Stimulation (hrs)

FIG. 1. **Egr-1 mRNA expression is rapidly and transiently upregulated following stimulation of normal splenic B cells with anti-μ.** Murine B cells were isolated and purified as previously described. Final preparations were centrifuged through Lympholyte-M (Cedarlane Labs. Limited, Hornby, Ontario, Canada) with a bouyant of 1.088 which allows recovery of dense, resting B cells (9). B cells were rested for 2 hours in media before stimulation. Total cellular RNA was isolated from B cells stimulated with anti-μ (10 μg/ml) for the indicated lengths of time and subjected to Northern analysis (10 μg total RNA/lane). Zero time points indicate unstimulated cells. The blot was analyzed for *egr-1* expression using the OC68 probe (49). Tick marks denote positions of 28S and 18S rRNA . (Adapted from ref. 26).

Egr-1 is inducible in mature B cells during activation.

Crosslinking of sIgM on mature murine resting B cells, with anti-immunoglobulin antibodies (anti-μ) induces activation of the B cell manifest as egress from the G_0 phase of the cell cycle and progression into the G_1 phase (8-13). Once in G_1, B cells are competent to receive further progression signals required to drive the cells through cell cycle and to differentiate into antibody secreting plasma cells (9,11,14). In some cases, anti-μ in high concentrations, can induce proliferation by a fraction of mature resting B cells (9,11). A number of biochemical and molecular events have been shown to be associated with anti-μ induced B cell activation (14,39), including increases in intracellular calcium (40), hydrolysis of inositol phospholipids (41-43), activation of protein kinase C (PKC) (44), and the induction of immediate early genes such as *c-fos* and *c-myc* (39,45,46). Several of these events have been shown to be critical for transducing sIgM signals into the appropriate B cell response (47,48), in this case proliferation. As in other cells, the role of immediate early genes in the B cell activation response is believed to be one of signal translation, linking biochemical "second messengers" in the cytoplasm with the expression of genes that typify the nature of the B cell response to sIg-generated signals. Many immediate/early genes encode transcriptional regulatory factors (16,17), and thus can regulate other genes necessary for the B cell activation response. We studied the regulation *egr-1* during anti-μ-induced B cell activation in order to determine if this gene functioned as a "third messenger or translator" for sIgM-mediated signals. *Egr-1* mRNA levels are rapidly (by 15 minutes) and transiently (returning to basal levels after 6 hours) increased following anti-μ stimulation (Figure 1). The increase in *egr-1* mRNA is a primary response to sIgM signaling since it is induced in the presence of cycloheximide but is blocked by actinomycin D (24). The observation that anti-μ can induce transcription of a CAT reporter gene driven by the *egr-1* promoter is consistent with the increase in message level as a result of increased transcription (24).

Analysis of EGR-1 protein expression in anti-μ stimulated mature resting B cells showed that protein induction is prolonged in comparison with *egr-1* mRNA increases. EGR-1 protein is first detected within 1 hour after stimulation and has declined but still detectable after 24 hours (Figure 2). The prolonged expression of EGR-1 protein is consistent with involvement of this molecule in maintaining the B cell in an activated state "primed" for secondary signals which then allow progression into the S phase of cell cycle. Further support for a role of *egr-1* in the B cell proliferative response is the correlation between the concentration of anti-μ required to induce *egr-1* expression and the concentration of anti-μ required for B cell G_0 to G_1 transition and S phase transition (Figure 3). Finally, preliminary studies using antisense *egr-1* to block EGR-1 expression have shown that antisense but not sense oligonucleotide inhibit anti-μ induced activation of

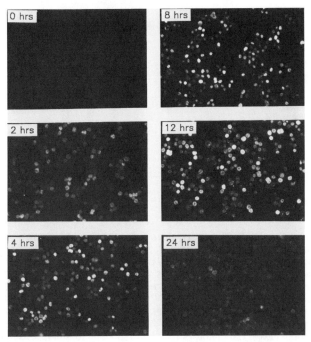

FIG 2.   **Kinetics of EGR-1 protein expression in B cells stimulated with anti-μ antibodies.** Murine splenic B cells were isolated as in Fig. 1. and stimulated for the indicated time with 10 μg/ml of anti-μ antibody. EGR-1 protein expression was detected by indirect immunofluorescence using a rabbit anti-Egr-1 primary antisera (50) followed by F(ab')$_2$ fragments of an affinity-purified FITC-conjugated goat anti-rabbit immunoglobulin antibody. (From ref. 24)

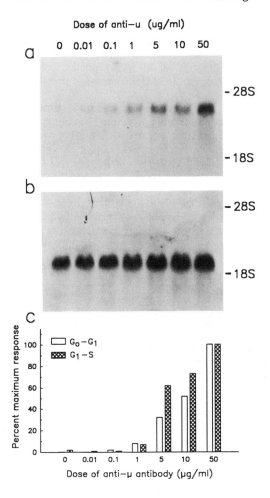

FIG. 3. *Egr-1* **mRNA is upregulated by anti-μ at doses which cause entry into cell cycle and S phase transition.** Upper panel shows Northern blot analysis performed on murine splenic B cells as in Fig. 1. B cells were stimulated for 45 minutes with the indicated concentrations of anti-μ antibody. Zero time point indicates unstimulated cells. The blot was analyzed for *egr-1* expression using the OC68 probe (49) (Panel a). Panel b shows the same blot hybridized to the beta-actin probe to control for equal loading. Positions of the 28S and 18S rRNA are indicated by tick marks to the side of panel. Lower panel shows relative responses to anti-μ as determined by G_0 to G_1 transition ($[^3H]$uridine incorporation at 24 hours) (open bars) or S phase transition ($[^3H]$thymidine incorporation at 48 hours) (hatched bars). Results are expressed as percent maximum responses; B cells were maximally stimulated at 50 μg/ml anti-μ (100%). Values represent means of three replicate cultures. (Figure from ref. 25).

mature B cells (Seyfert, VL and Monroe, JG; in preparation). These studies, combined with the observation that *egr-1* encodes a transcriptional regulatory factor, support a role for *egr-1* in regulating positive growth responses during B cell activation.

With respect to known sIgM signaling pathways, we have established that *egr-1* mRNA upregulation is coupled to the PKC pathway of B cell activation (24). These conclusions are based on our studies using pharmacologic agonists and antagonists to mimic or block the activation of the PKC or Ca^{2+} signaling components. We found that A23187 mediated $[Ca^{2+}]_i$ increases had no effect on *egr-1* expression, when used at doses relevant to ant-μ-mediated B cell activation. In contrast, PKC agonists SC-9 or PMA caused marked increases in *egr-1* mRNA levels (24). Staurosporine, sangivamycin, and H-7, each an inhibitor of PKC (51,52), were able to block anti-μ induced expression of *egr-1*, further supporting our conclusion that sIgM-mediated *egr-1* induction in mature murine B cells is coupled to the PKC pathway (24). Finally, promoter mapping studies showed that anti-μ- and PMA-induced *egr-1* transcription map to the same region within the *egr-1* promoter (McMahon, SB and Monroe, J G; unpublished results). The means by which PKC activation and *egr-1* expression are coupled remains to be defined.

Negative growth responses to sIgM-generated signals are not associated with *Egr-1* induction.

To further correlate *egr-1* induction with positive B cell growth responses, we evaluated *egr-1* expression in cases where sIgM-crosslinking results in a negative growth response or growth inhibition. For these studies we utilized B lymphoma cell lines which were known to be growth inhibited following sIgM crosslinking with anti-μ antibodies. B cell lines were chosen since they are homogenous populations and since they can be manipulated in culture. We initially analyzed *egr-1* expression in the B lymphoma cell line, CH31, which exhibits a decrease in proliferation in response to sIgM crosslinking (53,54). Our results indicated that

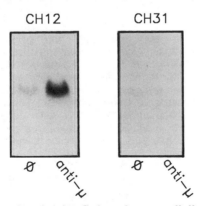

FIG 4. ***Egr-1* expression in the B lymphoma cell lines CH12 and CH31.** Northern blots of total cellular RNA (20 μg/lane) isolated from CH12 cells (left panel) or CH31 cells (right panel). RNA was extracted from unstimulated cells (Lane 1 in each panel) or cells stimulated for 1 hour with 10 μg of anti-μ antibody/ml (Lane 2 in each panel). Blots were probed for *egr-1* expression using the OC68 probe (49).

under conditions of maximum inhibition, there was no detectable change in *egr-1* mRNA levels (Figure 4). For comparison we studied *egr-1* expression in another B cell line in the CH B lymphoma series, CH12 (53). In contrast to the CH31 cells, anti-μ caused upregulation of *egr-1* mRNA levels in the CH12 cells. Importantly, sIgM crosslinking does not cause growth inhibition of CH12 cells,

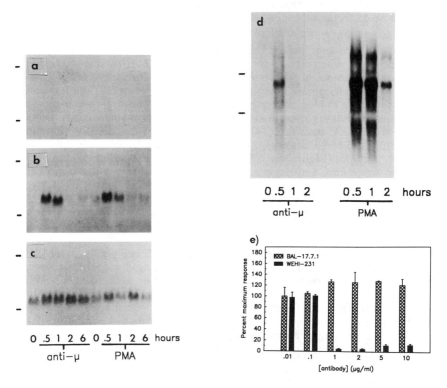

FIG. 5.   **Anti-μ induced *egr-1* expression and proliferative responses in WEHI-231 and BAL-17 cells.** Shown in left panel are Northern blots of total cellular RNA from WEHI-231 cells stimulated with anti-μ (10 μg/ml) or PMA (10 ng/ml) for the indicated times. The same blot was analyzed for expression of *egr-1* (a), *c-fos* (b), and beta-actin (c). Twenty μg of total RNA was loaded in each lane. Upper right panel (d) shows Northern blots of total cellular RNA isolated from BAL-17 cells stimulated with anti-μ (10 μg/ml) or PMA (10 ng/ml) for the indicated times. Tick marks in both panels indicate positions of 28S and 18S rRNAs. In the lower right panel (e) is shown proliferation analysis of BAL-17 or WEHI-231 cells cultured in the presence of the indicated amounts of anti-μ antibody for 48 hours. Cells were pulsed 16 hours prior to harvesting with [3H] thymidine. Results are expressed as percent proliferative response in counts per minute of incorporated [3H] thymidine relative to incorporation by cultures receiving no anti-μ stimulation (100%). Values represent means of three replicate cultures ± standard errors of the means. Values for unstimulated cultures were 171,054 ± 11,626 for BAL-17 cells and 133,072 ± 9,031 for WEHI-231 cells. (Adapted from ref. 25)

rather, ligand induced receptor crosslinking in the presence of appropriate secondary signals induces maturation of these cells to secrete IgM. These findings demonstrated an association between *egr-1* expression and a positive activation response to sIgM-generated signals.

Our observations in CH31 cells were further substantiated by studies using another B lymphoma cell line, WEHI-231. Constitutive proliferation of WEHI-231 cells is inhibited by anti-μ or by PMA (Figure 5) (55-58). *Egr-1* is not induced in WEHI-231 cells following sIgM crosslinking with anti-μ or after stimulation with PMA (Figure 5). This result is somewhat surprising since it is known that sIgM crosslinking on WEHI-231 cells leads to the same cascade of membrane and cytoplasmic signaling events as in mature splenic murine B cells (58-62). Furthermore, the inability of anti-μ or PMA to induce *egr-1* is not due to a general effect on immediate/early gene expression, as results in Figure 5 show that *c-fos* induction is coupled to both anti-μ and PMA stimulation. Thus the differential regulation of *egr-1*, in this case, could not be attributed to different second messenger systems in *egr-1* expressing versus non-expressing cells. Again for comparison, we evaluated *egr-1* expression in the B lymphoma, BAL-17 (63) (Figure 5), and a number of other cell lines (not shown) which are not growth inhibited by anti-μ treatment. As shown in Figure 5, *egr-1* is induced in BAL-17 cells by both anti-μ and PMA, and anti-μ does not inhibit the growth of BAL-17 cells, rather it appears to slightly augment their growth. Thus, in all cases where we have observed proliferation following anti-Ig stimulation, we have observed *egr-1* induction .

In addition to substantiating the association between *egr-1* expression and a positive B cell growth respons, our findings in WEHI-231 and BAL-17 cells are interesting with regards to B cell development. WEHI-231 cells express a phenotype characteristic of an immature B cell (64); they express sIgM but not sIgD. In addition, their negative growth response to sIgM-generated signals (59-62) has been postulated to be analogous to tolerance induction observed in immature B cells (3,4,57,58). Furthermore, crosslinking of sIgM on WEHI-231 cells not only causes growth inhibition of these cells, it eventually causes cell death (57). Therefore, WEHI-231 cells may represent an immature B cell susceptible to tolerance induction by a mechanism analogous to clonal deletion (6,7). BAL-17 cells, on the other hand, express a mature phenotype (63). They are sIgM[+], sIgD[+], and their growth is not inhibited by anti-μ (Figure 5). The different maturational states of B cells represented by BAL-17 and WEHI-231 cells, mature and immature B cells respectively, are by necessity committed to different growth responses to signaling via sIgM. In the case of mature cells, sIgM signaling results in B cell activation to allow their clonal expansion for mounting an immune response to foreign antigen. Alternatively, to ensure that self-reactive B cells are not activated and expanded, crosslinking of sIgM on immature B cells leads to the induction of tolerance or unresponsiveness either by deletion of self-reactive clones or by rendering cells anergic (3-7).

Based on our finding that *egr-1* is differentially regulated in immature versus mature B cells and because of the correlation between *egr-1* expression and a positive growth response to sIgM-generated signals, we hypothesized that sIgM-induced expression, or lack thereof, of *egr-1* and/or other growth-associated genes determines the growth response of a B cell and that differential regulation of these genes during B cell development accounts for the different growth responses of B cells at each stage. Our first approach for studying the developmental regulation of *egr-1* was to evaluate the effects of inducing maturation of an immature B cell, in

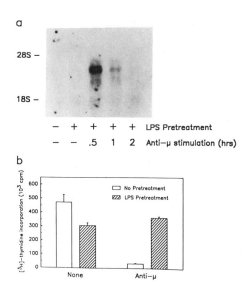

FIG. 6. Pretreatment of WEHI-231 cells with LPS allows induction of *Egr-1* expression by anti-μ and reverses the antiproliferative response. (a) WEHI-231 cells were cultured for 48 hours in the presence of 50 μg of LPS per ml (except for the negative control, leftmost lane). After the initial pretreatment period, the cells were washed and recultured at 37°C in the presence of anti-μ (5 μg/ml) for the indicated times; the cells were then harvested, and total cellular RNA was isolated and analyzed by Northern blotting. Unstimulated cultures were harvested at time zero. Relative Egr-1 expression was determined by hybridization to the OC68 probe (49). (b). WEHI-231 cells were pretreated for 24 hours in the presence of 50 μg/ml of LPS. At the end of the LPS treatment, cells were washed twice to remove LPS, and equal numbers (2×10^4 per well) of either LPS-pretreated or untreated cells were placed in culture (in 96-well plates) with or without anti-μ antibody (5 μg/ml). Relative thymidine incorporation was determined as described in FIG. 5. (Reproduced from ref. 26).

this case WEHI-231 cells, on *egr-1* expression. Specifically, we analyzed whether exposure of WEHI-231 cells to lipopolysaccharide (LPS) would allow anti-μ to induce *egr-1* expression. The rationale of this approach is based on previous studies showing that LPS prestimulation overcame the anti-μ-induced unresponsiveness of immature B cells from neonatal mice (65, Yellen, AJ, Wechsler, R and Monroe, JG; unpublished results). This effect of LPS is attributed to the ability of LPS to cause maturation of the immature B cells. In addition, it was prevously shown that costimulation of WEHI-231 cells with LPS and anti-μ blocked the anti-μ-induced growth inhibition of these cells (56). In our inducibility studies, we pretreated WEHI-231 cells for 48 hours with of LPS, washed away the LPS, and then treated the cells with anti-μ. As shown in Figure 6, LPS treatment alone had no detectable effect on *egr-1* message levels.

However, LPS pretreatment affected the cells in such a way that anti-μ was now able to induce *egr-1* expression. This change in *egr-1* expression was accompanied by a reversal of the anti-μ induced growth inhibition of WEHI-231 cells (Figure 6). Thus, under conditions where sIgM-generated signals can induce *egr-1* expression in WEHI-231 cells, the subsequent proliferative response is altered in a way consistent with our tenet that expression of this gene is involved in the translation of sIgM signals into positive versus negative activation responses.

In order to further assess the developmental regulation of *egr-1* in B cells, we analyzed *egr-1* expression in immature non-transformed B cells. These studies were necessary since our observations showing a lack of *egr-1* expression in immature cells were all derived from studies in transformed cell lines. We examined *egr-1* expression in immature sIgM$^+$, sIgD$^-$ B cells isolated from adult mouse bone marrow, a population of cells belonging to the tolerance-sensitive B cell population as defined by Teale and Klinman (3) and Metcalf and Klinman (66). In addition, we have found that these cells are not activated to enter cell cycle by anti-μ stimulation (Yellen, AJ and Monroe, JG; unpublished results). By analyzing EGR-1 protein expression on a single cell basis we found that neither anti-μ nor PMA was able to induce EGR-1 protein expression in sIgM$^+$, sIgD$^-$ bone marrow B cells (25). These findings demonstrated the presence of a non-transformed population of B cells that behave analogously to WEHI-231 cells with respect to *egr-1* expression and their growth growth response to sIgM crosslinking. The lack of anti-μ inducible *egr-1* expression in immature bone marrow cells supports the hypothesis that *egr-1* is developmentally regulated. Furthermore, these observations suggest that the inability of anti-μ to upregulate *egr-1* expression, in immature bone marrow derived B cells, may contribute to or be responsible for their tolerance susceptibility.

Further molecular studies to define the causative mechanisms responsible for the lack of *egr-1* expression in immature non-transformed B cells were impeded by the heterogenous nature of the bone marrow B cells and by the inability to maintain these cells in culture. It was therefore necessary to use WEHI-231 cells as a model system for these molecular studies. In approaching the question as to why anti-μ does not induce *egr-1* expression in WEHI-231 cells we considered the following possibilities: a) point mutation/s within the *egr-1* promoter in WEHI-231 cells; b) Lack of *trans*-activating factors necessary for the induction of *egr-1* by anti-μ or PMA; c) a trans-mediated repression of *egr-1* in these cells or; d) *cis*-acting DNA modifications of the gene, i.e., DNA methylation. Deletion of the *egr-1* gene or gross structural defects of the gene, such as translocation, in WEHI-231 cells were ruled out by Southern analysis (not shown) and because *egr-1* was expressed in WEHI-231 cells under some experimental conditions i.e., LPS pretreated cells. Point mutations of the *egr-1* gene or its promoter were also decided to be unlikely mechanisms to account for the lack of its expression in WEHI-231 cells since such a change would be recessive unless both alleles were affected.

To address the possible absence of *trans*-activating factors in WEHI-231 cells, we evaluated *egr-1* promoter function in these cells using the CAT gene as a reporter for *egr-1* promoter activity (68). The *egr-1* promoter region containing 935 base pairs 5' of the transcription start site was inserted 5' of a CAT gene to make the plasmid, p935, used in our CAT assays (24). CAT assays were performed using WEHI-231 cells transfected with p935 and then either left unstimulated for 24 hours or stimulated with anti-μ or PMA for 24 hours. CAT activity was observed with both anti-μ and PMA stimulated cells (Figure 7)

p935

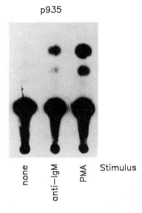

none   anti–IgM   PMA   Stimulus

FIG. 7.   The *egr-1* promoter is transcriptionally active in WEHI-231 cells. WEHI-231 cells were transfected with 30 µg of p935 using the DEAE-dextran method (67) with DMSO shock. After 24 hours in medium, the cells were stimulated with goat anti-µ (10 µg/ml) or PMA (10 ng/ml) for 24 hours and assayed for CAT activity (67,68). (Reproduced from ref. 25).

demonstrating that WEHI-231 cells contain the factors necessary for the *trans*-activation of the *egr-1* gene.

Our results, combined with the immature phenotype of WEHI-231 cells led us to investigate DNA methylation as a mechanism for inactivating *egr-1* in these cells. This idea was predicated upon the argument that DNA hypermethylation can silence genes in developing cells. The methylation status of the *egr-1* gene in the immature WEHI-231 cells was compared with that of mature BAL-17 cells using Southern analysis of genomic DNA digested with the restriction enzyme isoschizomers for CCGG, HpaII (an enzyme sensitive to cytosine methylation) and MspI (methylation insensitive) (Figure 8). Digestion of BAL-17 genomic DNA with either Hpa II or Msp I yielded the same restriction pattern (Figure 8) demonstrating that this gene is not methylated at these sites. Isolated mature splenic B cells from adult BALB/c mice gave results consistent with lack of methylation at these sites (Seyfert, VL and Monroe, JG; unpublished results). Genomic digests of WEHI-231 DNA with Msp I revealed a pattern similar to that seen in BAL-17; however, Hpa II digestion of WEHI-231 DNA produced a markedly different pattern, mainly consisting of larger bands than those of Msp I. Thus, *egr-1* in WEHI-231 cells is more methylated than in BAL-17 cells. The larger (2.0 kb) band observed in the Msp I digest of BAL-17 but not WHEI-231 DNA may reflect a polymorphism of *egr-1* in cell lines derived from these distinct inbred strains of mice.

The relevance of hypermethylation of the *egr-1* gene to the lack of anti-µ or PMA inducibility in WEHI-231 cells was evaluated by studying the effect of culturing these cells in the presence of 5-azacytidine, a non-specific inhibitor of DNA methylation (38). Northern analysis of RNA from WEHI-231 cells cultured with 5-azacytidine demonstrated that 5-azacytidine pretreatment allows both anti-µ and PMA to cause induction of *egr-1* expression, with kinetics similar to those of

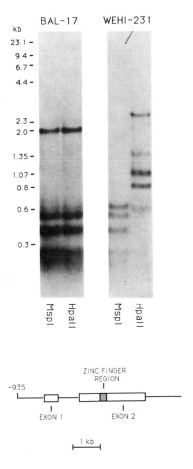

FIG 8.   **DNA methylation analysis of *Egr-1* in WEHI-231 versus BAL-17 cells.** Shown are Southern blots of genomic DNA (20 µg/lane) isolated from BAL-17 and WEHI-231 cells and digested with either Hpa II (lanes 2 and 4) or Msp I (lanes 1 and 3). The blot was probed with the Egr-1 genomic clone, p357 (27), shown in the bottom panel. (Reproduced from ref. 25).

mature B cells (Figure 9). Analysis of the methylation status of *egr-1* in the 5-azacytidine treated cells as performed for Figure 8 indicated that the gene was hypomethylated compared to untreated cells (Seyfert, VL and Monroe, JG; not shown). Importantly, this upregulation of *egr-1* appears to be a direct effect of 5-azacytidine on the gene itself, since WEHI-231 cells cultured with 5-azacytidine and then treated with anti-µ in the presence of cycloheximide still had marked increases of *egr-1* mRNA levels (Figure 9). The fact that *egr-1* is induced in 5-azacytidine treated WEHI-231 cells without *de novo* protein synthesis indicates that 5-azacytidine is causing demethylation of the *egr-1* gene itself and not a gene for some other protein that in turn facilitates *egr-1* expression. In addition, Southern analysis performed on genomic DNA isolated from untreated versus 5-azacytidine treated WEHI-231 cells revealed that this agent did in fact cause demethylation of the gene (not shown). These results established the functional importance of DNA methylation in the regulation of *egr-1* in WEHI-231 cells and, furthermore, established that *trans*-repression was unlikely to be responsible for the silencing of *egr-1* transcription in the immature B cells.

The GC-rich nature of the *egr-1* promoter and its correspondingly large number of restriction sites for methylation sensitive restriction enzymes (25) made it impossible to determine the relative methylation of the *egr-1* promoter in WEHI-231 and BAL-17 cells by Southern analysis. Therefore, to assess the effects of methylation of the *egr-1* promoter we examined the effect of *in vitro* cytosine methylation on inducible *egr-1* promoter activity. The plasmid p935 (the *egr-1* promoter coupled to the CAT reporter gene) was methylated at its HpaII or HhaI sites by the respective bacterial methylases. Methylation at all of these sites was verified by digestion of the methylase treated plasmids and resolution of the bands by polyacrylamide gel electrophoresis. In both cases where methylated p935 was tested, PMA inducible CAT activity, observed with the unmethylated construct, was abrogated. The same result was observed following anti-μ stimulation (not shown). The lack of CAT activity observed with the methylated

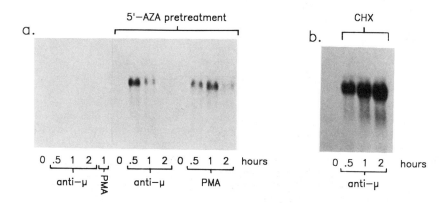

FIG. 9.   **Anti-μ or PMA induced *egr-1* in WEHI-231 cells treated with 5-azacytidine.** (A) Northern blot of total cellular RNA (20 μg/lane) from untreated WEHI-231 cells unstimulated (lane 1), stimulated with 5 μg/ml of anti-μ (lanes 2-4) or stimulated with 10 ng/ml PMA (lane 5) for the indicated times. On the same blot is total cellular RNA (20 μg/lane) from WEHI-231 cells pretreated 48 hours with 3 μM 5-azacytidine and then left unstimulated (lanes 6 and 10) or stimulated with 5 μg/ml anti-μ (lanes 7-9) or 10 ng/ml PMA (lanes 11-13). The blot was probed with an *Egr-1* cDNA clone OC68 (49), and the molecular weight of the complementary transcript indicated on the right. (B) Northern blot analysis of total cellular RNA (20 μg/lane) isolated from WEHI-231 cells cultured for 48 hours with 3 μM 5-azacytidine. After treatment with 5-azacytidine cells were cultured with cycloheximide (5 μg/ml) (except zero lane) beginning 15 minutes before stimulation with anti-μ (5 μg/ml) for the indicated times. This blot was probed as in (A). (Reproduced from ref. 25).

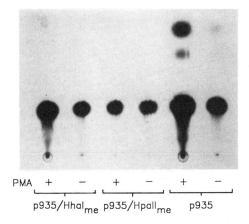

PMA + − + − + −

p935/Hhal$_{me}$ p935/Hpall$_{me}$ p935

FIG. 10. *In vitro* **methylation of the** *Egr-1* **promoter by Hha I or Hpa II bacterial methylases blocks inducible transcription activity.** The 5' flanking region of *egr-1* (up to -935) was used to make the plasmid p935 (Figure 7). Shown is CAT PMA (10 ng/ml; 24 hours) induced CAT activity in WEHI-231 cells transfected as in Figure 7 with unmethylated (p935) or Hha I or Hpa II methylase treated p935.

p935 was not due to methylation of the CAT gene itself, since it has been demonstrated that methylation of the CAT gene coding region does not block transcription of the gene (69). Importantly, methylation of p935 also blocks CAT activity in the permissive BAL-17 cells transfected with this construct and stimulated similarly (Seyfert, VL, McMahon, SB, Monroe, JG; unpublished results). Thus, methylation is capable of blocking transcription mediated through the *egr-1* promoter.

Taken together our results argue that methylation of *egr-1*, particularly of its promoter, is a mechanism used to silence this gene in WEHI-231 cells. This observation is unique because it is a case where methylation is regulating/silencing an immediate/early gene that is not associated with tissue specificity, as has been described, but rather, it is associated with a defined developmental stage within a single cell lineage.

DISCUSSION

A requirement for *egr-1* expression in the B cell activation response.

Our findings in B cells support a role for the immediate/early gene and transcription factor, *egr-1*, in the activation response of mature B cells to sIgM crosslinking. We have found that both anti-μ and PMA transiently induce *egr-1* mRNA and EGR-1 protein (for a more extended time period). Importantly, this induction occurs only at concentrations of anti-μ which also cause cell cycle progression. Increases in *egr-1* mRNA levels are strictly correlated with sIgM-induced B cell activation and are a primary response to sIgM-generated signals. Upregulation of the gene is probably mediated by increased transcription, as we have found that both anti-μ and PMA can cause transcriptional activation of the CAT reporter gene driven by the *egr-1* promoter. More compelling evidence supporting a role for *egr-1* in the B cell activation response is our preliminary finding, showing that *egr-1* antisense blocks the proliferative response of B cells to anti-μ (Seyfert, VL and Monroe, JG; in preparation).

We have also established that *egr-1* mRNA upregulation is coupled to the PKC pathway of B cell activation and is not affected by sIg induced increases in intracellular Ca^{2+} (24). Blocking PKC activity by PKC antagonists (24) also blocks anti-μ induced expression of *egr-1*, precluding involvement of other sIgM-linked signaling pathways (70-72) for induction of this gene. We hypothesize that the *egr-1* gene is regulated by a substrate of PKC that becomes phosphorylated and then causes upregulation of this gene by indirect or direct interaction with an as yet undefined enhancer element present between -388 and -395 in the *egr-1* promoter. PKC mediated control of *egr-1* transcription could be analogous to the regulation of genes containing the cAMP response element (CRE) by the CRE binding protein and transcription factor (CREB) which has been shown to be a substrate of protein kinase C and A (73). Although the *egr-1* promoter contains two AP-1 binding sites (27), these do not function as transcription enhancers for either anti-μ or PMA in B cells, in contrast to other PKC coupled genes. We are currently studying which DNA proteins are involved in anti-μ and PMA induction of *egr-1* and how these proteins are activated by PKC.

The importance of *egr-1* in B cell growth responses to sIgM-mediated signals was also demonstrated by our studies showing a lack of *egr-1* expression associated with growth inhibition following sIgM crosslinking. Our analyses of *egr-1* gene expression in the two immature murine B lymphoma cell lines, WEHI-231 and CH31 (both of which exhibit negative responses to sIgM crosslinking) (55-58), showed that both of these cell lines fail to exhibit detectable levels of *egr-1* mRNA with anti-μ stimulation. We have also obtained similar results in primary human B cell tumors that are growth inhibited by anti-μ (not shown). The lack of *egr-1* expression, in the the B cell lines is not due to a nonspecific effect of B cell transformation since we observed anti-μ and PMA induction of *egr-1* in BAL-17 cells and other mature B lymphoma cell lines (not shown). Finally, our recent findings (25) have established a lack of *egr-1* inducibility to anti-μ or PMA stimulation among immature B cells from murine bone marrow. Importantly, these B cells do not undergo a proliferative response or G_0 to G_1 transition following sIgM crosslinking by anti-μ (Yellen, AJ and Monroe, JG; unpublished observations).

The observation that crosslinking sIgM on WEHI-231 did not cause induction of *egr-1* gene expression was initially surprising to us. It has been shown that sIgM signal transduction in these cells is virtually identical to that of mature splenic B cells (39,46,62,74-77). These signaling similarities not only include utilization of the same second messengers and PKC activation (39,46,62), but also apply to the induction of other immediate/early genes (39). For example, it has been shown that anti-μ and PMA cause increases in *c-myc* (74,75) and *c-fos* (46) mRNA levels with the same kinetics as reported for splenic B cells, and in the case of *c-fos*, there are a number of enhancer elements in common with *egr-1* (27,28). We have also compared anti-μ and PMA induced expression of other genes in WEHI-231 cells and splenic B cells and have not found any other genes which are differentially expressed in these cells (*c-fos*, shown here) It appears then that immediate/early genes, such as *c-fos*, are induced by sIgM-generated signals but are not specifically associated with either a positive or negative growth response.

Thus, the lack of inducible *egr-1* expression appeared to be unique to this gene and as such, further supports the importance of this gene in determining a positive growth response of a B cell. Consistent with this interpretation are our studies showing that under conditions where *egr-1* can be induced in WEHI-231 cells, these cells no longer exhibit a negative growth response to sIgM signaling. Together these data support a relationship between the induction of *egr-1* expression and subsequent positive activation responses and cell growth.

WEHI-231 cells as a model for immature B cells.

Our results concerning *egr-1* expression in immature versus mature B cells has also led us to hypothesize that *egr-1* is developmentally regulated. Initially, our studies involved sIgM+, sIgD- B cells from the bone marrow as our source of immature stage B cells. However, as discussed before, subsequent molecular studies necessitated the identification of a homogeneous tissue culture model representing this population. We chose the WEHI-231 B lymphoma cell line as a model system for immature B cells. This approach was predicated upon the argument that WEHI-231 cells are representative of an immature B cell population that is present *in vivo*. The validity of the WEHI-231 cell line as a model system for immature B cells *in vivo* is based on their phenotype, sIgM+, sIgD-. In addition, their observed inability to express *egr-1* in response to sIgM crosslinking or PKC activation confirmed their similarity to their nontransformed bone marrow counterparts. Functionally, immature sIgM+, sIgD- B cells within the bone marrow do not exhibit positive growth responses to sIgM signaling (Yellen, AJ and Monroe, JG; unpublished results) and have been reported to be susceptible to tolerance, as defined by their inability to secrete immunoglobulin in response to antigen stimulation (66). These properties contrast those of mature B cells which proliferate and differentiate into antibody secreting plasma cells in response to crosslinking of sIgM and appropriate T cell help (8-14). The negative growth response associated with WEHI-231 to sIgM signaling is believed to be analogous to these responses in immature stage B cells (57,58). As such, by the criteria established by these and other studies (25,57,58), WEHI-231 appears to be an appropriate model system to study mechanisms of *egr-1* gene regulation that operate in immature, bone marrow-derived B cells.

Developmental Regulation of *egr-1* Expression.

The WEHI-231 system has suggested a mechanism, for the silencing of *egr-1*, that is consistent with developmental regulation of this gene in B cells. We found that the *egr-1* gene is more methylated in the phenotypically immature WEHI-231 cells than in the mature *egr-1* expressing cells and that methylation of *egr-1*, particularly of its promoter, can silence this gene in these cells. These results are consistent with the general premise that DNA methylation is a mechanism used for silencing of genes during development and differentiation. Our findings are also consistent with those of others showing; 1) an association between increased methylation of cytosine residues and gene inactivation in eucaryotic cells (35,78); and 2) hypomethylation associated with transcriptional activation of a gene (35,68,78,79).

Exactly how methylation interferes with transcriptional activation of *egr-1*, or other genes in general, is not understood, but there are two likely possibilities; one is that essential transcription factors are unable to bind at sites containing or proximal to methylated CpG's. The other is that methylated CpG's interact with nuclear proteins which in turn block transcription factor binding. Regarding the first possibility, there is conflicting evidence about whether transcription factors can or cannot bind to sites containing methylated CpG's. Two studies using the adenovirus late promoter and the DNA binding protein, USF, have shown that methylation of a CpG in the center of the USF binding site substantially reduces the amount of USF binding and transcription in *in vitro* assays (80). In this case, methylation of a CpG located only six bases away from the USF site had no affect on USF binding or on transcription (81). Iguchi-Ariga et al. (82) have shown that CpG methylation of the cAMP response element abolishes transcription factor binding as well as transcriptional activation. In contrast, Harrington et al. (83) and Hoeller et al. (84), have shown that methylation of the central CpG of the Sp1 transcription factor binding site has no affect on binding or transcription *in vitro*. Methylation of this CpG also had no affect on Sp1 mediated transcriptional activation *in vivo*.

The mechanisms by which cytosine methylation can affect protein binding may explain these conflicting results. A methyl group can provide a new hydrophobic contact for a protein, induce small local changes in DNA structure, or sterically hinder contacts between proteins and nearby functional DNA groups. These effects of cytosine methylation are all rather limited and specific changes in a given protein binding site, which could explain why different effects of methylation are observed, depending on the individual protein. Recent evidence has led to the suggestion that certain proteins exist which specifically bind methylated CpG's (85). A protein, MeCP, was found to specifically bind methylated multiple CpG's in a wide variety of unrelated DNA sequences and thus could function by directly interfering with DNA binding proteins or by inducing an inaccessible DNA conformation. In addition, studies by Antequerou et al. (86) have suggested that the binding of proteins like MeCP may explain the observed resistance of methylated CpG's to certain nucleases. As more information about CpG binding proteins is elucidated, it will be interesting to determine if they bind the *egr-1* promoter.

While it is known that transcriptional activation is associated with hypomethylation of a gene, it is not known whether transcriptional activation precedes demethylation or whether demethylation must occur before a gene can be transcribed. Studies showing that 5-azacytidine can cause transcriptional activation of genes, including *egr-1* (as shown here), suggest that demethylation of a gene precedes transcription (38,78,87-89); . However, studies in other systems provide evidence for transcriptional activation preceding demethylation (69,79). Using the muscle-specific alpha-actin gene, Yisraeli et al. (69) have shown that this gene is inhibited by DNA methylation when introduced into fibroblasts, but not when it is transfected into myoblasts. The myoblast, which expresses endogenous alpha-actin, is able to activate the transfected alpha-actin gene transcription in the presence of methyl moieties. Then the alpha-actin gene appears to undergo site-specific demethylation which resembles that of the active endogenous gene.

More evidence for transcription preceding demethylation is derived from studies of the IgG kappa gene (89). The IgG kappa gene is highly methylated and inactive in pre-B cells, but it can be activated in these cells by LPS. Interestingly, no detectable demethylation accompanies this activation, suggesting that expression of the gene in pre-B cells relies on enhancer dependent mechanisms, presumably the interaction of an LPS inducible factor with a specific kappa enhancer. As the B cell matures, kappa expression becomes constitutive and the gene becomes demethylated. Kelley et al. (89) have proposed that expression of kappa in mature B cells is enhancer independent and thus requires demethylation. These findings are in some ways analogous to what we have observed in our studies. For example, we know that LPS allows anti-μ and PMA to induce *egr-1* in the non-expressing immature WEHI-231 cells. Our analyses of the methylation status of the *egr-1* gene in the LPS pretreated cells has been inconclusive. As such, these cells which can now express *egr-1* may do so by a methylation-independent mechanism. This question is currently being addressed in our laboratory.

As before mentioned, the *egr-1* promoter is extremely GC rich and contains many CpG islands. CpG island promoters are generally associated with genes maintained in a state of continuous transcriptional activation and thus, are found predominantly associated with housekeeping type genes which are constitutively and ubiquitously expressed (35). It is believed that CpG islands are more resistant to DNA methylation, thereby preventing the gene from becoming inactivated by cytosine methylation. While *egr-1* expression appears to be ubiquitous, it is not constitutively expressed, rather it is an inducible gene. There are several other immediate/early type genes which contain CpG islands, including *c-fos* (35), but none of them have been found to be inactivated by methylation. We do know that there are a number of CpG clusters in the *egr-1* promoter, and that there are two CpG's between -388 to -359; the region known to contain the response element required for *egr-1* transcriptional activation by anti-μ or PMA in B cells. By comparing the methylation status of *Egr-1* CpG's, both within and proximal to this region, in immature and mature B cells, we will be able to better evaluate if CpG methylation is able to interfere with the binding of transcriptional regulatory factors to their sites in the *egr-1* promoter.

Recently, Antequera, et al. (90) have suggested that *de novo* methylation of CpG island genes is an artifact of maintaining cell lines *in vitro*. This study brought into question our results in WEHI-231 cells, since these cells are an *in vitro* tissue culture cell line. In the Antequera et al. study (90), it was discovered that CpG island tissue specific genes, but not housekeeping genes, were methylated in many human and mouse cell lines. While *egr-1* in not a classic housekeeping gene, it is ubiquitously expressed and does appear to be associated with promoting growth. Hence, methylation of *egr*-1 would probably result in a growth disadvantage in culture. In addition we evaluated *egr-1* expression and methylation in a variety of other B cell lines and found the gene to be expressed and unmethylated. Most importantly, we have identified a population of primary B cells in adult mouse bone marrow which behave like the WEHI-231 cells with respect to *egr-1* expression. This is also in direct conflict with the study by Antequera et al. (90), who showed that methylation and expression of CpG island genes in the cell lines was not reflected in the primary cells from which they were derived. For these reasons, we believe that methylation of the *egr-1* gene in WEHI-231 cells is not due to a tissue culture artifact but reflects a regulatory mechanism that is used during B cell development *in vivo*.

Developmental regulation of B cell activation and tolerance.

The silencing of the *egr-1* gene in the immature versus mature B cell begs the question of whether or not this gene is silenced as part of the mechanism to prevent activation responses in immature B cells. Our studies to date have focused on the molecular mechanisms accounting for the observed unresponsiveness of the immature B cell to sIgM signaling. In this respect, methylation of *egr-1* at the immature stage of B cell development provides a mechanistic explanation for the unresponsiveness associated with this population. If one accepts that entry into cell cycle is a prerequisite to differentiation into an antibody secreting plasma cell, then lack of expression of a gene critical to a positive growth and proliferative response would ensure immune unresponsiveness.

To what extent the observed unresponsiveness of immature B cells to sIgM signaling reflects the more active process of tolerance induction is unclear. In this regard, our studies in the bone marrow-derived immature B cells may be more informative than the WEHI-231 cells, where the relevance of sIgM-mediated inhibition of cell growth and proliferation is unclear. Signals generated through sIgM in the unresponsive cells may be abortive and result in cell death, thereby eliminating the affected clone. Alternatively, signals generated may be incomplete and result in paralysis of the B cell until generation of secondary signals that can complement sIgM-generated signals. It remains a possibility that the unresponsiveness observed in bone marrow-derived immature B cells may be transitional and not reflect induction of sustained unresponsiveness to subsequent antigen challenge. Although still preliminary, results from our studies of *egr-1* developmental regulation and function, as well as our recent studies of transmembrane signaling in various immature B cell populations (discussed below), may provide some insight.

Bone marrow-derived immature B cells do not express *egr-1* in response to either anti-μ or phorbol diester stimulation; however, this does not appear to apply to all immature B cell populations, specifically immature B cells from the day 1-4 neonatal mouse spleen. Like bone marrow immature B cells, these neonatal splenic B cells are unresponsive to anti-μ stimulation as determined by lack of entry into cell cycle (91). It is these cells which others have established to be highly sensitive to antigen receptor-induced tolerance (2). Consistent with the paradigm developed above, these cells also do not express *egr-1* following sIgM crosslinking. However, in contrast to bone marrow-derived immature B cells and WEHI-231, these B cells do express *egr-1* in response to phorbol diester stimulation; a result inconsistent with methylation of the *egr-1* gene in this population. Thus, the underlying signaling defect in the neonatal B cells does not appear to be inability to induce expression of the *egr-1* gene. Our recent studies (91) have established that in immature B cells from neonatal mouse spleen, sIgM is uncoupled from phospholipase C catalyzed inositol phospholipid hydrolysis, and all subsequent PKC-linked signaling events; including *egr-1* induction. The molecular basis for this uncoupling is still undefined, but we have found it to be correlated with a lack of association with a specific sIgM-associated protein (92). Thus, inability to induce *egr-1* expression in immature B cells from neonatal mice in response to sIgM-crosslinking is not due to *cis*-inactivation of the gene as was concluded for bone marrow-derived immature B cells, but rather an earlier signaling defect which confers an inability to activate PKC. Subsequent studies in the bone marrow system have confirmed that uncoupling of sIgM from inositol phospholipid hydrolysis in this population as well. Thus, these latter cells appear to possess two signaling defects which may account for their unresponsiveness to sIgM-generated signals.

Our observations of developmentally-regulated B cell unresponsiveness in each of the systems described in this review have suggested to us the model depicted in Figure 11 for the development of B cell responsiveness to antigen. B cells first acquire the ability to bind and respond to antigen during the immature stage of development, at which point they express the membrane form of IgM on their surface. We maintain that this occurs in the bone marrow of adult mammals. In the normal, non-diseased individual, the only antigens encountered by these newly receptor-positive cells would be self-antigens. We believe these cells maintain *egr-1* in its methylated state and, in addition, are defective in coupling sIgM to inositol phospholipid hydrolysis. As such, these cells are unresponsive to sIgM-mediated activation under all conditions (see below). Somewhat later in development, as the sIgM+,sIgD- cell leaves the bone marrow, *egr-1* becomes unmethylated and can be induced under certain conditions. However, these cells still maintain the receptor-associated signaling defect and are, therefore, still unresponsive to antigen. It is these cells which we believe comprise the majority of the sIgM+ cell population of the day 1-4 neonatal spleen. These cells clearly differ from the immature cells in the bone marrow because we have found that co-stimulation with IL-4 and anti-μ drives the neonatal B cells into cell cycle (Yellen, AJ and Monroe, JG; unpublished observations). In view of this observation, it is interesting that IL-4 will stimulate induction of *egr-1* (Seyfert, VL and Monroe, JG; unpublished observations), suggesting that its mechanism of action may be to bypass the specific sIgM signaling defect in these cells. This possibility may be analogous to studies where the B cell tolerant state has been broken or sensitivity to tolerance abrogated by provision of T cell help. These results then, suggest that mechanistic differences may exist for tolerance induction at these two substages of the immature B cell. Whether these mechanisms account for the multiple mechanisms suggested by historical and more recent (5,7) cellular studies remains to be evaluated.

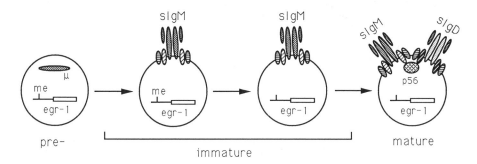

Figure 11. Development of B cell activation signaling.

It is likely that these more developed immature B cells also reside in the peripheral lymphoid organs of the adult animal, but are present in very low frequency because they quickly proceed to the next stage of development shown in Figure 11. At the fully mature, immunocompetent stage of development, *egr-1* remains unmethylated and the mechanism by which sIgM is coupled to inositol phospholipid hydrolysis becomes functional. In this figure, this is depicted by the expression of a new sIgM-associated protein, although the actual mechanism may involve be entirely different. At this stage of development, sIgM-generated signals are transduced across the plasma membrane by a process which includes hydrolysis of inositol phospholipids and activation of PKC. This process, among other critical activation events, leads to the induction of primary activation-related genes, of which *egr-1* is a member. Subsequent alterations in the physiology of the B cell and the expression of secondary response genes coordinate to drive the B cell into cell cycle whereupon it is then competent to receive the secondary signaling necessary to stimulate clonal expansion and the eventual differentiation into an antibody secreting effector cell.

The above scenario is discussed only as our current working model to attempt to formulate a unifying hypothesis, incorporating the findings of our laboratory and others with regard to developmental-regulated responsiveness and tolerance in the B lymphocyte. Clearly, many of these hypotheses remain to be formally tested, and there remains many solid data that are not adequately explained as yet by the model discussed above. For example, it it not clear to what extent the observed differences in responsiveness to anti-μ stimulation in WEHI-231, adult bone marrow, neonatal, and adult B cells may reflect lineage or subpopulation differences in these different B cell pools. A significant proportion of the neonatal B cells are Ly1+ while others have established that the majority of the cells in the adult bone marrow are Ly1- and cannot give rise to a Ly1+ population (93). Rothstein and Kolber (94) have shown very different responses to sIgM crosslinking when comparing Ly1+ and Ly1- B cells. To date, we have not observed any difference in the signaling or functional responses of the Ly1+ versus Ly1- populations in the neonatal spleen. Furthermore, WEHI-231 cells (which are Ly1+) behave in a manner identical to adult bone marrow-derived B cells with respect to *egr-1* gene expression. Thus, while lineage differences could suggest an alternative interpretation of our data, our studies to date are all consistent with our proposed model. However, much work is still required to definitively assess the relevance of the observed unresponsiveness of these populations to the regulation of the humoral immune response, as well as to define the developmental progression of activation signaling in B cells. The model systems described in this review will be very useful to address these issues and our interpretations of our results to date, will hopefully suggest experimental approaches to manipulate these systems to provide useful information.

ACKNOWLEDGEMENTS

The authors wish to acknowledge Drs. Vikas Sukhatme and Xinman Cao for sharing ideas and unpublished data on the *egr-1* gene as well as providing us with critical reagents; both of which facilitated our progress in these studies. We also thank our colleagues who were helpful in a number of these studies as well as providing helpful suggestions and criticism. They include: Steve McMahon, William Glenn, Amy Yellen, and Drs. Michael Cancro, Tom Kadesch, and Rebecca Taub

REFERENCES

1.   Burnet, FM. The clonal selection theory of acquired immunity, New York: Cambridge University Press, 1959.

2.   Metcalf, ES, Klinman, NR. *In vitro* tolerance induction of neonatal murine B cells. J. Exp. Med. 1976;143:1327-1340.

3.   Teale, JM, Klinman, NR. Tolerance as an active process. Nature 1980;288:385-387.

4.   DeFranco, AL. Tolerance: a second mechanism. Nature 1989;342:340-341.

5.   Goodnow, CC, Crosbie, J, Adelstein, S, et al. Altered immunoglobulin expression and functional silencing of self-reactive B lymphocytes in transgenic mice. Nature 1988;334:676-682.

6.   Nossal, GJV and Pike, BL. Evidence for the clonal abortion theory of B-lymphocyte tolerance. J. Exp. Med. 1975;141:904-914.

7.   Nemazee, DA and Burki, K. Clonal deletion of B lymphocytes in a transgenic mouse bearing anti-MHC class I antibody genes. Nature 1989;337:562-566.

8.   Julius, MH, von Boehmer, H, Sidman, C. Dissociation of two signals required for activation of B cells. Proc. Natl. Acad. Sci., USA 1982;79:1989-1993.

9.   DeFranco, AL, Raveche, ES, Asofsky, R, Paul, W. Frequency of B lymphocytes responsive to anti-immunoglobulin. J. Exp. Med. 1982; 155:1523-1536.

10. Howard, M and Paul, WE. Regulation of B-cell growth and differentiation by soluble factors. Ann. Rev. Immunol. 1983;1:307-333.

11. Cambier, JC and Monroe, JG. B cell activation. V. Differentiation signaling of B cell membrane depolarization, increases I-A expression, G$_0$ to G$_1$ transition, and thymidine uptake by anti-IgM and anti-IgD antibodies. J. Immunol. 1984;133:576-581.

12. LoCascio, NJ, Haughton, G, Arnold, LW, Corley, RB. Role of cell sruface immunoglobulin in B-lymphocyte activation. Proc. Natl. Acad. Sci., USA. 1984;81:2466-2469.

13. DeFranco, AL, Raveche, ES, Paul, WE. Separate control of B lymphocyte early activation and proliferation in response to anti-IgM antibodies. J. Immunol. 1985;135:87-94.

14. Cambier, JC and Ransom, JT. Molecular mechanisms of transmembrane signaling in B lymphocytes. Ann. Rev. Immunol. 1987;5:175-199.

15. Denhardt, DT, Edwards, DR, Parfett, CLJ. Gene expression during the mammalian cell cycle. Biochim. Biophys. Acta. 1986;65:83-125.

16. Dynan, WS and Tjian, R. Control of eukaryotic messenger RNA synthesis by sequence-specific DNA-binding proteins. Nature 1985;316:774-778.

17. Mitchell, PJ and Tjian, R. Transcriptional regulation in mammalian cells by sequence-specific DNA binding proteins. Science 1989;245:371-378.

18. Sukhatme, VP, Cao, X, Chang, LC, et al. A zinc finger-encoding gene coregulated with *c-fos* during growth and differentiation, and after cellular depolarization. Cell 1988;53:37-43.

19. Lau, LF and Nathans, D. Expression of a set of growth-related immediate early genes in BALB/c 3T3 cells: coordinate regulation with *c-fos* or *c-myc*. Proc. Natl. Acad. Sci., USA 1987;84:1182-1186.

20. Milbrandt, J. A nerve growth factor-induced gene encodes a possible transcriptional regulatory factor. Science 1987;238:797-799.

21. Chavrier, P, Jannssen-Timmen, U, Mattei, MG, Zerial, M, Bravo, R, Charnay, P. Structure, chromosome location, and expression of the mouse zinc finger gene *Krox-20*: Multiple gene products and coregulation with the protooncogene *c-fos*. Mol Cell. Biol. 1989;9:787-797.

22. Christy, B and Nathans, D. DNA binding site of the growth factor-inducible protein Zif268. Proc. Natl. Sci., USA. 1989;86:8737-8741.

23. Lamaire, P, Vesque, C, Schmit, J, Stunnenberg, H, Frank, R, Charnay, P. The serum-inducible mouse gene *Krox-24* encodes a sequence-specific transcriptional activator. Mol. Cell. Biol. 1990;10:3456-3467.

24. Seyfert, VL, McMahon,SB, Glenn,WD, Cao, X, Sukhatme, VP, Monroe, JG. *Egr-1* expression in surface immunoglobulin-mediated B cell activation: kinetics and association with protein kinase C activation. J. Immunol. 1990;145:3647-3653.

25. Seyfert, VL, McMahon, SB, Glenn, WD, Yellen, AJ, Sukhatme, VP, Cao, X, Monroe, JG. Methylation of an immediate/early inducible gene as a mechanism for B cell tolerance induction. Science 1990;250:797-800.

26. Seyfert, VL, Sukhatme, VP, Monroe, JG. Differential expression of a zinc finger-encoding gene in response to positive versus negative signaling through receptor immunoglobulin in murine B lymphocytes. Mol. Cell Biol. 1989;9:2083-2088.

27. Tsai-Morris, C, Cao, X, Sukhatme, VP. 5' flanking sequence and genomic structure of *Egr-1*, a murine mitogen inducible zinc finger encoding gene. Nuc. Acids Res. 1988;16:8835-8846.

28. Treisman, R. Identification of a protein-binding site that mediates transcriptional response of the *c-fos* gene to serum factors. Cell 1986;46:557-574.

29. Bohmann, D, Bos, TJ, Admon, A, Nishimura, T, Vogt, PK, Tjian, R. Human proto-oncogene *c-jun* encodes a DNA binding protein with structural and functional properties of transcription factor AP-1. Science 1987;238:1386-1392.

30. Halazonetis, TD, Georgopoulos, K, Greenberg, ME, Leder, P. *c-Jun* dimerizes with itself and with *c-fos* forming complexes of different DNA binding affinities. <u>Cell</u> 1988;55:917-924.

31. Kouzarides, T and Ziff, E. The role of the leucine in the fos-jun interaction. <u>Nature</u> 1988;336646-651.

32. Sassone-Corsi,P, Ransone,LJ, Lamph, WW, Verma, IM. Direct interaction between fos and jun nuclear oncoproteins: role of the leucine zipper domain. <u>Nature</u> 1988;336:690-695.

33. Kadonaga, JT, Carner, KR, Masiarz, FR, Tjian, R. Isolation of cDNA encoding transcription factor Sp1 and functional analysis of the DNA binding domain. <u>Cell</u> 1987;51:1079-1090.

34. Montminy, MR and Bilezikijian, LM. Binding of a nuclear protein to the cyclic-AMP response element of the somatostatin gene. <u>Nature</u> 1987;328:175-178.

35. Bird, AP. CpG-rich islands and the function of DNA methylation. <u>Nature</u> 1986;321:209-213.

36. Keshet, I, Yisraeli, J, Cedar, H. Effect of regional DNA methylation on gene expression. <u>Proc. Natl. Acad. Sci., USA</u>. 1985;82:2560-2564.

37. Lock, LF, Melton, DW, Caskey, T, Martin, G. Methylation of the mouse hprt gene differs on the active and inactive X chromosomes. <u>Mol. Cell. Biol.</u> 1986;6:914-924.

38. Jones, PA. Altering gene expression with 5-azacytidine. <u>Cell</u> 1985;40:485-486.

39. Monroe, JG and Seyfert, VL. Studies of surface immunoglobulin-dependent B cell activation. <u>Immunol. Res.</u> 1988;7:136-151.

40. Ransom, JT, Digiusto, DL, Cambier, JC. Single cell analysis of calcium mobilization in anti-immunoglobulin-stimulated B lymphocytes. <u>J. Immunol.</u> 1986;136:54-57.

41. Maino, VC, Hayman, MJ, Crumpton, MJ. Relationship between enhanced turnover of phosphatidylinositol and lymphocyte activation by mitogens. <u>Biochem. J.</u> 1975;146:247-252.

42. Coggeshall, KM and Cambier, JC B cell activation. VIII. Membrane immunoglobulins transduce signals via activation of phosphatidylinositol hydrolysis. <u>J. Immunol.</u> 1985;133:3382-3386.

43. Bijsterbosch, MK, Meade, CJ, Turner, GA, Klaus, GGB. B lymphocyte receptors and polyphosphoinositide degradation. <u>Cell</u> 1985;41:99-1006.

44. Hornbeck, P and Paul, WE. Anti-immunoglobulin and phorbol ester induce phosphorylation of proteins associated with the plasma membrane and

cytoskeleton in murine B lymphocytes. J. Biol. Chem. 1986;31:14817-14824.

45. Kelly, K, Cochran, BH, Stiles, CD, Leder, P. Cell-specific regulation of the c-myc gene by lymphocyte mitogens and platelet-derived growth factor. Cell 1977;35:603-610.

46. Monroe, JG. Up-regulation of *c-fos* expression is a component of the mIg signal transduction mechanism but is not indicative of competence for proliferation. J. Immunol. 1988;140:1454-1460.

47. Monroe, JG and Kass, MJ. Molecular events in B cell activation. I. Signals required to stimulate G_0 to G_1 transition of resting B lymphocytes. J. Immunol. 1985;135:1674-1682.

48. Klaus, GGB, O'Garra, A, Bijesterbosch, MK, Holam, M. Activation and proliferation of mouse B cells. VIII. Induction of DNA synthesis in B cells by a combination of calcium ionophores and phorbol myristate acetate. Eur. J. Immunol. 1986;16:92-97.

49. Sukhatme, VP, Kartha, S, Toback, GG, Taub, R, Hoover, RG, Tsai-Morris, C. A novel early growth response gene rapidly induced by fibroblast, epithelial cell and lymphocyte mitogens. Oncogene Res. 1987;1:343-355.

50. Cao, X, Koski, RA, Gashler, A, et al. Identification and characterization of the *Egr-1* gene product, a DNA-binding zinc finger protein induced by differentiation and growth signals. Mol. Cell. Biol. 1990;10:1931-1939.

51. Loomis, CR and Bell, RM. Sangivamycin, a nucleoside analogue, is a potent inhibitor of protein kinase C. J. Biol. Chem. 1988;263:1682-1692.

52. Hidaka, H, Inagaki, M, Kawamoto, S, Sasaki, Y. Isoquinoline-sulfonamides, novel and potent inhibitors of cyclic nucleotide dependent protein kinase and protein kinase C. Biochem. 1984;23:5036-5040.

53. Haughton, G, Arnold, LW, Bishop, GA, Mercolino, TJ. The CH series of murine B cell lymphomas: neoplastic analogues of Ly-1+ normal B cells. Immunol. Rev. 1986;93:35-51.

54. Pennell, CA and Scott, DW. Lymphoma models for B cell activation and tolerance IV. Growth inhibition by anti-Ig of CH31 and CH33 B lymphoma cells. Eur. J. Immunol. 1986;16:1577-1581.

55. Boyd, AW and Schrader, JW. The regulation of growth and differentiation of a murine B cell lymphoma. II. The inhibition of WEHI-231 by anti-immunoglobulin antibodies. J. Immunol. 1981;126:2466-2469.

56. Jakway, JP, Usinger, WR, Gold, MR, Mishell, RI, DeFranco, AL. Growth regulation of the B lymphoma cell line WEHI-231 by anti-

immunoglobulin, lipopolysaccharide, and other bacterial products. J. Immunol. 1986;

57. Scott, DW, Tuttle, J, Livnat, D, Haynes, W, Cogswell, JP, Keng, P. Lymphoma models for B-cell activation and tolerance. II. Growth inhibition by anti-μ of WEHI-231 and the selection and properties of resistant mutants. Cell. Immunol. 1985;93:124-131.

58. Scott, DW, Livnat, D, Pennell, CA, Keng, P. Lymphoma models for B cell activation and tolerance. III. Cell cycle dependence for negative signaling of WEHI-231 B lymphoma cells by anti-μ. J. Exp. Med. 1986;164:156-164.

59. Monroe, JG and Haldar, S. Involvement of a specific guanine nucleotide binding protein in receptor immunoglobulin stimulated inositol phospholipid hydrolysis. Biochim. Biophy. Acta. 1989;1013:273-278.

60. Harnett, MM and Klaus, GGB. G protein coupling of antigen receptor-stimulated polyphosphoinositide hydrolysis in B cells. J. Immunol. 1988;140:3135-3139.

61. Gold, MR, Jakway, JP, DeFranco, AL. Involvement of a guanine nucleotide-binding component in membrane IgM-stimulated phosphoinositide breakdown. J. Immunol. 1987;139:3604-3613.

62. Fahey, KA and DeFranco, AL. Crosslinking membrane IgM induces production of inositol trisphosphate and inositol tetrakisphosphate in WEHI-231 B lymphoma cells. J. Immunol. 1987;138:8935-8942.

63. Mizuguchi, J, Tsang, W, Morrison, SL, Beaven, MA, Paul, W. Membrane IgM, IgD, and IgG act as signal transmission molecules in a series of B lymphomas. J. Immunol. 1986;137:2162-2167.

64. Lanier, LL, Warner, NL, Ledbetter, JA, Herzenberg, LA. Quantitative immunofluorescent analysis of surface phenotypes of murine B cell lymphomas and plasmacytomas with monoclonal antibodies. J. Immunol. 1981;127:1691-1696.

65. Hammerling, U, Chin, AF, Abbott, J. Ontogeny of murine B lymphocytes: sequence of B-cell differentiation from surface-immunoglobulin-negative precursors to plasma cells. Proc. Natl. Acad. Sci., USA 1976;73:2008-2012.

66. Metcalf, ES and Klinman, NR. In vitro tolerance induction of bone marrow cells: a marker for B cell maturation. J. Immunol. 1977;118:2111-2116.

67. Ausubel, FM, Brent, R, Kingston, RE, Moore, DD, Seidman, JG, Smith, JA, Struhl, K. Current Protocols in Molecular Biology John Wiley and Sons, New York, 1989.

68. Gorman, CM, Moffat, LF, and Howard, BM. Recombinant genomes which express chloramphenicol acetyltransferase in mammalian cells. Mol. Cell. Biol. 1982;2:1044-1051.

69. Yisraeli, J, Adelstein, RS, Melloul, D, Nudel, U, Yaffe, D, Cedar, H. Muscle-specific activation of a methylated chimeric actin gene. Cell 1986;46:409-416.

70. Gold, MR, Law, DA, DeFranco, AL. Stimulation of protein tyrosine phosphorylation by the B-lymphocyte antigen receptor. Nature 1990;345:810-813.

71. Campbell, M and Sefton, BM. Protein tyrosine phosphorylation is induced in murine B lymphocytes in response to stimulation with anti-immunoglobulin. EMBO J. 1990;9:2125-2131.

72. Mond, JJ, Feuerstein, N, Finkelman, FD, Huang, F, Huang, K, Dennis, G. B-lymphocyte activation mediated by anti-immunoglobulin antibody in the absence of protein kinase C. Proc. Natl. Acad. Sci. USA 1987;84:8588-8592.

73. Yamomoto, KK, Gonzale, GA, Biggs III, WH, Montminy, M. Phosphorylation-induced binding and transcriptional efficacy of nuclear factor CREB. Nature 1988;334:494-498.

74. Snow, EC, Fetherston, JD, Zimmer, S. Induction of the c-myc protooncogene after binding to hapten-specific B cells. J. Exp. Med. 1986;164:944-949.

75. McCormack, JE, Pepe, VH, Kent, RB, Dean, M, Marshak-Rothstein, A, Sonnenshein, GE. Specific regulation of *c-myc* oncogene expression in a murine B cell lymphoma. Proc. Natl. Acad. Sci. (USA) 1984;81:5546-5550.

76. LaBaer, J, Tsien, RY, Fahey, KA, DeFranco, AL. Stimulation of the antigen receptor on WEHI-231 B lymphoma cells results in a voltage-independent increase in cytoplasmic calcium. J. Immunol. 1989;143:1032-1039.

77. Monroe, JG and Seyfert, VL. Negative signaling through surface immunoglobulin. Molecular mechanisms and relevance to induced B cell unresponsiveness. In: Receptors and Signal Transduction in Regulation of Lymphocyte Function. Ed. by J.C. Cambier. American Association for Microbiology Press, Washington, D.C. 1990;51-66.

78. Cedar, H. DNA methylation and gene activity. Cell 1988;53:3-4.

79. Enver, T, Zhang, J, Papayannopoulou, T, Stamatoyannopoulos, G. DNA methylation: a secondary event in globin gene switching? Genes Dev. 1988;2:698-706.

80. Watt, F and Molloy, PL. Cytosine methylation prevents binding to DNA of a HeLa cell transcription factor required for optimal expression of the adenovirus major late promoter. Genes Dev. 1988;2:1136-1143.

81. Jove, R, Sperber, DE, Manley, JL. Transcription of methylated eukaryotic viral genes in a soluble in vitro system. Nuc. Acids Res. 1984;12:4715-4730.

82. Iguchi-Ariga, SMM and Schaffner, W. CpG methylation of the cAMP-responsive enhancer/promoter sequence TGACGTCA abolishes specific factor binding as well as transcriptional activation. Genes Dev. 1989;2:612-619.

83. Harrington, MA, Jones, PA, Imagawa, M, Karin, M. Cytosine methylation does not affect binding of transcription factor Sp1. Proc. Natl. Acad. Sci., USA 1988;85:2066-2070.

84. Hoeller, M, Westin, G, Jiricny, J, Schaffner, W. Sp1 transcription factor binds DNA and activates transcription even when the binding site is CpG methylated. Genes Dev. 1988;2:1127-1135.

85. Meehan, RR, Lewis, JD, McKay, S, Kleiner, EL, Bird, AP. Identification of a mammalian protein that binds specifically to DNA containing methylated CpGs. Cell 1989;58:499-507.

86. Antequera, F, Macleod, D, Bird, AP. Specific protection of methylated CpGs in mammalian nuclei. Cell 1989;58:509-517.

87. Jones, PA, Taylor, SM, Mohandas, T, Shapiro, LJ. Cell cycle-specific reactivation of an inactive X-chromosome locus by 5-azadeoxycytidine. Proc. Natl. Acad. Sci., USA 1982;79:1215-1219.

88. Michalowsky, LA and Jones, PA. Gene structure and transcription in mouse cells with extensively demethylated DNA. Mol. Cell. Biol. 1989;9:885-892.

89. Kelley, DE, Pollok, B, Atchison, ML, Perry, RP. The coupling between enhancer activity and hypomethylation of K immunoglobulin genes is developmentally regulated. Mol. Cell. Biol. 1988;8:930-937.

90. Antequera, F, Boyes, J, Bird, AP. High levels of de novo methylation and altered chromatin structure at CpG islands in cell lines. Cell 1990;62:503-514.

91. Yellen, AJ, Glenn, W, Sukhatme, VP, Cao, X, and Monroe, JG. Signalling through surface IgM in tolerance-susceptible immature murine B lymphocytes. Developmental regulated differences in transmembrane signalling in splenic B cells from adult and neonatal mice. J. Immunol. 1991;146:1446-1454.

92. Yellen, AJ and Monroe, JG. A tyrosine phosphoprotein associated with sIgM and implicated in the developmental regulation of transmembrane signaling through this receptor. Submitted for publication.

93. Herzenberg, LA, Stall, AM, Lalor, PA, Sidman, C, Moore, WA, Parks, DR, and Herzenberg, LA. The Ly-1 B cell lineage. Immunol. Rev. 1986;93: 81-102.

94. Rothstein, TL and Kolber, DL. Anti-immunoglobulin antibody inhibits the phorbol ester induced stimulation of peritoneal B cells. J. Immunol. 1988;141:4089.

Advances in Regulation of Cell Growth, Volume 2;
Cell Activation: Genetic Approaches, edited by
James J. Mond, John C. Cambier, and Arthur Weiss.
Raven Press, Ltd., New York © 1991.

11

Transgenic Systems for the Analysis of *src*-Family Kinase Function

Kristin M. Abraham, Steven D. Levin, Michael P. Cooke, and Roger M. Perlmutter

Howard Hughes Medical Institute and the Departments of Biochemistry, Immunology and Medicine (Medical Genetics), University of Washington School of Medicine, Seattle, Washington 98195

INTRODUCTION

The role of *src*-family protein tyrosine kinases in lymphocyte activation

Lymphocyte activation is regulated through an intricate series of biochemical events intiated at antigen specific receptors. In many respects, the molecular events that follow engagement of the antigen-specific receptor resemble protoypical activation pathways that have been documented in other cell systems, including the augmentation of phospholipase C activity, and the resultant generation of diacylglycerol and inositol 1,4,5-trisphosphate (IP_3) second messengers (1,2). Thus, many features of the antigen-specific activation of lymphocytes can be mimicked through the use of pharmacologic agents such as phorbol esters and Ca^{++} ionophores (3).

Kinetic studies indicate that prior to these phospholipase C-dependent activation events, engagement of the T cell antigen receptor (TCR) provokes rapid phosphorylation of several proteins on tyrosine (4,5). Increases in the accumulation of these phosphotyrosine-containing proteins can be observed within 60 seconds following TCR stimulation (4,6,7). Although several lines of evidence indicate that the TCR and its associated CD3 complex mediate these biochemical changes following activation, neither the TCR nor the CD3 complex possess any known intrinsic enzymatic activity. However, the specific association of a member of the *src*-family of protein tyrosine kinases, $p59^{fyn}$, with CD3 components provides the TCR/CD3 complex with a signal transducing element capable of generating the observed increases in tyrosine phosphorylation (8). Evidence suggesting that the association of $p59^{fyn}$ with this TCR/CD3 complex is functionally responsible for several of the biochemical activation events associated with TCR engagement has been provided by transgenic experiments in which overexpression of $p59^{fyn}$ within developing thymocytes increases responsiveness of these cells to TCR-mediated stimuli (7).

Although the T cell antigen receptor/CD3 complex clearly mediates the primary signal transduction response during antigen recognition, stimulation through ancillary receptors on the lymphocyte surface can critically influence the outcome of these responses both in vitro and in vivo. CD4 and CD8, the TCR "coreceptors", influence the activation state of T cells by binding to non-polymorphic determinants of Class II and Class I MHC molecules concomitant with TCR binding to antigen (9-11). As in the case of the TCR/CD3 complex, cross-linking of CD4 and CD8 activates an endogenous tyrosine kinase activity in lymphocytes (12). The array of phosphotyrosine containing proteins which accumulate following stimulation of CD4 and CD8 coreceptors is, however, distinct from that observed following TCR/CD3 crosslinking (13). This unique phosphotyrosine pattern must reflect in part the substrate specificity of the *src*-family kinase associated with CD4 and CD8, $p56^{lck}$ (14-16). Thus, the activation state of lymphocytes has been intimately associated with the activities of at least two specific *src*-family protein tyrosine kinases, $p59^{fyn}$ and $p56^{lck}$.

Protein tyrosine kinases are an important class of regulatory enzymes that assist in controlling cellular proliferation in several systems. Many protein tyrosine kinases were first identified as the products of virally-encoded oncogenes, and several have subsequently been identified as receptors for known growth factors (17). Nevertheless, a substantial fraction of the more than 50 currently recognized protein tyrosine kinases cannot be explicitly associated with a growth-stimulation pathway (18,19). For example, the *src*-family of protein tyrosine kinases includes 8 evolutionarily conserved elements (*src, hck, lck, blk, fyn, lyn, yes,* and *fgr*) that appear to participate in specialized signal transduction circuits (20,21). Although expression of altered forms of the *src*-family subgroup of protein tyrosine kinases can result in neoplastic transformation, in general, expression of *src* family kinases increases as cells differentiate. Indeed, many members of the *src* family are maximally expressed in normal cells that have little replicative potential (20).

Among members of the *src* family kinases, $p56^{lck}$ is especially amenable to experimental manipulation. Products of the *lck* gene accumulate exclusively in lymphocytes and particularly in T cells (22). Lymphocyte populations can be easily obtained in large quantities, and the phenotypic characteristics and activation requirements of these cells have been widely studied. By experimental manipulation of levels of $p56^{lck}$ activity in normal cells, we have endeavored to define the signalling pathways to which $p56^{lck}$ contributes. At the same time, our experiments provide potential insights into the physiologic function of *src* family kinases in general.

Structure and regulation of $p56^{lck}$

The structure of a typical *src*-family kinase, as exemplified by $p56^{lck}$, is depicted in FIG. 1. Unlike their growth factor receptor counterparts, *src*-family kinases do not possess extracellular ligand binding domains nor membrane-spanning regions. Nonetheless, they are preferentially localized to the inner face of the cytoplasmic membrane, an association that is probably facilitated by amino-terminal myristylation (23-25). The 70 amino acids encoded within this amino terminal region also contain motifs that are unique to each kinase, and which are postulated to promote specific interactions between the kinases and their substrates or ligands. Indeed, within this region of the *lck*-encoded kinase are residues that are critical for the interaction of $p56^{lck}$ with its associated receptor molecules CD4 and CD8 (Figure 1, 26-28). Two cysteine residues at positions 20 and 23 in $p56^{lck}$ are essential for its association with complementary cysteine residues contained in the carboxy terminus of the CD4 (and CD8α) molecule; replacement of these residues by site-directed mutagenesis abolishes the CD4:$p56^{lck}$ interaction (27). Since *o*-phenanthroline treatment can

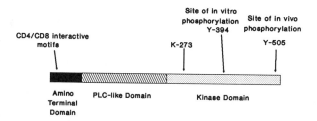

FIG. 1. Diagrammatic representation of p56^{lck}.

inhibit the p56^{lck}:CD4 interaction in vitro, the association facilitated by these four cysteine residues may occur through the coordination of a metal ion. Although p56^{lck} becomes specifically associated with CD4 and CD8 in cells that express these molecules, it is important to note that CD4/CD8 and p56^{lck} are not coexpressed in lymphocytes in an obligate fashion. Cells that express neither CD4 nor CD8 (such as NK cells, $\gamma\delta$ TCR-bearing T cells or B cells) nevertheless express p56^{lck} (20). Presumably other ligands must exist within these cells for p56^{lck}.

Adjacent to the unique amino terminal region is a central domain containing short stretches of sequence similarity with other *src* family kinases and with a set of other molecules including phospholipase C-γ, *ras*-GAP and p47$^{gag-crk}$ (29-31). Interactions among phosphotyrosine-containing proteins and proteins bearing these SH2 and SH3 (*src*-homology-2 & 3) domains have been demonstrated directly via co-precipitation (32). It has been postulated that binding of phosphotyrosine-containing substrates to the SH2 domain of *src*-family kinases may alter kinase activity (33). However, whether *src*-family kinases in turn phosphorylate these substrates or alter their activity is unknown.

The carboxy-terminal half of p56^{lck} encodes a catalytic unit whose sequence is highly conserved among members of the *src* family (18). Within this region are common structural motifs including a lysine residue critical for nucleotide binding (K273), an in vitro tyrosine autophosphorylation site (Y394 in p56^{lck}), and a tyrosine residue that becomes phosphorylated in vivo (Y505 in p56^{lck}). These sites of tyrosine phosphorylation appear to assist in regulating tyrosine kinase activity. For example, mutations that alter the in vivo autophosphorylation sites of several of the kinases (including p60^{src}, p56^{lck} and p59^{hck}) convert each of these enzymes into a transforming protein (34-38). Hence phosphorylation of this carboxy-terminal site appears to negatively regulate kinase activity, a phenomenon that can be directly observed in the case of p60^{src} (39). In contrast, phosphorylation at the in vitro tyrosine autophosphorylation site appears to augment kinase activity, albeit modestly (40).

p56^{lck} and lymphocyte activation

Expression of p56^{lck} is limited to lymphoid cells among normal cell populations, and is greatest in developing thymocytes. In mature T-lymphocytes, expression of the *lck* gene is transcriptionally regulated in response to mitogenic activation (41). In addition, mitogenic stimulation or phorbol ester treatment of T-lymphocytes leads to modification of the protein product of the *lck* gene, p56^{lck}. This modification results in a shift in apparent molecular size from 56 kDa to 60 kDa (41-43). Although the molecular basis of this post-translational modification remains undefined, multiple sites of serine phosphorylation have been identified in the amino terminal domain of $p56^{lck}$ that may play a role in the conversion (44). Similar alterations are provoked in T lymphocytes and NK cells within minutes after exposure to IL-2 (45-47).

Perhaps the most compelling evidence for the involvement of $p56^{lck}$ in lymphocyte activation comes from the demonstration that the majority of $p56^{lck}$ molecules in T-lymphocytes are found in non-covalent association with the surface "coreceptors" CD4 and CD8 (14-16,27,48). This association is both physical and functional, in that cross-linking of surface CD4 molecules results in enhanced $p56^{lck}$ activity (12). Thus, CD4 (and CD8) provide an extracellular domain structure that couples to the *lck*-encoded kinase domains, in effect reconstituting a receptor-linked protein tyrosine kinase. While expression and engagement of CD4 and CD8 are capable of affecting T cell activation in vitro, the biologic consequences of these CD4- and CD8- mediated signalling events have been difficult to discern. A wealth of evidence has accumulated that can support both positive and negative roles for these molecules in lymphocyte activation (9-11,49,50). For example, Glaichenhaus and coworkers have demonstrated the requirement for CD4 in the antigen-specific responses of two lymphoma cell lines (51). In these studies, antigen-induced IL2 production by two lines of MHC class II-restricted hen egg lysozyme-specific T cells was dramatically enhanced by the introduction of a transfected CD4 gene into the line. This enhancement was apparently dependent on the ability of CD4 to productively associate with $p56^{lck}$, as introduction of mutants with altered $p56^{lck}$:CD4 binding motifs produced only modest enhancement of IL2 production. Although the demonstration of a physical link between $p56^{lck}$ and CD4/CD8 has provided a framework for the analysis of $p56^{lck}$ function in these specific processes, the biological consequences of $p56^{lck}$ activation within normal lymphocyte populations as well as its potential for participation in alternative signalling pathways in these cells remains to be elucidated.

That $p56^{lck}$ can participate in biological responses initiated at alternative receptors has been illustrated by the recent demonstration of the physical association of $p56^{lck}$ with the IL2R β chain. This interaction is distinct from that previously characterized for $p56^{lck}$:CD4/CD8, and has been mapped to regions within the catalytic domain of $p56^{lck}$ (46). Several observations indicate that this physical association may be biologically relevant. For example, IL2 treatment of a factor-dependent cell line (CTLL) results in conversion of $p56^{lck}$ to its p60 form (52), and IL2 treatment of human PBL results in both the $p56^{lck}$ to $p60^{lck}$ conversion and activation of $p56^{lck}$ kinase activity (45,46). Association of $p56^{lck}$ with the IL2R might explain the presence of $p56^{lck}$ within cells that express neither CD4 or CD8 coreceptors. Based on these results, it is plausible that $p56^{lck}$ will participate in signal transduction pathways emanating from multiple receptors, and as such its activity will be dependent upon the array of surface receptors being expressed in the T-lymphocyte.

Since tyrosine phosphorylation plays a central role in lymphocyte activation, mechanisms that regulate the level of cellular phosphotyrosine may influence lymphocyte stimulability. Indeed, several independent lines of evidence support the notion that protein tyrosine phosphatases also regulate lymphocyte activation responses. Lymphocytes contain a variety of protein tyrosine phosphatases, the best characterized being the CD45 molecule in its multiple isoforms (53-55). The cytoplasmic portion of CD45 contains two prototypical protein tyrosine phosphatase domains, and expression of this molecule affects the responsiveness of lymphocytes (56-58). For example, in several T cell lines, loss of CD45 expression correlates with an inability of the cells to respond to antigenic stimulation (59), and signalling responses generated after cross-linking of several lymphocyte surface receptors can be influenced by co-crosslinking the receptors with the CD45 molecule (60,61). The protein tyrosine phosphatases thus add an additional level of complexity to the regulation of lymphocyte activation via tyrosine phosphorylation.

MODEL SYSTEMS FOR THE ANALYSIS OF P56^{lck} FUNCTION: AUGMENTING P56^{lck} ACTIVITY IN VIVO.

The *lck* transgenic mouse - experimental design

To examine the role of p56^{lck} in T-lymphocyte biology, we sought to overexpress the kinase in cells which normally express p56^{lck} using transgenic technology (62). The advantages of this system are threefold. First, by overexpressing the kinase in transgenic animals, we are able to assess the effects of p56^{lck} simultaneously on all relevant subsets of cells. Secondly, this system removes the necessity for using cultured cell lines or clones, which as a result of their propagation may have been exposed to undefined selection pressures and hence may not retain normal signalling responses. Finally, by using a transgenic system we were able to analyze the effects of p56^{lck} overexpression in developing thymocytes, which cannot readily be examined using in vitro methods.

In order to mimic the normal pattern of *lck* expression as precisely as possible, we chose to generate transgenic mice in which *lck* genes were expressed under the control of cis-acting regulatory elements from the *lck* gene itself. In both mouse and man, expression of transcripts is controlled by either of two promoter elements (proximal and distal); the promoter positioned most distal to the structural gene is active in both thymocytes and peripheral T cells, whereas the proximal promoter is active primarily in thymocytes (63,64). In preliminary experiments, we defined sequences necessary for *lck* proximal promoter function using heterologous reporter genes (7,65-67). These studies indicated that as little as 584 bp of 5' flanking region sequence was sufficient to direct lymphocyte-specific expression in transgenic mice. Armed with this information, two constructs were generated in which wild type (pLGY) or mutant (pLGF) versions of the *lck* structural gene were reconstructed to include the proximal promoter (FIG. 2).

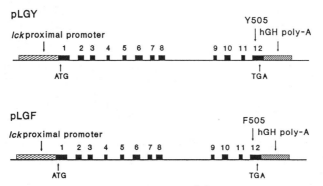

FIG. 2. Transgenes encoding wild-type p56^{lck}(Y505:pLGY) and the activated, mutant version of p56^{lck} (F505:pLGF). (Adapted from ref. 68).

The constructs differ in that a portion of exon 12 sequence encoding the in vivo phosphorylation site (Y505) in the pLGY transgene encoding wild type p56^{lck} was substituted with a mutated version of the gene encoding phenylalanine (F505) in pLGF (35,68,69). This strategy permits the analysis of both wild-type and activated p56^{lck} overexpression in vivo. The remainder of the transgene sequence includes a polyadenylation signal provided by sequences obtained from the human growth hormone gene (hGH; 70). The inclusion of these heterologous hGH sequences provides a useful marker sequence for transgene integration and expression. We employed these constructs to generate more than two dozen independent mouse lines in which p56^{lck} expression levels have been systematically increased (Table 1).

Table 1

Transgene Construct	Founder	#[a]	Transgene Expression[b]	Tumor Formation[c]	CD3 (%) Thymus[d]
pLGF	1127	23	33.8	+	1.0
	2949	7	35.0	+	4.0
	2943	1	23.4	+	9.5
	2954	13	11.7	−	13.0
	2964	5	10.0	−	12.0
	3082	7	ND	−	10.6
	3122	1	9.0	−	33.0
	3073	9	6.0	−	36.5
	2961	3	ND	−	62.0
	701	13	1.4	−	70.3
	2957	5	ND	−	76.5
	795	6	ND	−	78.8
pLGY	4220	12	75.0	+	11.9
	7233	5	ND	+	35.9
	1610	1	11.0	ND	68.6
	7240	1	ND	−	88.1
	1627	1	ND	ND	98.6
	1570	1	1.5	ND	99.2
	7246	1	ND	−	123.9
	1592	4	ND	−	92.2
	1572	3	ND	−	101.3

[a]Number of individual animals analyzed in each founder line
[b]Transgene mRNA expression in pg/μg total cellular RNA
[c]Tumor formation by 8 weeks of age
[d]Total number of CD3$^+$ cells in thymus as a percentage of normal littermate control value

Table 1. Characteristics of pLGF and pLGY transgenic animals (Adapted from refs. 68,69).

In these transgenics, the promoter activity of the transgenes faithfully reflects the tissue distribution of proximal promoter activity in vivo. That is, transgene expression is easily detectable in thymocytes, but undetectable in other organs (68,69). As expected, thymocytes from animals expressing the transgenes also detectably overexpress p56^{lck} protein when analyzed by immunoblotting using anti-p56^{lck} antisera (FIG. 3).

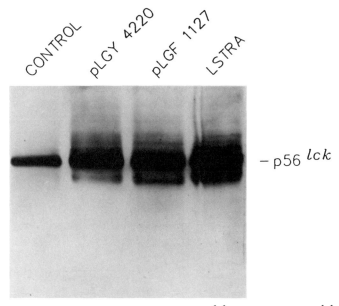

FIG. 3. Detection of transgene-derived p56^{lck} using anti-p56^{lck} specific antisera. The migration position of p56^{lck} is indicated at the right of the figure (Adapted from ref. 68).

Dose-related disturbances in thymopoiesis in pLGF animals

The mature thymus in mice consists of distinct subpopulations of thymocytes that can be distinguished by the array of surface markers which they express. Using in vitro culture techniques and in vivo repopulation studies it has been possible to assign to these cells a developmental sequence which is defined by the acquisition of CD4, CD8 and CD3 surface markers (71,72). The most immature thymocyte subpopulation, comprising <5% of the total thymocyte number, does not express any of these molecules. This subpopulation is localized to the subcapsular region of the thymus, is mitotically active, and is capable of reconstituting all thymocyte populations following adoptive transfer. Such "triple negative" cells during differentiation first gain expression of low levels of the CD8 molecule (CD4$^-$CD8^{lo}), and subsequently acquire surface CD4. These double positive (CD4$^+$CD8$^+$) cells, constitute the most numerous thymocyte type (>85%). During this stage of development antigen receptor rearrangements are completed and expression of the clonotypic TCR and its associated CD3 chains becomes detectable at low levels. The density of TCR and CD3 expression subsequently increase during the latest stages of development, coincident with the loss of expression of either the CD4 or CD8 molecule. Thus, the most mature thymocytes express high levels of TCR/CD3 in the presence of either CD4 or CD8.

The proximal *lck* promoter used in the transgenes is known to be active in all thymocyte subsets defined on the basis of CD4 or CD8 expression (63,67). In generating multiple independent transgenic founder animals expressing the pLGF transgene, we have been able to assess in semi-quantitative terms the effects of p56^{lck} overexpression on this developmental process. Such an analysis reveals that the predominant effect of p56^{lck} overexpression is to inhibit thymocyte maturation.

This maturational inhibition is readily assessed by monitoring the surface phenotypes of lymphocytes in the transgenic animals. The expression of CD3 in the pLGF thymus as compared with controls is dramatically altered, with the percentage of thymocytes expressing the TCR/CD3 complex reduced in the most severe cases to 1.0% of the normal percentage of CD3^{+} cells. This reduction in CD3^{+} cell numbers is correlated with a dramatic alteration in the CD4 and CD8 profiles of the transgenic thymus (FIG. 4).

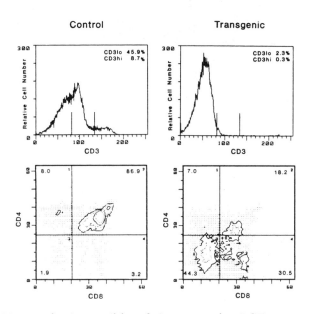

FIG. 4. Phenotypic abnormalities of thymocytes in pLGF transgenic mice. Thymocytes recovered from pLGF transgenic animals of the 1127 line and matched littermate controls were analyzed by flow cytometry for the percentage of cells expressing surface CD3 (represented as single parameter histograms in the upper panels) and similarly for CD4 (x-axis) and CD8(y-axis) in the dual parameter histograms in the lower panels (Adapted from ref. 69).

Strikingly, the magnitude of the observed inhibition in thymocyte development is strictly a function of the amount of p56^{lck} present within the developing thymus. That is, as levels of p56^{lck} activity are systematically increased, the severity of the abnormality increases correspondingly (Table 1). This observation implies that levels of p56^{lck} in the thymus may be limiting in vivo, and must be maintained below a critical threshold level to avoid severe developmental abnormalities.

The pLGF/pLGY phenotypic abnormality is a unique characteristic of $p56^{lck}$ overexpression

Substrate specificities among the members of the *src* family kinases have in general been difficult to demonstrate. Thus, it might well be imagined that overexpression of any *src* family kinase in this transgenic system could significantly affect thymocyte development. Subsequent experiments revealed, however, that the developmental inhibition observed in pLGF and pLGY transgenic animals is a unique characteristic of $p56^{lck}$ overexpression. Transgenic animals overexpressing the activated version of human *hck* (*hck*F501) exhibit no thymic phenotype, even when $p59^{hck}$ is expressed at levels where comparable pLGF expression produces dramatic abnormalities (K.M.A., S.D.L., and R.M.P., unpublished results). In addition, overexpression of wild-type $p59^{fyn}$, a kinase that, like $p56^{lck}$, is normally expressed within the developing thymus has a very different effect. Detailed studies of transgenic mice bearing a *fyn* expression construct under the control of the *lck* promoter revealed that although $p59^{fyn}$ overexpression profoundly affects the efficiency of signalling from the TCR complex, thymocyte development is essentially normal (7). Hence the capacity of $p56^{lck}$ to induce this developmental abnormality must reflect specialization of the *src* family kinases with respect to substrates and/or receptors.

$p56^{lck}$ activity alters the kinetics of development

One advantage of using a transgenic system to analyze gene function is the ability to examine gene dosage effects on developmental processes. Thymocyte ontogeny in particular is readily monitored by assessing the temporal acquisition of CD8, CD4 and CD3 surface markers during fetal and neonatal development. The progression of the *lck* transgenic phenotype during this developmental window was assessed in pLGF and pLGY transgenic animals (FIG. 5).

Animals overexpressing pLGF and pLGY during defined stages of fetal and postnatal development appear to be undergoing a relatively normal developmental progression. That is, CD3⁻CD4⁻CD8⁻ cells first acquire CD8 at a low density, followed by CD4. The kinetics of this development however, are profoundly delayed. Thus, animals overexpressing wild-type $p56^{lck}$ at four-times normal levels (pLGY4220) maintain a fetal developmental profile at adulthood. The observation that the *lck* transgenic developmental abnormality is evident during ontogeny at time periods preceding the expression of high levels of the previously characterized $p56^{lck}$ ligands, CD4 and CD8, implies that the developmental abnormality may not be primarily the result of heightened responsiveness via these molecules or pathways.

An in vivo estimate of the relative potency of $p56^{lckF505}$ as compared with wild type $p56^{lck}$

Since the extent of disruption of thymocyte maturation in *lck* transgenic mice is strictly dependent on *lck* kinase activity, it is possible to make an estimate of the increase in *lck* activity afforded by the Y505→F505 activating mutation in vivo. Thus, by comparing the phenotypes associated with overexpression of pLGY as versus pLGF, one can make some statements regarding the effects of

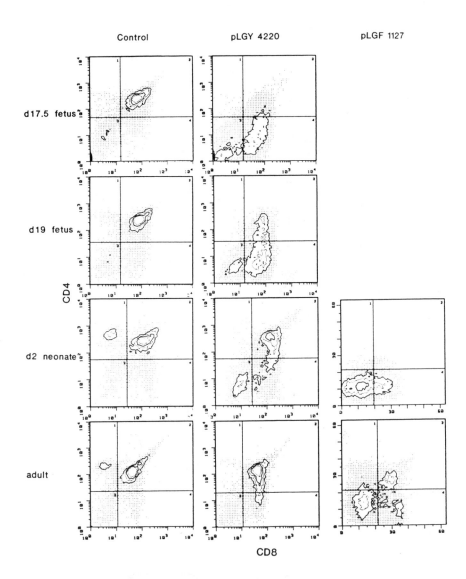

FIG. 5. Developmental kinetics of the pLGF and pLGY thymocyte abnormality.
Thymocytes recovered from transgenic and control animals at the ages
indicated were analyzed by flow cytometry as in FIG. 4 for the presence
of surface CD4 and CD8 (Adapted from ref. 69).

introduction of the Y→F mutation at the in vivo autophosphorylation site on kinase activity and function.

First, it appears that the pLGF and pLGY constructs function in essentially the same manner, each generating similar phenotypes in vivo. Both constructs are capable of producing developmental abnormalities with the same characteristics, and of increasing suceptibility to thymoma formation with similar kinetics (see below). Thus by these criteria the Y→F mutation does not grossly alter the specificity of the kinase.

However, the efficiencies with which the transgenes generate developmental abnormalities differ when pLGF and pLGY are directly compared. For example, a simple doubling of the steady state level of $p56^{lck}$ in developing thymocytes (as in pLGY #1610) appreciably alters thymopoiesis. This alteration, however, is of a magnitude similar to that seen in pLGF animals expressing approximately seven-fold less $p56^{lckF505}$ (compare pLGF #701). Examined in this way, pLGF functions with approximately seven-fold higher efficiency than pLGY.

Both wild-type and activated forms of $p56^{lck}$ function as oncogenes when overexpressed in vivo

The *lck* gene coding sequence is structurally quite similar to retroviral protein tyrosine kinase genes that behave as oncogenes. Indeed, as already mentioned, substitution of phenylalanine for tyrosine at position 505 of $p56^{lck}$ unmasks the transforming potential of this kinase in fibroblast assays (34,35). The *lck* gene is positioned at a site of chromosomal abnormalitites in about 10% of non-Hodgkins lymphomas (73,74), and is implicated in the pathogenesis of at least two retrovirally-induced lymphomas in which Moloney murine leukemia virus has inserted immediately adjacent to *lck* (75,76).

In light of this information, it is perhaps not surprising that overexpression of the activated form of $p56^{lck}$ in transgenic mice using the pLGF transgene construct leads to thymoma development. These tumors are lymphoid in origin, develop with rapid kinetics (6-8 weeks), and are capable of metastasis. The surface phenotype characteristic of these tumors is invariably CD3⁻CD4⁻CD8⁻ or CD3⁻CD4⁻CD8^{lo}, reflecting the dominant phenotype of thymocytes in high expressor *lck*-transgenics (Table 1). The tumors can be readily propagated in vitro as factor-independent cell lines, where they retain an immature phenotype (68).

Interestingly, tumor formation was observed in pLGY transgenic animals overexpressing the wild-type version of $p56^{lck}$ as well (Table 1). These tumors are strikingly similar to those found in pLGF animals with respect to their kinetics of development, cell surface phenotypes, and their ability to be propagated in vitro. Thus, the wild-type version of $p56^{lck}$ can act as a potent oncogene when overexpressed in thymocytes. The ability of wild-type $p56^{lck}$ to transform thymocytes but not fibroblasts presumably results from its ability to interact with particular substrates and/or receptor structures found in lymphoid cells.

ALTERNATIVE GENETIC STRATEGIES FOR ALTERING p56^{lck} ACTIVITY IN VIVO.

Expression of altered forms of p56^{lck} with mutant receptor binding motifs

The primary focus of the previous experiments has been to elevate levels of p56^{lck} activity in cells, with the goal of discerning the role of this *src* family kinase during thymocyte development. In order to clearly delineate the physical components of the kinase that are critical determining factors in the pLGF and pLGY phenotypes, additional mutant versions of these transgenes can be tested within transgenic contexts.

For example, as p56^{lck} is known to be tightly associated with CD4 and CD8 in lymphocytes, and the motifs responsible for mediating this association are known, it is possible to overexpress mutant versions of p56^{lck} in transgenics that are incapable of interacting with these surface components. By analyzing animals overexpressing such altered forms of *lck*, we can assess to what extent the phenotypes associated with *lck* overexpression can be attributed to increased coupling of *lck* to CD4 and CD8. Preliminary results indicate that the developmental abnormality observed in *lck* transgenic animals is not primarily mediated by p56^{lck} interactions involving the cysteine residues that are critical for CD4 and CD8 binding. In a similar vein, additional mutant versions of the kinase that are altered within the IL2R-interactive domain may be tested, in order to determine the requirement for binding via these motifs for generation of the phenotype.

The *lck* transgenic phenotype is a unique characteristic of p56^{lck} overexpression, in that expression of closely-related *src*-family kinases cannot readily recapitulate the *lck*-transgenic phenotype. Given the evolutionary conservation of functional domains within the *src* family kinases, it should be possible to dissect the physical components of the kinase that are responsible for this unique attribute of p56^{lck} by replacing *lck* domains with similar regions obtained from related *src*-family kinases. Such an experimental design would facilitate analysis of functional regions of the kinase by incorporating a phenotypic readout, and may ultimately lead to dissection of previously uncharacterized receptor binding pathways, or to the identification of unique p56^{lck} substrates.

Reducing p56^{lck} activity using dominant negative mutations

Although overexpression strategies can be extremely informative in dissecting the functional capability of proteins, several caveats are inherent in such an experimental design. By its very nature, overexpression can potentially lead to interactions which would not take place in the presence of physiologic concentrations of the protein under study. In order to develop alternative strategies to aid in elucidating the role of p56^{lck}, we have initially focussed on trying to decrease p56^{lck} function in vivo. One experimental approach to achieve this goal is to introduce a dominant negative p56^{lck} mutation into transgenic mice. This can be accomplished by interfering with the kinase activity of p56^{lck} through the introduction of a specific mutation at the codon encoding the ATP binding residue within the *lck* gene (K273 - see Figure 1). This mutational strategy has been used to produce "dead" kinase versions of several *src*-family members including *src*, *hck* and *fyn* (7,36,77). Overexpression of these "dead" kinases in normal cells allows the catalytically inactive version of the kinase to compete with the functional endogenous kinase for substrates or receptors. Such cells should exhibit disturbances in responses involving $p56^{lck}$-specific signal transduction pathways. This strategy has been used successfully to inhibit $p59^{fyn}$ function in transgenic thymocytes, and may ultimately prove extremely informative in the study of $p56^{lck}$ as well.

Production of lymphocytes lacking $p56^{lck}$ via homologous recombination in embryo-derived stem cells

As an alternative to producing cells that are phenotypically $p56^{lck}$ negative through the use of dominant negative mutations, cells can be produced by genetic means that lack functional *lck* genes. Such cells can be produced by homologous recombination using gene knockout techniques (for review, see 78). By such techniques, a genetically-induced interruption of *lck* is produced in embryonal stem cells, which can then be used to produce strains of animals which possess only non-functional *lck* alleles. Cells obtained from these animals would be unable to mount signalling responses that require the $p56^{lck}$ molecule. This technique will undoubtedly open new avenues of experimental design and ultimately provide extremely valuable information in the future.

A SPECULATIVE VIEW OF $p56^{lck}$ FUNCTION DURING THYMOCYTE DEVELOPMENT.

By overexpressing wild-type and activated versions of $p56^{lck}$ in transgenic animals, it has been possible to assess the role of this signal transduction molecule during all phases of thymocyte development. As a result of this analysis it seems likely that $p56^{lck}$ plays a role in thymocyte biology apart from its interaction with CD4 and CD8 coreceptor molecules. The phenotypic abnormality characteristic of $p56^{lck}$ overexpression, and the neoplastic transformation associated with extremely elevated levels of $p56^{lck}$ both become evident in cells that express little to undetectable levels of the previously characterized $p56^{lck}$ ligands (CD4 or CD8). The balance of the evidence would therefore suggest that $p56^{lck}$ is interacting with alternative ligands under these circumstances. Thus, the function of $p56^{lck}$ is likely being determined by the particular set of $p56^{lck}$ interactive receptors being expressed within the cell at any particular time during development.

Upon expression of CD8, and subsequently CD4, $p56^{lck}$ becomes recruited into the T cell receptor signal transduction apparatus. This permits two individual *src* family members to participate in regulating thymocyte selection, a process which is dependent upon both CD3/TCR and CD4/CD8 signalling pathways. Thus, the balance of signals being generated through $p56^{lck}$ and $p59^{fyn}$ kinases may ultimately prove to be a critical factor influencing the outcome of repertoire selection processes within the thymus.

But what can be said of the function of *src*-family kinases in general, and of $p56^{lck}$ in particular? First, although structural considerations argue that all *src*-family kinases will mediate similar functions in their respective cell types (20), data derived from the *lck*, *fyn* and *hck* transgenic mice suggests that individual *src*-family kinases may have acquired quite specialized functions. Thus, although all of these proteins may act as components of signal transduction complexes, the nature of the signalling pathways to which each contributes may be very distinct.

Secondly, we find it intriguing that the developmental arrest observed in *lck* transgenic mice occurs prior to the time when $V\beta$ gene segments rearrange (69) and in fact precludes the production of cells bearing both CD4 and CD8 surface proteins. During the past two years we have evaluated transgenic animals bearing a variety of signalling molecules under the control of the proximal *lck* promoter, and each system has exhibited a distinct phenotype. However, the *lck* transgenic mice resemble in some respects animals expressing SV40 large T antigen during thymocyte development (65). The *lck*-SV40 transgene elicits thymoma development in mice, albeit with a long latency, but it is provocative that thymocytes from *lck*-SV40 animals exhibit a developmental arrest in the preneoplastic period that phenotypically resembles the pattern seen in pLGF and pLGY mice. We suspect that this finding may indicate that the substrates for $p56^{lck}$ participate in cell regulation pathways that are also disrupted by SV40 large T antigen, perhaps including pathways involving the RB-1 protein. It may be the case that $p56^{lck}$ disrupts thymocyte maturation by perturbing the correct sequencing of commitment events during thymocyte development. These speculations lead us to consider the nucleus as a primary site of action of $p56^{lck}$. Viewed in this context, it is perhaps not surprising that $p56^{lck}$ interacts with components of the T lymphocyte antigen receptor complex, and with growth factor receptors. In each case, $p56^{lck}$ may be acting primarily by altering cell cycle control. More detailed analysis of this process will require systems in which lymphocyte cell growth can be synchronized, and would benefit tremendously from the availability of pharmacologic reagents that specifically inhibit $p56^{lck}$. Steady progress is being made in each of these areas.

SUMMARY

By genetic manipulation of *src*-family protein tyrosine kinase activity in vivo, we have produced unique model systems in which developmentally regulated protein tyrosine kinase signalling pathways can be analyzed. Using such a strategy it has been possible to uncover the participation of $p56^{lck}$ in what appear to be multiple signal transduction responses carried out through independent stages of thymocyte development. The phenotypic abnormality associated with $p56^{lck}$ overexpression is a consequence of increased $p56^{lck}$ activity early in thymocyte ontogeny, well before the acquisition of CD4 and CD8 coreceptor expression. As such, the abnormality may reflect hyper-responsiveness through alternative $p56^{lck}$-coupled pathways during this developmental window. In the future the study of $p56^{lck}$ and other *src*-family kinase expressing transgenic animals promises to provide further insights into the function of what clearly are complex and multi-functional signal transducing elements.

REFERENCES

1. Imboden, J.B., Weiss, A., and Stobo, J.D. The antigen receptor on a human T cell line initiates activation by increasing cytoplasmic free calcium. J. Immunol. 1985; 134:663-665.

2. Gelfand, E.W., Mills, G.B., Cheung, R.K., Lee, J.W.W., and Grinstein, S. Transmembrane ion fluxes during activation of human T lymphocytes: role of Ca^{2+}, Na^+/H^+ exchange and phospholipid turnover. Immunol. Rev. 1987; 95:59-87.

3. Weiss, A., Imboden, J., Hardy, K., Manger, B., Terhorst, C., and Stobo, J. The role of T3/antigen receptor complex in T cell activation. Ann. Rev. Immunol. 1986; 4:593-619.

4. Samelson, L.E., Patel, M.D., Weissman, A.M., Harford, J.B., and Klausner, R.D. Antigen activation of murine T cells induces tyrosine phosphorylation of a polypeptide associated with the T cell antigen receptor. Cell 1986; 46:1083-1090.

5. June, C.H., Fletcher, M.C., Ledbetter, J.A., and Samelson, L.E. Increases in tyrosine phosphorylation are detectable before phospholipase C activation after T cell receptor stimulation. J. Immunol. 1990; 144:1591-1599.

6. Baniyash, M., Garcia-Morales, P., Luong, E., Samelson, L.E., and Klausner, R.D. The T cell antigen receptor ζ chain is tyrosine phosphorylated upon T cell activation. J. Biol. Chem. 1988; 263:18225-18230.

7. Cooke, M.P., Abraham, K.M., Forbush, K.A., and Perlmutter, R.M. Regulation of thymocyte signal transduction by a non-receptor protein tyrosine kinase, p59$^{fyn(T)}$. Cell 1991; in press.

8. Samelson, L.E., Philips, A.F., Loung, E.T., and Klausner, R.D. Association of the *fyn* protein tyrosine kinase with the T cell antigen receptor. Proc. Natl. Acad. Sci. USA 1990; 87:4358-4362.

9. Eichmann, K., Johnson, J.-I., Falk, I., and Emmrich, F. Effective activation of resting mouse T lymphocytes by cross-linking submitogenic concentrations of the T cell antigen receptor with either Lyt-2 or L3T4. Eur. J. Immunol. 1987; 17:643-650.

10. Emmrich, F., Strittmatter, U., and Eichmann, K. Synergism in the activation of human CD8 T cells by cross-linking the T-cell receptor complex with the CD8 differentiation antigen. Proc. Natl. Acad. Sci. USA 1986; 83:8298-8302.

11. Anderson, P., Blue, M.-L., Morimoto, C., and Schlossman, S.F. Cross-linking of T3 (CD3) with T4 (CD4) enhances the proliferation of resting T lymphocytes. J. Immunol. 1987; 139:678.

12. Veillette, A., Bookman, M.A., Horak, E.M., Samelson, L.E., and Bolen, J.B. Signal transduction through the CD4 receptor involves the activation of the internal membrane tyrosine-protein kinase p56^{lck}. Nature (London) 1989; 338:257-259.

13. Veillette, A., Bolen, J.B., and Bookman, M.A. Alterations in tyrosine protein phosphorylation induced by antibody-mediated cross-linking of the CD4 receptor of T lymphocytes. Mol. Cell. Biol. 1989; 9:4441-4446.

14. Rudd, C.E., Trevillyan, J.N., Wong, L.L., Dasgupta, J.D., and Schlossman, S.F. The CD4 receptor is complexed to a T-cell specific tyrosine kinase (pp58) in detergent lysates from human T lymphocytes. Proc. Natl. Acad. Sci. USA 1988; 85:5190-5194.

15. Veillette, A., Bookman, M.A., Horak, E.M., and Bolen, J.B. The CD4 and CD8 T cell surface antigens are associated with the internal membrane tyrosine-protein kinase $p56^{lck}$. Cell 1988; 55:301-308.

16. Barber, E.K., Dasgupta, J.D., Schlossman, S.F., Trevillyan, J.M., and Rudd, C.E. The CD4 and CD8 antigens are coupled to a protein-tyrosine kinase ($p56^{lck}$) that phosphorylates the CD3 complex. Proc. Natl. Acad. Sci. USA 1989; 86:3277-3281.

17. Hunter, T., and Cooper, J.A. Protein tyrosine kinases. Ann. Rev. Biochem. 1985; 54:897-930.

18. Hanks, S.K., Quinn, A.M., and Hunter, T. The protein kinase family: conserved features and deduced phylogeny of the catalytic domains. Science 1988; 241:42-52.

19. Hunter, T. Cooperation between oncogenes. Cell 1991; 64:249-270.

20. Perlmutter, R.M., Marth, J.D., Ziegler, S.F., Garvin, A.M., Pawar, S., Cooke, M.P., and Abraham, K.M. Specialized protein tyrosine kinase proto-oncogenes in hematopoietic cells. Biochim. Biophys. Acta 1988; 948:245-262.

21. Dymecki, S.M., Niederhuber, J.E., and Desiderio, S.V. Specific expression of a tyrosine kinase gene, *blk*, in B lymphoid cells. Science 1990; 247:332-336.

22. Marth, J.D., Peet, R., Krebs, E.G., and Perlmutter, R.M. A lymphocyte-specific protein-tyrosine kinase is rearranged and overexpres sed in the murine T cell lymphoma LSTRA. Cell 1985; 43:393-404.

23. Buss, J.E., and Sefton, B.M. Myristic acid, a rare fatty acid, is the lipid attached to the transforming protein of Rous sarcoma virus and its cellular homolog. J. Virol. 1984; 53:7-12.

24. Marchildon, G.A., Casnellie, J.E., Walsh, K.A., and Krebs, E.G. Covalently bound myristate in a lymphoma tyrosine kinase. Proc. Natl. Acad. Sci. USA 1984; 81:7679-7682.

25. Pellman, D., Garber, E.A., Cross, F.R., and Hanafusa, H. Fine structural mapping of a critical NH_2-terminal region of $p60^{src}$. Proc. Natl. Acad. Sci. USA 1985; 82:1623-1628.

26. Shaw, A.S., Amrein, K.E., Hammond, C., Stern, D.F., Sefton, B.M., and Rose J.K. The *lck* tyrosine protein kinase interacts with the cytoplasmic tail of the CD4 glycoprotein through its unique amino-terminal domain. Cell 1989; 59:627-636.

27. Turner, J.M., Brodsky, M.H., Irving, B.A., Levin, S.D., Perlmutter, R.M., and Littman, D.R. Interaction of the unique amino-terminal region of the tyrosine kinase $p56^{lck}$ with the cytoplasmic domains of CD4 and CD8 is mediated by cysteine motifs. Cell 1990; 60:755-765.

28. Vega, M.A., Kuo, M.C., Carreroy, A.C., and Strominger, J.L. Structural nature of the interaction between T lymphocyte surface molecule CD4 and the intracellular protein tyrosine kinase *lck*. Eur. J. Immunol. 1990; 20:453-456.

29. Matsuda, M., Mayer, B.J., Fukui, Y., and Hanafusa, H. Binding of a transforming protein $p47^{gag-crk}$ to a broad range of phosphotyrosine-containing proteins. Science 1990; 248:1537-1539.

30. Moon, K.H., Suh, H.W., and Rhee, S.G. Inositol phospholipid-specific phospholipase C: complete cDNA and protein sequences and sequence homology to tyrosine kinase-related oncogene products. Proc. Natl. Acad. Sci. USA 1988; 85:5419-5423.

31. McCormick, F. *ras*GTPase activating protein: signal transmitter and signal terminator. Cell 1989; 56:5-8.

32. Mayer, B.J., Jackson, P.K., and Baltimore, D. The noncatalytic *src* homology region 2 segment of *abl* tyrosine kinase binds to tyrosine phosphorylated cellular proteins with high affinity. Proc. Natl. Acad. Sci. USA 1991; 88:627.

33. Cantley, L.C., Auger, K.R., Carpenter, C., Duckworth, B., Graziani, A., Kapeller, R. and Soltoff, S. Oncogenes and signal transduction. Cell 1991; 64:281-302.

34. Amrein, K.E., and Sefton, B.M. Mutation of a site of tyrosine phosphorylation in the lymphocyte-specific tyrosine protein kinase, $p56^{lck}$, reveals its oncogenic potential in fibroblasts. Proc. Natl. Acad. Sci. USA 1988; 85:4247-4251.

35. Marth, J.D., Cooper, J.A., King, C.S., Ziegler, S.F., Tinker, D.A., Overell, R.W., Krebs, E.G., and Perlmutter, R.M. Neoplastic transformation induced by an activated lymphocyte-specific protein tyrosine kinase ($p56^{lck}$). Mol. Cell. Biol. 1988; 8:540-550.

36. Ziegler, S.F., Levin, S.D., and Perlmutter, R.M. Transformation of NIH3T3 fibroblasts by an activated form of $p59^{hck}$. Mol. Cell. Biol. 1989; 9:2724-2727.

37. Cartwright, C.A., Eckhart, W., Simon, S., and Kaplan, P.L. Cell transformation by $pp60^{c-src}$ mutated in the carboxy-terminal regulatory domain. Cell 1987; 49:83-91.

38. Kmiecik, T.E., and Shalloway, D. Activation and suppression of $pp60^{c-src}$ transforming ability by mutation of its primary sites of tyrosine phosphorylation. Cell 1987; 49:65-73.

39. Cooper, J.A., and King, C.S. Dephosphorylation or antibody binding to the carboxy terminus stimulates $pp60^{c-src}$. Mol. Cell. Biol. 1986; 6:4467-4477.

40. Abraham, N., and Veillette, A. Activation of p56^{lck} through mutation of a regulatory carboxy-terminal tyrosine residue requires intact sites of autophosphorylation and myristylation. Mol. Cell. Biol. 1990; 10:5197-5206.

41. Marth, J.D. Lewis, D.B., Wilson, C.B., Gearn, M.E., Krebs, E.G., and Perlmutter, R.M. Regulation of pp56^{lck} during T-cell activation: functional implications for the src-like protein kinases. EMBO 1987; 6:2727-2734.

42. Casnellie, J.E., and Lamberts, R.J. Tumor promoters cause changes in the state of phosphorylation and apparent molecular weight of a tyrosine protein kinase in T lymphocytes. J. Biol. Chem. 1986; 261:4921-4925.

43. Casnellie, J.E. Sites of in vivo phosphorylation of the tyrosine protein kinase in LSTRA cells and their alteration by tumor-promoter phorbol esters. J. Biol. Chem. 1987; 262:9859-9864. 1987

44. Veillette, A., Horak, I.D., and Bolen, J.B. Post-translational alterations of the tyrosine kinase p56^{lck} in response to activators of protein kinase C. Oncogene Res. 1988; 2:385-392.

45. Horak, I.D., Gress, R.E., Lucas, P.J., Horak, E.M., Waldmann, T.A., and Bolen, J.B. T-lymphocyte interleukin 2-dependent tyrosine protein kinase signal transduction involves the activation of p56^{lck}. Proc. Natl. Acad. Sci. USA 1991; 88:1996-2000.

46. Hatakeyama,. M., Kono, R., Kobayashi, N., Kawahara, A., Levin, S.D., Perlmutter, R.M., and Taniguchi, R. IL-2 receptor interacts with a src-family kinase p56^{lck}; identification of novel intermolecular association. 1991; manuscript submitted.

47. Einspahr, K.J., Abraham, R.T., Dick, C.J., and Leibson, P.J. Protein tyrosine phosphorylation and p56^{lck} modification in IL-2 or phorbol ester-activated human natural killer cells. J. Immunol. 1990; 145:1490-1497.

48. Shaw, A.S., Amrein, K.E., Hammond, C., Stern, D.F., Sefton, B.M., and Rose, J.K. The lck tyrosine protein kinase interacts with the cytoplasmic tail of the CD4 glycoprotein through its unique amino-terminal domain. Cell 1989; 59:627-636.

49. Bank, J., and Chess, L. Perturbation of the T4 molecule transmits a negative signal to T cells. J. Exp. Med. 1985; 162:1294-1303.

50. Tite, J.P., Sloan, A., and Janeway, C.A. The role of L3T4 in T cell activation: L3T4 may be both an Ia-binding protein and a receptor that transduces a negative signal. J. Mol. Cell. Immunol. 1986; 2:179-190.

51. Glaichenhaus, N., Shastri, N., Littman, D.R., and Turner, J.M. Requirement for association of p56^{lck} with CD4 in antigen-specific signal transduction in T cells. Cell 1991; 64:511-520.

52. Salzman, E.M., Thom, R.R., and Casnellie, J.E. Activation of a tyrosine protein kinase is an early event in the stimulation of T lymphocytes by interleukin-2. J. Biol. Chem. 1988; 263:6956-6959.

53. Trowbridge, I.S., Ralph, P., and Bevan, M.J. Differences in the surface proteins of mouse B and T cells. Proc. Natl. Acad. Sci. USA 1975; 72:157-161.

54. Omary, M.B., Trowbridge, I.S., and Battifera, J.A. Human homologue of murine T200 glycoprotein. J. Exp. Med. 1980; 152: 842-852.

55. Komuro, K., Hakura, K., Boyse, E.A., and John, M. Ly-5: A new T-lymphocyte antigen system. Immunogenetics 1975; 1:452-456.

56. Tonks, N.K., Charbonneau, H., Diltz, C.D., Fischer, E.H., and Walsh, K.A. Demonstration that the leukocyte common antigen CD45 is a protein tyrosine phosphatase. Biochemistry 1988; 27:8695-8701.

57. Charbonneau, H., Tonks, N.K., Walsh, K.A., and Fischer, E.H. The leukocyte common antigen (CD45): a putative receptor-linked protein tyrosine phosphatase. Proc. Natl. Acad. Sci. USA 1988; 85:7182-7186.

58. Ledbetter, J.A., June, C.H., Grosmaire, L.S., and Rabinovitch, P.S. Cross-linking of surface antigens causes mobilization of intracellular ionized calcium in T lymphocytes. Proc. Natl. Acad. Sci. USA 1987; 84:1384-1388.

59. Koretzky, G.A., Picus, J., Thomas, M.L., and Weiss, A. Tyrosine phosphatase CD45 is essential for coupling T cell antigen receptor to the phosphatidylinositol pathway. Nature 1990; 346:66-68.

60. Ledbetter, J.A., Tonks, N.K., Fischer, E.H., and Clark, E.A. CD45 regulates signal transduction and lymphocyte activation by specific association with receptor molecules on T or B cells. Proc. Natl. Acad. Sci. USA 1988; 85:8628-8632.

61. Ledbetter, J.A., Schieven, G.L., Uckun, F.M., and Imboden, J.B. CD45 cross-linking regulates phospholipase C activation and tyrosine phosphorylation of specific substrates in CD3/Ti-stimulated T cells. J. Immunol. 1991; 146:1577-1583.

62. Palmiter, R.D., and Brinster, R.L. Germ-line transformation of mice. Ann. Rev. Genet. 1986; 20:465-499.

63. Wildin, R.S., Garvin, A.M., Pawar, S., Lewis, D.B., Abraham, K.M.,, Forbush, K.A., and Perlmutter, R.M. Developmental regulation of *lck* gene expression in T lymphocytes. J. Exp. Med. 1991; 173:383-393.

64. Reynolds, P.J., Lesley, J., Trotter, J., Schulte, R., Hyman, R., and Sefton, B.M. Changes in the relative abundance of Type I and Type II *lck* mRNA transcripts suggest differential promoter usage during T cell development. Mol. Cell. Biol. 1990; 10:4266-4270.

65. Garvin, A.M., Abraham, K.M., Forbush, K.A., Peet, R., Farr, A.G., and Perlmutter, R.M. Disruption of thymocyte development by SV40 T-antigen. Intern. Immunol. 1990; 2:173-180.

66. Chaffin, K.E., Beals, C.R., Forbush, K.A., Wilkie, R.M., Simon, M.I., and Perlmutter, R.M. Dissection of thymocyte signaling pathways by in vivo expression of pertussis-toxin ADP ribosyltransferase. EMBO 1990; 9:3821-3829.

67. Teh, H-S., Garvin, A.M., Forbush, K.A., Carlow, D.A., Davis, C.B., Littman, D.R., and Perlmutter, R.M. Participation of CD4 coreceptor molecules in T-cell repertoire selection. Nature (London) 1991; 349:241-243.

68. Abraham, K.M., Levin, S.D., Marth, J.D., Forbush, K.A., and Perlmutter, R.M. Thymic tumorigenesis induced by overexpression of p56^{lck}. Proc. Natl. Acad. Sci. USA 1991; in press.

69. Abraham, K.M., Levin, S.D., Marth, J.D., Forbush, K.A., and Perlmutter, R.M. Delayed thymocyte development induced by augmented expression of p56^{lck}. J. Exp. Med. 1991; in press.

70. Seeburg, P.H. The human growth hormone gene family: nucleotide sequences show recent divergence and predict a new polypeptide hormone. DNA 1982; 1:239-249.

71. von Boehmer, H. The developmental biooblogy of T lymphocytes. Ann. Rev. Immunol. 1988; 6:309-326.

72. Fowlkes, B.J., and Pardoll, D.M. Molecular and cellular events of T cell development. Advances in Immunology 1989; 44:207-264.

73. Levine, E.G., Arthur, D.C., Frizzera, G., Peterson, B.A., Hurd, D.D., and Bloomfield, C.D. There are differences in cytogenetic abnormalities among histologic subtypes of the non-Hodgkin's lymphomas. Blood 1985; 66:1414-1422.et al.,

74. Marth, J.D., Disteche, C., Pravtcheva, D., Ruddle, F., Krebs, E.G., and Perlmutter, R.M. Localization of a lymphocyte-specific protein tyrosine kinase gene (lck) at a site of frequent chromosomal abnormalities in human lymphomas. Proc. Natl. Acad. Sci. USA 1986; 83:7400-7404.

75. Garvin, A.M., Pawar, S., Marth, J.D., and Perlmutter, R.M. Structure of the murine lck gene and its rearrangement in a murine lymphoma cell line. Mol. Cell. Biol. 1988; 8:3058-3064.

76. Adler, H.T., Reynolds, P.J., Kelley, C.M., and Sefton, B.M. Transcriptional activation of lck by retrovirus promoter insertion between two lymphoid-specific promoters. J. Virol. 1988; 62:4113-4122.

77. Cooper, J.A. The src-family of protein tyrosine kinases. Peptides and Protein Phosphorylation. 1989; CRC press.

78. Frohman, M.A., and Martin, G.A. Cut, paste, and save: new approaches to altering specific genes in mice. Cell 1989; 56:145-147.

Advances in Regulation of Cell Growth, Volume 2;
Cell Activation: Genetic Approaches, edited by
James J. Mond, John C. Cambier, and Arthur Weiss.
Raven Press, Ltd., New York © 1991.

12

Segment Polarity Genes and Intercellular Communication in *Drosophila*

John Klingensmith and Norbert Perrimon

Howard Hughes Medical Institute, Dept. of Genetics, Harvard Medical School
25 Shattuck St. Boston, MA 02115

Cell-cell interactions play a crucial role in pattern formation of all multicellular organisms, from early cell divisions onward (1). Such intercellular communication activates genetic mechanisms directing cells down particular routes of differentiation, such that the end product is an organism composed of an enormous complexity of cell types.

The genetic circuitry regulating development of higher organisms, such as vertebrates, has proven difficult to dissect for technical reasons; thus developmental biologists have turned to the straightforward genetics afforded by many invertebrate species. The fruit fly *Drosophila melanogaster* provides an excellent model system in which to investigate the molecular genetic basis of cell-cell communication in development, allowing characterization of these processes at the organismic, cellular, molecular and biochemical levels.

The first genes mediating intercellular communication in the developing *Drosophila* embryo appear to be the segment polarity genes, which are involved in pattern formation of cells within the segmental units of which the embryo is composed (2). One of these genes, *wingless* (*wg*), encodes a protein which is made by a small domain of cells in each segment but is required by other cells to achieve their proper fates (3,4). As such, *wg* protein behaves as a paracrine signal, and in fact is taken up directly by those cells which require *wg* activity (5). Our research aims to understand segment polarity gene function in cell-cell communication, and particularly to identify the genes encoding components of the *wg* signalling pathway. This paper reviews current understanding of segment polarity genes in this context.

Although our primary objective is to understand segment polarity gene function in the pattern formation of *Drosophila*, this work may be directly relevant to the development of higher organisms as well. For example, *wg* is the *Drosophila* homolog of *Wnt-1* (formerly called *int-1*), a proto-oncogene which when misexpressed can result in mammary tumors in mice (6). Absence of expression of *Wnt-1* results in deletion of a large region of the brain in fetal mice (7,8). The *Wnt-1* gene is conserved in a broad evolutionary range of animals, and in each case appears to encode a secreted product (9). Thus it seems likely that molecules interacting with *wg* in *Drosophila* might be conserved in the *Wnt-1* pathway of mice and other species.

DROSOPHILA DEVELOPMENTAL GENETICS

The basic segmental body pattern of *Drosophila* is established during embryogenesis. Systematic screens for mutants with altered pattern elements on the embryonic cuticle have identified about fifty genes, acting maternally and/or zygotically, that are involved in embryonic segmentation (10-16). Initially, embryonic polarity is set up along the anteroposterior and dorsoventral egg axes by maternal gene products, which permits the subsequent establishment of restricted domains of zygotic gene expression (2,17-19).

The genes which specifically direct the process of segmentation can be grouped into three classes: gap, pair-rule and segment polarity (20). Detailed genetic and molecular analysis of many of these genes has shown that the process of segmentation results from a dynamic pattern of genetic interactions, under precise spatial and temporal control (2). In response to positional cues provided by maternal gene products, a few restricted domains of gap gene expression occur along the syncitium of nuclei in the precellular blastoderm. When mutant for a gap gene, an embryo lacks derivatives of the corresponding domain, creating a "gap" in its body plan. The gap genes in turn set up expression boundaries of the pair-rule genes, which serve to further subdivide the blastoderm into regions consisting of approximately two segmental primordia each. Lesions in the pair-rule genes result in the absence of structures at two-segment intervals. Finally, the segment polarity genes begin to be expressed around the time of cellularization. In some cases defined by the borders of expression domains of different pair-rule genes, the expression pattern of a given segment polarity gene is reiterated in every segment. These genes form pattern within single segments; in mutant embryos, each segment is present but organization within it is abnormal. The segment polarity genes are thus the first pattern formation genes to act after the cellularization of the blastoderm.

After generation of the segmental body plan of the embryo, a variety of genes act to bring about the particular differentiation of the cells in each segment. This process is controlled by the homeotic genes, each of which is specific to a subset of segments. A homeotic mutation results in the expression of incorrect segmental identities in regions where the gene is required, transforming these segments into others (homeosis). Mostly homeobox-containing genes, the homeotic genes are thought to be involved in regulation of the cascades of genes bringing about the characteristic differentiation of each segment.

After a day, development of the embryo is complete, and the animal hatches into the larval phase of the life cycle. The three larval stages are primarily a period of rapid growth, with little pattern formation. However, even before hatching a subset of cells have been set aside to form the imaginal discs, from which the adult structures are formed. These cells remain diploid and divide rapidly, while the cells of the larval body become large and polyploid.

After a few days the larvae become quiescent and enter pupation. The imaginal discs undergo complex morphogenetic changes in response to hormones. Meanwhile, the polyploid larval cells begin to die as the disc and neural cells proliferate and differentiate. Gradually the adult body pattern begins to take form within the pupal case. Ultimately an animal bearing no resemblance to its precursor emerges, displaying a complexity of body parts, each decorated with a vast, precise array of bristles and hairs.

The metamorphosis of the *Drosophila* larva into the adult fly is developmentally tantamount to a second program of pattern formation, in which each disc develops independently of the others to form a particular part of the adult body. As in embryonic development, mutations have been isolated which affect various stages and aspects of imaginal pattern formation. Most of these genes are specific to imaginal growth, but some are also involved in embryogenesis. Several segment polarity genes can be mutated to reveal an imaginal role. As in embryogenesis, these defects tend to suggest failures of normal cell-cell interaction within the disc. The phenotypes, which range from major structural abnormalities to subtle patterning defects of individual cells, will be discussed following an overview of the embryonic phenotypes.

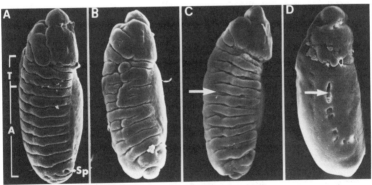

Fig. 1. Scanning electron micrographs of segment polarity phenotypes. These panels show for comparison surface views of wildtype (A), *naked* (B), *patched* (C) and *wingless* (D) embryos at 10 - 11 hours of development. These are lateral perspectives, with anterior at top and ventral to the left. The thoracic (**T**) and abdominal segments (**A**) are indicated in the wildtype embryo, each segment delimited by a segmental furrow. The posterior spiracle is indicated (**S p**). In *naked* (B) segment borders are often incomplete and irregular, with partial fusion of some segments. *patched* (C) embryos have the normal set of segmental furrows, but also an ectopic second furrow within each segment. This results in narrow outfoldings of epidermis between the segments (arrow). In *wingless* embryos (D) segmental furrows are completely absent. The tracheal pits of each segment, normally closed by this stage, fuse into large holes (arrow). All segmental units appear merged, except the maxillary appendage (open arrowhead). These embryos also lack posterior spiracles (compare to panel A).

SEGMENT POLARITY PHENOTYPES

Segment polarity genes act during embryogenesis to define and maintain the cell states that are required for proper organization of individual segmental units (3). The juxtaposition of these different cell states leads to the establishment of boundaries delimiting the segmental primordia. In wildtype embryos, the ventral cuticle of each segment is covered by a belt consisting of rows of tooth-like denticles in the anterior region, followed by naked cuticle to the posterior (Fig. 4A). Deep segmental furrows delimit each segment (Fig. 1A). The denticle belts, naked cuticle and segmental furrows provide morphological landmarks by which to compare the affects of various mutants on segmental organization. Mutations in any of the segment polarity genes cause deletions in each segment of various cuticular structures, often accompanied by ectopic occurrence of others.

Group	Locus	Location	Function M	E	I	Deduced Stucture	Refs
I	*zeste-white 3*	3B	+	+	+	serine/threonine kinase	21,54,55
	naked	75F-76B	?	+	?	unknown	10,21
	lines	44F-46D	?	+	?	unknown	11
II	*patched*	44D	-	+	+	transmembrane	20,22,61
	costal- 2	43B-43C	+	+	+	unknown	22,33
III	*armadillo*	2B	+	+	+	plakoglobin	58,60,74,86
	dishevelled	10B	+	+	+	unknown	24,75,82
	porcupine	17A	+	+	+	unknown	16,75,82
	fused	17C	+	+	+	serine/threonine kinase	24,56
	wingless	28A	-	+	+	secreted factor	5,65,67,71
	smooth	(2-4)	?	+	?	unknown	11
	gooseberry	60F	-	+	?	homeobox (nuclear)	20,51,52
	hedgehog	94E	?	+	+	unknown	20,31
	cubitus-interruptus Dominant	102B	-	+	+	zinc-finger (nuclear)	25,53
IV	*engrailed*	48A	-	+	+	homeobox (nuclear)	20,26,45,46

TABLE I. Properties of the Segment Polarity Genes.
The loci are listed in groups reflecting the position within the segment the lesions primarily affect, from anterior (I) to posterior (IV), as detailed in the text. Location refers to the cytological location of the loci on polytene chromosomes. In the case of *smooth* only the meiotic map position is known. The developmental phases at which the genes function are indicated - maternal (M), embryonic (E) and imaginal (I). A plus (+) indicates that a gene functions at a given phase, while a minus (-) indicates lack of function. In some cases function has not been assayed, denoted (?). The deduced structure of the protein is indicated for those characterized. References include the most comprehensive phenotypic and, where appropriate, molecular descriptions, numbered as in the text.

Four groups of segment polarity genes can be distinguished by their mutant phenotypes (Table I), in that each mutation affects a specific portion of the segment, considered here from anterior to posterior (Fig. 2). Three genes cause pattern disturbances in the anterior of the segment (Group I). In *naked* (*nkd*) and *zeste-white 3* (*zw3*), mutant embryos lack almost all denticles, with only naked cuticle remaining (10,21). Segment borders are incomplete in these mutants (Fig. 1B). The *lines* (*lin*) lesion results in a more limited deletion of anterior pattern elements, leaving about half the normal number of denticle rows (11).

Central pattern elements are deleted from each segment in two mutations (Group II). In *patched* (*ptc*) embryos, the posterior of each denticle belt is deleted, along with some adjacent naked cuticle, as schematized in Fig. 2. This results in a mirror-image duplication of the anterior part of each denticle belt flanked by an ectopic segment furrow (20,22). The extra segment border manifests itself as a narrow outfolding of cuticle between segments, shown in Fig. 1C. A similar but less precise denticle pattern is seen in *costal-2* (*cos-2*) embryos (23).

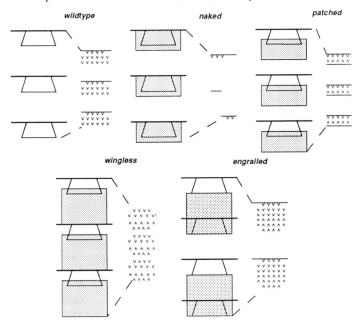

Fig. 2. Portions of the segment affected in mutants and the resulting phenotypes. This drawing schematizes the embryonic segment and its patterning defects in the various segment polarity phenotypes, as described in the text. The left side of each schematic signifies the primordia of two full segments (three for *engrailed*), and the right represents the pattern of denticles generated in each case. Rows of v characters represent the extent and polarity of denticles, although in all cases except *naked* more rows of denticles occur than diagrammed. Horizontal lines represent segment borders. The shaded boxes indicate the portion of the segment affected in each mutant. A representative locus of each phenotypic group (I-IV) is shown. Other loci in a given group result in qualitatively similar defects, as described in the text.

In the largest group of segment polarity mutants, the naked cuticle of the posterior region is largely replaced by denticles (Group III). Instead of being oriented toward the posterior as are the denticles of the anterior belt, these ectopic denticles are often of inverse or chaotic polarity (Fig.4). The members of this class are *armadillo* (*arm*), *cubitus-interruptus Dominant* (*ci-D*), *dishevelled* (*dsh*), *fused* (*fu*), *gooseberry* (*gsb*), *hedgehog* (*hh*), *porcupine* (*porc*), *smooth* (*smo*), and *wingless* (*wg*) (12,16,20,24,25). The status of segmental borders depends on the mutant locus in question; for example, they are normal in *gsb* but absent in *wg* embryos (Fig. 1D). Similarly, the presence of naked cuticle varies among these loci. Many embryos, such as *gsb* mutants, have nearly mirror-image duplications of each denticle belt, but still retain some naked cuticle along the posterior margin of the segment. An extreme segment polarity phenotye is seen in *wg* embryos, in which the ventral epidermis is covered by a lawn of denticles, with no naked cuticle remaining (Fig. 4B).

Mutations in *engrailed* (*en*) constitute a fourth group. This locus affects the segment border and flanking cells (Fig. 2). Lesions result in deletion of the most posterior region, along with the most anterior rows of denticles in the next segment as well as the intervening segment border (20,26). Superimposed on this pattern is a pair-rule phenotype in which alternating segments are missing; this complication may reflect the regulatory role of pair-rule genes on *en* expression (27,28).

Because different regions of the segment are affected in various segment polarity mutations, the phenotypes must arise by different mechanisms. Genes within a group all show similar phenotypes, which probably have a related or sometimes identical cellular basis. Thus among the segment polarity genes there appear to be subsets of genes with related roles in embryonic cell patterning.

IMAGINAL ROLES OF SEGMENT POLARITY GENES

Many segment polarity genes function in pattern formation during imaginal disc development as well as during embryonic segmentation (Table I). In most cases, the imaginal role of a given gene has been revealed by the phenotypes of mutants in alleles that permit survival through adult differentiation. Often the defective structures do not interfere with the viability of the fly. When no such allele exists, small clones of cells homozygous for a lethal allele can be induced in an otherwise wildtype animal (29). Such clones are induced, via mitotic recombination, after embryogenesis but before imaginal development; thus the embryonic function of the gene is completed yet any imaginal role of the gene is revealed by a phenotype of the clone in the adult.

As in embryonic development, the imaginal phenotypes of the segment polarity genes are quite varied, from severe to subtle. In *wg* pupal-lethal alleles, which permit survival through imaginal development, the distal derivatives of many imaginal discs are deleted. Thus antennae, lower legs, and wings are absent (30). Clones of *hh* cells can also result in deletions within these structures (31). A number of loci which later were found to have segment polarity phenotypes were initially known for their wing patterning defects in viable alleles: *fu*, *ci-D* and *cos-2* (32,33). *ptc* is now known to have a wing defect also (34). Mutant clones for *zw3* (also known as *shaggy*) differentiate bristles rather than hairs (21,35). In *dsh* viable-allele flies, the polarity of hairs and bristles on many body parts is chaotic relative to wildtype (24,36,37). Any imaginal roles of *gsb*, *nkd*, *smo*, *lin*, *arm*, and *porc* are yet to be characterized. However, because all alleles of *porc* and some alleles of *arm* result in death at the larval-pupal interface, these two genes do have some role in imaginal development (16).

Thus, all segment polarity genes tested to date are used again in development, during the pattern formation of imaginal discs. However, the relationships of embryonic and imaginal functions of these genes are completely unknown. Only in the case of *en* does a connection seem clear. Throughout development, *en* appears to be required for organization of the posterior of segments. In the adult, *en* mutations result in replacement of posterior disc structures by those of the anterior (38).

MOLECULAR ANALYSIS OF SEGMENT POLARITY GENES

Several segment polarity genes have now been isolated and analyzed, revealing much about their functions. Some general features of this analysis provide clues as to what these genes are doing as a group. Their zygotic expression begins around the time the blastoderm cellularizes and continues through late stages of embryogenesis. During this time the major cellular pattern-forming events are occuring. As the embryo gastrulates in the hours following cellularization, transient parasegmental borders and then segmental borders form in each segment. The parasegmental borders divide each segment into anterior and posterior compartments (39). Because denticles are not secreted until much later in embryogenesis, segmental and parasegmental furrows provide the best morphological landmarks to compare gene expression patterns.

Each segment polarity gene analyzed to date is expressed in a pattern reiterated in every segment during gastrulation (Figure 3). In some cases expression is restricted to a narrow stripe of cells, while in others it occurs across the entire segment. However, the exact spatial and temporal pattern of expression of each gene is unique. This suggests that the different developmental potentials of cells within the segment are conferred, at least in part, by the particular combination of genes expressed in those cells at a given time in development. The deduced structures of segment polarity proteins suggest involvement at several levels of cellular regulation (Table I). Below we consider genes analyzed thus far according to their deduced molecular structures.

Genes Encoding Putative DNA Binding Products

Like many genes involved in *Drosophila* embryonic development, some of the segment polarity genes contain homeoboxes. These are structurally-related domains, initially identified in homeotic genes, that are known to bind genetic regulatory sequences, the specificity of which is dependent upon the nature of the particular homeobox in question (40). The *en* homeobox was among the first to be shown to bind specific DNA sequences (41). *en* appears to function via transcriptional regulation of downstream genes, in at least some cases acting as a repressor (42,43,44). The *en* transcipts and protein are expressed by gastrulation in the posterior-most row of cells in each segment (45-48). These cells coincide with the posterior compartment of the segment, in that they are bounded by the transient parasegmental border to the anterior and the segmental border to the posterior (Fig. 3). The *en* gene is also expressed exclusively in the posterior compartment of the imaginal discs (47,49,50). This pattern of expression coincides with the genetic requirement for *en*, in that it is required by cells in posterior compartments to achieve their proper fate (38). In the embryo, the defects of *en* mutations extend beyond the posterior compartment (Fig. 2), presumably reflecting extensive cell-cell interaction (see below)

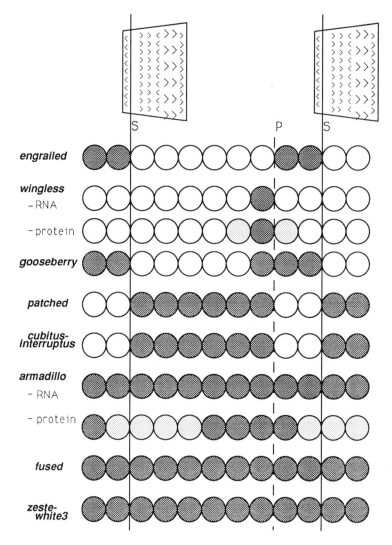

Fig. 3. Expression patterns of segment polarity genes. The molecular expression patterns of genes cloned to date are shown relative to the parasegmental (P) and segmental (S) borders in mid-gastrulation embryos. The relationship of these landmarks to the denticle belts of the mature embryo is shown by the parallelograms above, in which > characters indicate denticle rows. The diagram schematizes longitudinal sections through the ventral epidermis, circles representing cells. Dark circles indicate transcript expression. The *en* protein pattern is identical to that of the transcript. *wg* protein is found in the cells expressing the transcript but also at lower levels in flanking cells, due to secretion and uptake (light shading). *arm* transcript occurs uniformly, as does a low level of protein (light shading). In cells around the parasegmental border *arm* protein occurs at high levels (dark circles). Protein expression patterns of the remaining loci are unknown.

Two other segment polarity genes also encode putative DNA-binding proteins. The *gsb* locus, which consists of two transcription units, contains a homeobox similar to that of the pair-rule gene *paired* (51,52). It appears to be expressed in the *en*-expressing cells (the *en* domain) and the cells immediately to the anterior, at a width of approximately two cells as gastrulation begins (Fig. 3). The *ci-D* gene appears to encode a product with a zinc-finger motif, found in many transcription factors and known to bind DNA (53). *ci-D* is unusual in that its expression pattern is initially uniform and does not resolve into a periodic pattern until late stages of gastrulation. It is then expressed in the anterior three-quarters of every segment, excluded from the *en* domain. In the absence of *en*, *ci-D* is expressed throughout the segment, suggesting direct repression (43).

Putative Serine-Threonine Kinases

Two segment polarity genes appear to encode serine-threonine kinases: *zw3* (54,55) and *fu* (56). Both of these genes have maternal as well as zygotic requirements: neither maternal nor zygotic expression of either gene is sufficient to fulfill its embryonic function (16). Transcripts of both genes are uniformly distributed in the embryo throughout embryogenesis (54-56). Protein distribution patterns are not yet reported. The developmental significance of the apparent kinase function of these genes is not clear. However, because some serine-threonine kinases are known to be involved in signal transduction (57), it is possible that *zw3* and *fu* are involved in transduction of extracellular stimuli from the plasma membrane to the nucleus.

Putative Membrane Proteins

Peripheral

Antibody localization and cell fractionation experiments reveal that the product of the *arm* gene is peripherally associated with the cell membrane (58). Although the transcript is expressed uniformly throughout development (59), the protein is expressed at high levels in a restricted part of each segment (58). Comparison to the domains of cells expressing *wg* and *en* show that the cells expressing high levels of *arm* protein center on the *wg* domain but extend beyond it into the anterior part of the *en* domain (Figure 3). As will be discussed below, this enhanced *arm* protein expression is dependent upon the *wg* product. Although initial databank searches revealed no homologies of *arm* to known proteins, it has recently been determined that *arm* is the *Drosophila* homolog of the mammalian desmosome constituent plakoglobin (60). Desmosomes are a type of cell-cell junction in epithelial cell sheets. This homology has important implications for understanding the mechanism by which segment polarity genes function.

Integral

The *ptc* gene encodes a putative transmembrane protein, predicted to span the membrane seven times (22,61). The deduced peptide sequence bears no significant homolgy to other proteins, nor are any other functional motifs recognizable. However, it is structurally similar to the *Drosophila* gene *frizzled* (62), which is involved in positional signalling in the adult epidermis (63,64). The *ptc* message is expressed in a dynamic pattern: although uniform as it first appears during the cellular blastoderm stage, it is resolved into a segmental pattern in which it is expressed in all but the *en* domain as the embryo undergoes gastrulation (Figure 3). Finally, as gastrulation ends, it is expressed in two stripes per segment: along the anterior border and along the posterior border of the anterior compartment (i.e. in the *wg* domain) (44).

Secreted Products

The *wg* gene, which is homologous to the mammalian proto-oncogene *Wnt-1*, encodes a cystine-rich protein with a signal peptide (65-67). Immunolocalization experiments via electron microscopy reveal that the *wg* protein is indeed secreted (5). As mentioned, *wg* transcipt is expressed in the cells immediately anterior to the *en* domain, along the parasegmental border. The protein is expressed in these cells but is also taken up by adjacent cells (Figure 3). The molecular biology of *wg* as well as its developmental genetics suggest that it is a morphogenetic signalling molecule within the segment; this idea will be discussed in detail below.

The molecular natures of *dsh, porc, lin, hh, smo, nkd* and *cos-2* remain unknown, although characterization of each is underway in various laboratories. Although only about half the segment polarity genes have been cloned and sequenced, their deduced structures suggest a variety of points of action, from the nucleus to the extracellular matrix. Investigations of the remaining genes and use of antibodies to all these proteins in biochemical assays should fill in some of the missing links in our understanding of segment polarity gene function.

The Initiation of Segment Polarity Gene Expression

The expression pattern of each of the segment polarity genes is reiterated in every segment, whether expressed in a single cell or in all the cells of the segment (Figure 3). The transcripts of *arm, fu* and *zw3* are expressed maternally as well as zygotically and so are present from the outset of development. However, the timing of the onset of stripes among those genes restricted to subsets of cells in the segmental primordia varies in an important way. Flanking the parasegmental border are cells expressing *wg* (to the anterior) and *en* (to the posterior). The position of the parasegmental border is determined by certain pair-rule genes, and these genes also regulate the initiation of expression of *wg* and *en* (28,68). As mentioned above, *gsb* expression overlaps the *wg* and *en* stripes exactly (Figure 3). While these three genes are expressed in narrow stripes by the beginning of gastrulation, *ptc* and *ci-D* expression is not resolved into stripes until late germ-band extension, after the expression of pair-rule genes has ceased (22,53,61). Thus the pair-rule genes are necessary for the initiation of segment polarity gene expression at the borders of the parasegments, but do not appear to regulate expression in interior cells. Rather it appears that the refinement and maintenance of segment polarity gene expression depends on interactions among the segment polarity genes themselves. Many of these interactions among segment polarity genes appear to be mediated via cell-cell communication, as discussed below.

SEGMENT POLARITY GENES MEDIATE CELL INTERACTIONS

Segmental expression of the segment polarity genes begins around the time of cellularization of the blastoderm and continues through late embryogenesis. These genes appear to be the first pattern formation genes to act after cellularization, perhaps suggesting that their function is dependent on a cellular context. Phenotypic and molecular genetic analyses indicate that many of these genes have a role in processes involving cell-cell communication. Such evidence comes from both embryonic and imaginal studies.

Segment Polarity Genes and Cell Interactions in Imaginal Development

Although segment polarity loci have been considered together on the basis of their embryonic defects in intrasegmental patterning, many of them also exhibit similarities in their imaginal phenotypes, as discussed previously. Some of the phenotypes exhibited by segment polarity mutants involve failures in patterning processes known to involve cell-cell interactions and communication. Flies mutant for a viable allele of *dsh* exhibit a tissue polarity phenotype, in that the normally distally-aligned rows of bristles and hairs on many body parts instead appear chaotic (24,36,37). The organization of cell polarity in the adult epidermis involves positional information to which each cell responds, though the nature of such information remains unknown (63,64). The adult function of the *zw3* locus is also involved in the patterning of bristles. Although all alleles are lethal, the phenotype of small clones in an otherwise wildtype adult wing is striking: bristles occur in place of hairs (21,35). This locus perturbs the process of lateral inhibition among cells in the wing, in which innervated bristle cells inhibit many of their neighbors from similarly adopting a neural fate, leaving them to differentiate non-innervated hairs (69).

The induction of clones of cells mutant for other loci in adult flies has revealed another aspect of cell-cell communication affected by some segment polarity genes in imaginal development. When a genetically mutant cell exhibits the mutant phenotype of a gene, and surrounding wildtype cells are unaffected, the gene acts in a cell-autonomous manner. However, if the cell does not display the mutant phenotype, or if adjacent wildtype cells also show the phenotype, the gene functions non-autonomously. Non-autonomy indicates the involvement of cell-cell interactions in the process affected by the mutant gene. Clones of genetically-marked *wg* (70-72), *hh* (31) and *ptc* (34) cells all exhibit non-autonomy of function in their respective imaginal roles.

The expression patterns in imaginal discs of the segment polarity genes thus far cloned have yet to be reported in most cases. *en* transcripts occur throughout the posterior compartments of discs, the same regions affected by the lesion (47,49,50). However, it is clear that in at least two cases the regions of the disc affected by the mutation do not coincide with the spatial distribution of the genes' transcripts. *ptc* transcript appears throughout the anterior compartment of the wing disc, most intensely along the anterior-posterior compartment border; nevertheless it appears that cells along the border do not require *ptc* function (34). The *wg* transcript in wing discs occurs in a complex pattern with relatively few cells expressing the gene (30), yet the viable and pupal-lethal *wg* lesions eliminate most wing structures (30,70).

Intercellular Communication in the Embryonic Segment

A characteristic feature of the embryonic phenotypes of the various segment polarity genes is that, among genes cloned, the domain of expression of a given gene is not coincident with the affected portion of the segment. For example, *wg* is expressed in a narrow stripe of cells yet affects the entire posterior of the segment (5,65) while *zw3* is expressed throughout the segment but lesions affect only the anterior (21,54,55). The segment polarity phenotypes themselves reveal that these genes function in cell patterning within the segmental primordium, and the discrepancy between phenotypes and sites of expression shows that their overall function involves cell-cell interactions and communication.

Perhaps the most striking evidence that segment polarity function involves cell-cell communication is the relationship between *wg* and *en* expression. As discussed, *wg* and *en* are expressed in adjacent, mutually exclusive stripes of cells on either side of the parasegmental border. Although their expression is initiated via the pair-rule genes, maintenance of expression of each gene requires cell-cell interactions mediated by the other, as well as additional segment polarity genes. In *en* embryos, *wg* expression is initially normal, appearing in a stripe a single cell wide in each segment, just anterior to the *en* stripe. But it begins to fade and eventually disappears in the absence of *en* expression in the adjacent cell (3). Mutations in *ptc*, *nkd* (3), and *hh* (44) reportedly also affect the temporal and spatial patterns of *wg* transcript accumulation. Studies on the distribution of *wg* protein in these and the other segment polarity mutants suggests that regulation of *wg* expression by segment polarity genes is very complex, involving several elements of cell-cell interaction (73). The dependence of *wg* expression on *en* expression will be discussed further later in the text.

The nature of the requirement for *wg* in the maintenance of *en* expression is better understood than its reciprocal. Although *en* expression is initiated correctly in *wg* mutants, it begins to deteriorate by the time of germ-band extension (3,4). Several other segment polarity loci, such as *arm*, result in a similar disappearence of *en* expression (4). The deterioration of *en* expression has immediate consequences for the developing segment primordium. As soon as *en* disappears in *arm*, *wg* and other embryos with an extreme segment polarity phenotype (see below), cell death can be detected in the *en* expression domain (74,75). This cell death provides a link between changes in molecular expression patterns and the cellular basis of the resulting phenotype, but its rapidity complicates the analysis of the affects of mutations on expression patterns. Nonetheless, sufficient evidence has been gathered to begin to dissect the signalling pathway by which *wg* affects *en* expression, and as such reveal much about how segment polarity genes mediate intercellular communication in the developing segment.

THE *WINGLESS* SIGNALLING PATHWAY

The nonautonomy of *wg* in imaginal disc clones, its narrow stripe of expression relative to the broad region affected in mutant embryos, and the requirement for *wg* by cells adjacent to those expressing it collectively suggest that *wg* might function as an intrasegmental signal. The molecular structure of the deduced protein is consistent with such a role: it has a hydrophobic signal peptide and a potential secondary structure which suggests it might be secreted (65,67). Immunocytochemistry using an antibody to *wg* has confirmed this notion (5). The *wg* protein appears in diffuse stripes centering on those of the transcript, only broader. Electron microscopy reveals that *wg* antigen occurs not only in those cells expressing the transcript but also in neighboring cells and in the extracellular space between them. In these cells the antigen is contained within vesicles which are thought to represent endocytotic uptake of the protein (5). By double-staining with *en* antibody, van den Heuvel and colleagues observed that *en* cells take up *wg* protein. Cells anterior to the *wg* expressing cells probably take up *wg* protein also, but there is as yet no unambiguous way to identify these cells (5,76). Thus immunocytochemical data support the genetic evidence and suggest that *wg* protein does indeed function as a signal between domains of cells in the segmental primordium.

Identification of Pathway Components

The regulation of *en* expression by *wg* and the uptake of secreted *wg* protein by *en* cells constitute a paracrine regulatory interaction. As such this signalling mechanism provides an excellent *in vivo* model system for studying cell-cell communication in pattern formation. However, it is clear that many other molecules must be involved in the pathway in addition to the *wg* and *en* proteins. Since the *en* protein is a nuclear DNA binding function, it could not interact with *wg* protein directly - there must be molecular machinery for receiving the signal at the cell surface and for transducing the stimulus to the nucleus. The molecular and cell biology of the *wg* protein suggest that uptake by *en* cells might involve some form of receptor-mediated endocytosis (5,9). Moreover, it is likely that additional molecules might be required for presentation of an active *wg* signal to its target cells. Such activities might be involved in processing and secreting the *wg* protein from its source cells and transporting it through the extracellular matrix.

Determination of which molecules are components of a signalling pathway and the mechanism by which they function ultimately requires difficult biochemical experiments. As an initial means of identifying likely candidates in a genetically mutable pathway one can consider loci giving rise to similar phenotypes. This approach has proven quite productive in *Drosophila*. For example, a group of genes directing terminal development of the early embryo all mutate to very similar phenotypes; genetic and molecular experiments have demonstrated that these genes constitute a signal transduction mechanism and have allowed several to be ordered into a pathway (19,57). Similarly, in eye development the *sevenless* gene encodes a receptor tyrosine kinase required to receive positional signals (77,78). On the basis of identical phenotype, the *bride-of-sevenless* locus was identified successfully as encoding another key component of the signalling pathway (79,80). Thus it is likely that other genes mutating to segment polarity phenotypes encode elements of the *wg* signalling pathway, particularly those with defects similar to *wg*. Below we present a consideration of these phenotypes.

Phenotypic Implications for the *wingless* Pathway

The most obvious feature of the *wg* phenotype is that the naked cuticle of the posterior of the segment is deleted, replaced by ectopic denticles. Many of the segment polarity loci (group III of Table I) share this feature, and as such are readily distinguished from the other sorts of segment polarity phenotypes, groups I, II and IV, typified by *nkd*, *ptc* and *en* respectively (Figures 1 and 2). These other phenotypic groups involve different areas of the segment (schematized in Figure 2) and likely involve a different cellular basis for their defects. However, even among those loci superficially similar to *wg* there are several differences in phenotype. The *wg* phenotype can be considered the most extreme of these in that it differs most dramatically from wildtype. Below we consider the features of mutant *wg* embryos and the other loci which share these defects, with the purpose of determining those most likely to result from the same mechanisms and thus most likely to be in the *wg* signalling pathway.

Segmentation defects are most easily seen when soft tissue is dissolved, leaving behind the chitinous cuticle. Because the cuticle is secreted locally by cells immediately beneath, it reflects cell organization. Cuticle preparations of *wg* embryos exhibit a lawn of denticles, with no borders between segments, as shown in Figure 4 (24,71). Embryos mutant for *dsh* (24) and *porc* (16) appear identical (Figure 4). The strongest *hh* (31) and *arm* (74) phenotypes also look similar to *wg* embryos.

The apparent lack of segment borders in *wg* embryos is confirmed by scanning electron microscopy (SEM) of specimens throughout development (24), as can be seen in Figure 1D. This is also true of *dsh* embryos (24), and on the basis of Nomarski optical sections, of *porc* and *arm* embryos (75). Other loci appear to generate at least transient or partial segmental furrows.

In addition to failure in segmental delimitation, *wg* embryos exhibit fusion of tracheal pits, transiently occuring invaginations in the middle of each segment. This can be seen via SEM (Figure 1D) as well as by a molecular marker which is expressed in the embryonic tracheal system (81). Stainings of *dsh*, *porc* and *arm* embryos also reveal fusion of tracheal pits, while *hh* embryos do not exhibit this phenotype (75).

Finally, the pattern of cell death in *wg*, *dsh*, *porc* and *arm* embryos is identical (74,75). Cell death in the embryonic epidermis can be detected by methylene blue staining after fixation of tissue, in that dead cells fail to occlude the dye and appear

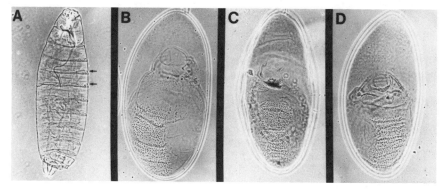

Fig. 4. The extreme segment polarity phenotype. The ventral cuticle of the wildtype embryo (A) shows segmental organization, each segment consisting of an anterior belt of denticles with naked cuticle to its posterior, delimited by segmental furrows (arrows). In the *wg* embryo (B), naked cuticle is deleted, denticles are disorganized, and segmental furrows are missing. This is termed the extreme segment polarity phenotype. In the absence of maternal and zygotic activity, *dsh* (C) and *porc* (D) embryos exhibit an identical phenotype. At least one allele of *arm* also shows this phenotype. All such cuticles are indistinguishable. Anterior is uppermost.

fragmented. Counter staining with X-gal of an *en-lacZ* fusion (50) show that this cell death occurs primarily in the *en* domain at the posterior of the segment (74). Counter staining with a *wg-lacZ* fusion reveals that the cell death progresses into the *wg* domain in these four mutants (75).

To summarize, the extreme segment polarity phenotype of *wg* is also exhibited by *dsh*, *porc* and *arm*. In addition to the cuticular defects, these loci are set off from others in that they also lack segmental furrows, show fusion of tracheal pits, and exhibit a characteristic pattern of cell death. In fact, these embryos are indistinguishable. Thus phenotypic analyses implicate *dsh*, *porc* and *arm* as being involved in the *wg* signalling pathway.

Regulation of Downstream Genes

Another means of identifying loci which might function in a regulatory pathway is to determine how candidate mutations affect the expression of genes known to be downstream of a given member of the pathway. In the case of *wg*, maintainance of *en* expression is a key function. If *arm*, *dsh* and *porc* are in the *wg* pathway and thus mediate the affects of *wg* on *en* expression, then mutations in these genes should affect *en* expression in the same manner. The strongest *arm* phenotype results in a similar pattern of deterioration of *en* expression, as judged by *en-lacZ* fusion stainings of *arm* embryos (4,74). *en* expression also decays in *dsh* embryos (24). Recent work comparing the cell biology of *en* disappearence reveals that *en* antigen staining decays in an identical spatial and temporal pattern in *wg*, *dsh*, *porc*, and *arm* embryos (82).

Although the *wg* signalling pathway was initially defined by its affects on *en* expression in adjacent cells, other genes also appear to be regulatory targets. One such affect is post-transcriptional regulation of *arm* (58). Although the *arm* transcript is expressed uniformly, the protein is expressed at high levels in cells overlapping those expressing *wg*. These enhanced stripes of expression rapidly follow the evolution of the *wg* pattern. In *wg* embryos only background levels of *arm* protein are detected, demonstrating that *wg* regulates this phenomenon. Because the domain of cells expressing high levels of *arm* includes not only the stripe of cells expressing *wg* transcript but also cells to either side, this modulation appears to be mediated by the paracrine signalling pathway of *wg*. Riggleman et al. (58) observed that among other segment polarity mutants, only *dsh* and *porc* are required for this affect on *arm* expression. This observation adds independent evidence that *dsh* and *porc* mediate *wg* signalling. The affect of extreme *arm* alleles on *arm* protein expression was not studied, although the above discussion suggests that they too should block modulation.

In addition to its regulatory affects on other segment polarity genes, *wg* also has a role in directing the expression of certain other genes. For example, the *Distal-less* gene (*Dll*) is expressed in the embryo at the sites of vestigal limbs in each trunk segment, in a manner dependent on *wg* function in adjacent cells (83). The *wg* gene also appears to have a function in non-epidermal cells of the embryo. Recent results suggest that *wg* provides a signal between mesodermal and endodermal cells as well, as judged by its synthesis in the visceral mesoderm but regulatory affect on *labial* expression in adjacent endodermal midgut cells (84). It will be important to determine if *dsh*, *porc* and *arm* also mediate these functions of *wg* as well.

Roles of Segment Polarity Genes in the *wingless* Pathway

The mechanisms by which *dsh*, *porc* and *arm* function in mediating the *wg* signal remain unknown. Molecular characterization of *dsh* and *porc*, underway in our laboratory, should provide some clues as to their modes of action. Nevertheless, genetic information will be required to determine their functional roles to support any hints revealed by their DNA sequences or expression patterns. For this it will be essential to determine the cellular requirements for *dsh* and *porc* function. By generating small genetically-marked clones of mutant cells and comparing their phenotype to adjacent wildtype cells, the requirement by individual cells for either gene product can be analyzed. If each cell requires the gene function to achieve its normal fate (i.e. if it is cell autonomous), the gene would most likely function in reception or interpretation of the *wg* signal in its target cells. Conversely, if a genetically mutant cell can be rescued by an activity supplied by nearby wildtype cells, a condition of non-autonomy, the gene would probably be involved in presentation of active *wg* signal to its target cells. The results of such clonal analyses indicate that *dsh* is cell autonomous, while *porc* is non-autonomous (82).

The *arm* gene has been the subject of extensive study and its potential roles in the *wg* pathway are better understood. Several studies have revealed that *arm* is cell autonomous (74,85,86), suggesting that *arm* functions in the response of target cells to the *wg* signal. The deduced sequence of the *arm* protein and its extraction properties suggest an abundant molecule with structural properties and a peripheral membrane association (58,59). The basis of these findings has become clear with the recent observation that *arm* is the *Drosophila* homolog of plakoglobin (60). Plakoglobin is a component of desmosomes, which are a type of junction between epithelial cells. Conventional wisdom suggested that desmosomes were largely inert structural linkages, but the pattern formation phenotypes of *arm* suggest that they may have a more active role in mediating intercellular communication. A possible model for the role of *arm* in the *wg* signalling pathway is that desmosomes may be the sites of relatively high concentrations of receptor molecules in the cell membrane, including those receiving the *wg* signal. Alternatively, *arm* may fulfill a more passive function in that desmosomes may bring cells into sufficiently close contact that short-range paracrine factors such as *wg* can be passed successfully between source and target (60).

An interesting feature of *dsh* and *porc* is that their embryonic (segment polarity) defects are seen only if the mutant embryo is derived from a mother lacking the gene activity in her germline (16). In other words, maternally contributed gene product can mask the affects of the mutation in embryonic development. This maternal product is not necessary however because a single wildtype copy of the gene from the father suffices for normal development. The allele of *arm* giving rise to the extreme segment polarity phenotype also exhibits this fully-rescuable maternal affect (74). This reveals that the functions encoded by *dsh*, *porc* and *arm* are present before fertilization but are not needed for embryonic development until after zygotic gene expression begins. Thus the components of the *wg* signalling pathway may be in place well before the signal is produced.

These loci, along with *wg*, are only known examples giving rise to the extreme segment polarity phenotype. It seems very unlikely however that they could encode all the elements necessary for the signalling pathway. It is likely that some components are not yet represented by mutations. Theoretical considerations argue that the genome should contain approximately fifteen loci mutable to an extreme segment polarity phenotype (16). Some of these may be in hand but have not been recognized because of maternal masking of zygotic phenotypes. It is also possible that some elements of the pathway also have earlier functions and any mutations in these genes would exhibit an earlier defect than segment polarity. Finally, some components of the *wg* pathway may exist among the other segment polarity genes. Because there is considerable cell-cell interaction apart from the *wg* to *en* pathway phenotypes might not reflect functional relationships in a straightforward manner.

Relationship of Other Segment Polarity Genes to the *wingless* Signalling Pathway

As mentioned, *wg* protein is secreted by its source cells and is taken up by adjacent *en* cells to the posterior. It also appears that *wg* protein is taken up by cells to the anterior (5,76). Yet these cells anterior to the *wg*-expressing cells do not respond in the same way as those to the posterior. One possibility is that they lack the receptor molecule(s) for the *wg* signal. A more likely explanation is revealed by the *ptc* mutation - in *ptc* embryos, en expression is observed on both sides of the *wg* domain (3). This suggests that these cells also have the *wg* reception apparatus, but that *ptc* inhibits its function in some way. Support for this idea is the observation that *ptc* transcript is expressed uniformly at first but becomes repressed in the *en* domain (22,44,61). The putative transmembrane structure of the *ptc* protein suggests that its inhibitory affects on the *wg* signalling pathway occur in the membrane, perhaps by competing with the *wg* receptor molecule(s) for ligand binding or by disruption of consequent signal transduction (61).

Another mode of intercellular communication mediated by the segment polarity genes is that from the *en* domain to the *wg* domain, the reciprocal of the pathway on which we have focussed. As mentioned, maintenance of *wg* expression requires *en* function in cells to the posterior (3). Since *en* is a nuclear factor it cannot act directly as a signal to the *wg* domain, but rather probably allows for the production of some signal or stimulus, whether it be a paracrine factor or a contact-mediated cue. A candidate gene for functioning in this pathway is *hh*. *wg* expression decays in strong *hh* mutants in a manner similar to that in *en* embryos (44,73). The non-autonomy of *hh* in imaginal development (31) is consistent with an involvement in the signalling aspect of this intercellular communication. Its molecular structure remains unknown.

The potential roles of many of the segment polarity genes in cell-cell interactions within the developing segment remain unclear. The apparent DNA-binding functions encoded by *gsb* and *ci-D* (51-53) may be the eventual targets of intercellular stimuli, but as yet have not been shown to be. Other genes, such as *cos-2*, *lin* and *smo* have not been studied in sufficient detail to reveal their possible functions. However, it seems likely that at least some of the other segment polarity genes do have a role in intercellular communication. As discussed, *zw3* is clearly involved in cell interactions in the patterning of cells in imaginal development, but any analogous role in embryonic development is as yet undetected. Given the serine-threonine kinase homology of its product, *zw3* may be involved in transduction of signal information from the cell surface to nuclear targets (54,55). Because *nkd* exhibits an identical phenotype and has similar affects on downstream gene expression, such as *en*, it seems likely that *nkd* is in the same pathway, perhaps as a substrate of *zw3* kinase activity (54,57). The *fu* product, which is also a putative serine-threonine kinase, might also be involved in signal transduction of paracrine stimuli (56).

SEGMENT POLARITY HOMOLOGS IN VERTEBRATES

One of the most remarkable aspects of the segment polarity genes is that many of them appear to be conserved in mammals and other vertebrates, which are evoutionarily quite distant from *Drosophila*. Few genes from earlier levels of the pattern formation heirarchy are conserved in higher species. This is probably because these genes function largely in an acellular system, the early fly blastoderm, whereas segment polarity genes are involved in patterning after cellularization of the embryo. In the embryos of higher organisms cell divisions accompany nuclear divisions, perhaps eliminating the need for a precellular pattern formation network (1).

In some cases segment polarity homologs in other species have known functions in pattern formation. The most conspicuous example of these is the *wg* homolog *Wnt-1* (previously known as *int-1*). Originally identified as a proto-oncogene implicated in virally-induced breast tumors in mice, *Wnt-1* is now known to be present in virtually all multicellular organisms, from nematodes to humans (9). In mouse, *Wnt-1* is expressed in the embryonic central nervous system (87,88), implying a role in neural development. This has been demonstrated recently via homologous-recombination mediated knock-out of both copies of the *Wnt-1* gene (7,8). Homozygous fetuses show absence of the midbrain and some adjacent tissue, evident by mid-gestation. The *Wnt-1* product has been shown to be secreted into the extracellular matrix, as is the *wg* product in flies (89,90), suggesting that it also functions as a paracrine factor. Although *Wnt-1* misexpression can result in breast tumors, the gene is not normally expressed in mammary tissue (87,88). Recent work indicates that the *Wnt-1* gene is one of a family of structurally-related genes, some of which are expressed in breast (91). Possibly ectopic *Wnt-1* protein interacts with the potential receptor(s) of these other products in generating breast tumors. There is as yet no information as to the nature of *Wnt* receptors, or whether there is a single receptor or a whole family of receptors interacting with *Wnt* gene products (92).

Genes homologous to *gsb* and *en* are also known to be expressed in mouse embryos. The *Pax 1* gene, highly homologous to *gsb*, is expressed in a segmental pattern during mouse development (93). Although a role in metamerism of the mouse embryo would be an exciting analogy to *gsb* function, no phenotypic evidence has been obtained as yet for *Pax 1*. Mice possess two genes homologous to *en*, *En-1* and *En-2*, which are expressed in the area of the embryonic CNS giving rise to the midbrain, as is *Wnt-1* (94,95). Because *Wnt-1* mutation deletes this region, it seems possible that *En* and *Wnt* might have a funtional interaction in mouse development as do their *Drosophila* homologs (8). Thus not only the genes but also their regulatory interactions may be conserved in evolution.

Several other segment polarity genes also have vertebrate homologs, but any involvement in pattern formation is unknown. The homology of *arm* to plakoglobin (60) has important implications for the function of desmosomes in development, inviting a new course of experimentation. Perhaps the homology of *zw3* to the mammalian gene for glycogen synthase kinase-3/ factor A (57) and of *ci-D* to the human transcription factor GLI (53) will lead to a better understanding of the developmental roles of these genes also.

SUMMARY AND PROSPECTUS

In this review we have discussed the segment polarity genes of *Drosophila* and the evidence that they mediate intercellular communication in pattern formation. Their phenotypes indicate that they function not only in embryonic pattern formation but in generation of the adult body plan as well. The molecular structures of the genes cloned thus far suggest that they function at several different levels of cell regulation, from the nucleus to the cell membrane and even in the extracellular space and adjacent cells. Future experiments should reveal whether the embryonic and imaginal requirements represent the same pathways used twice or the involvement of common factors in otherwise unrelated modes of cell interaction.

Phenotypic and *in situ* expression analyses suggest that these genes function in at least a few different pathways of cell communication. The best understood of these is the *wingless* signalling pathway, which regulates *en* transcription and *arm* protein levels in adjacent cells. At least three segment polarity genes - *dsh*, *porc* and *arm* - mediate the paracrine function of the *wg* protein. Molecular characterization and biochemical analysis of these gene products should elucidate the molecular mechanism by which *wg* signalling occurs. Further mutagenic screens are expected to reveal additional components of this pathway.

Many of the segment polarity genes, possibly all of them, have homologs in mice and other higher organisms. Future work will reveal whether these genes function analogously in these evolutionarily distant organisms. Some of these homologs are of special interest because of previously characterized roles in mammalian biology. For example, the *Wnt-1* product appears to act as a paracrine factor, as does its its fly homolog *wg*. It will be especially interesting to determine if the genes mediating the *wg* pathway also have a role in the *Wnt-1* pathway, in that their conservation could provide insight into the oncogenic properties of *Wnt-1* in breast tissue, which have proven difficult to address by conventional approaches.

ACKNOWLEDGMENTS

We are grateful to E. Siegfried for comments on the manuscript, to M. van den Heuvel and R. Nusse for communication of results prior to publication, and to all the above for helpful discussions. This work was supported by the Howard Hughes Medical Institute and N.I.H. grant HD23684.

REFERENCES

1. Davidson, E. How embryos work: a comparative view of diverse modes of cell fate specification. *Development* 1990; 108: 365-389.

2. Ingham, PW. The molecular genetics of embryonic pattern formation in *Drosophila. Nature* 1988; 335: 25-33.

3. Martinez-Arias, A Baker, N and Ingham, PW. Role of segment polarity genes in the definition and maintenance of cell states in the *Drosophila* embryo. *Development* 1988; 103: 157-170.

4. DiNardo, S, Sher, E, Heemskerk-Jongens, J, Kassis, JA, and O'Farrell, PH. Two-tiered regulation of spatially patterned *engrailed* gene expression during *Drosophila* embryogenesis. *Nature* 1988; 332: 604-609.

5. van den Heuvel, M, Nusse, R, Johnston, P and Lawrence, P. Distribution of the *wingless* gene product in *Drosophila* embryos: a protein involved in cell-cell

communication. *Cell* 1989; 59: 739-749.

6. van Ooyen, A and Nusse, R. Structure and nucleotide sequence of the putative mammary oncogene *int-1*; proviral insertions leave the protein-coding domain intact. *Cell* 1984; 39: 233-240.

7. Thomas, KR and Capecchi, MR. Targeted disruption of the murine *int-1* proto-oncogene results in severe abnormalities in midbrain and cerebellar development. *Nature* 1990; 346: 847-850.

8. McMahon, AP and Bradley, A The *Wnt-1* (*int-1*) proto-oncogene is required for development of a large region of the mouse brain. *Cell* 1990; 62: 1073-1085.

9. Nusse, R. The *int* genes in mammary tumorigenesis and in normal development. *TIG* 1988; 4: 291-295.

10. Jurgens, G, Wieschaus, E, Nusslein-Volhard, C and Kluding, H. Mutations affecting the pattern of the larval cuticle in *Drosophila melanogaster*. II. Zygotic loci on the third chromosome. *Roux's Arch. Dev. Biol.*1984;193: 283-295.

11. Nusslein-Volhard, C, Wieschaus, E and Kluding, H. Mutations affecting the pattern of the larval cuticle in *Drosophila melanogaster*. I. Zygotic loci on the second chromosome. *Roux's Arch. Dev. Biol.* 1984; 193: 267-282.

12. Wieschaus, E, Nusslein-Volhard, C and Jurgens, G. Mutations affecting the pattern of the larval cuticle in *Drosophila melanogaster*. III. Zygotic loci on the X chromosome and the fourth chromosome. *Roux's Arch. Dev. Biol.* 1984; 193: 296-307.

13. Perrimon, N, Mohler, JD, Engstrom, L, and Mahowald, A. X-linked female sterile loci in *Drosophila melanogaster*. *Genetics* 1986; 113: 695-712.

14. Schupbach, T and Wieschaus, E. Maternal effect mutations affecting the segmental pattern in *Drosophila*. *Roux's Arch. Dev. Biol.* 1986; 195: 302-307.

15. Nusslein-Volhard, C, Frohnhofer, HG, and Lehmann, R. Determination of anteroposterior polarity in *Drosophila*. *Science* 1987; 238: 1675-1681.

16. Perrimon, N, Engstrom, L and Mahowald, AP. Zygotic lethals with specific maternal effect phenotypes in *Drosophila melanogaster*: I. Loci on the X-chromosome. *Genetics* 1989; 121: 333-352.

17. Scott, MP and O'Farrell, PH. Spatial programming of gene expression in early *Drosophila* embryogenesis. *Ann. Rev. Cell Biol.* 1986; 2: 49-80.

18. Akam, M. The molecular basis for metameric pattern in the *Drosophila* embryo. *Development* 1987; 101: 1-22.

19. Manseau, LJ and Schupbach, T. The egg came first, of course. *TIG* 1989; 5: 400-405.

20. Nusslein-Volhard, C, and Wieschaus, E. Mutations affecting segment number and polarity in *Drosophila*. *Nature* 1980; 287: 795-801.

21. Perrimon, N and Smouse, D. Multiple functions of a *Drosophila* homeotic gene, *zeste-white 3*, during segmentation and neurogenesis. *Dev. Biol.* 1989; 135: 287-305.

22. Hooper, J and Scott, MP. The *Drosophila patched* gene encodes a putative membrane protein required for segmental patterning. *Cell* 1989; 59: 751-765.

23. Grau, Y and Simpson, P. The segment polarity gene *costal-2* in *Drosophila*. I. The organization of both primary and secondary embryonic fields may be affected. *Dev. Biol.* 1987; 122: 186-200.

24. Perrimon, N and Mahowald, AP. Multiple functions of segment polarity genes in *Drosophila*. *Dev. Biol.* 1987; 119: 587-600.

25. Orenic, T, Chidsey, J, and Holmgren, R. *Cell* and *cubitus-interruptus Dominant*: two segment polarity genes on the fourth chromosome in *Drosophila*.

Dev. Biol. 1987; 124: 50-56.
26. Kornberg, T. *Engrailed*: a gene controlling compartment and segment formation in *Drosophila. PNAS* 1981; 78: 1095-1099.
27. DiNardo, S and O'Farrell, P. Establishment and refinement of segmental pattern in the *Drosophila* embryo: spatial control of engrailed expression by pair-rule genes. *Genes and Dev.* 1987; 1: 1212-1225.
28. Ingham, PW, Baker, NE, and Martinez-Arias, A. Regulation of segment polarity genes in the *Drosophila* blastoderm by *fushi tarazu* and *even-skipped. Nature* 1987; 331: 73-75.
29. Lawrence, PA, Johnston, P and Morata, G. Methods of marking cells. in *Drosophila*: A Practical Approach (ed. D. Roberts) Oxford: IRL Press, 1984.
30. Baker, NE.Transcription of the segment polarity gene *wingless* in the imaginal discs of *Drosophila*, and the phenotype of a pupal-lethal *wg* mutation. *Development* 1988; 102: 489-498.
31. Mohler, J. Requirements for *hedgehog*, a segment polarity gene, in patterning larval and adult cuticle of *Drosophila. Genetics* 1988; 120: 1061-1072.
32. Lindsley, DL and Grell, EH. Genetic Variations of *Drosophila melanogaster*. Washington: Carnegie Institution of Washington, 1968.
33. Simpson, P and Grau, Y. The segment polarity gene *costal-2* in *Drosophila*. II. The origin of imaginal pattern duplications. *Dev. Biol.* 1987; 122: 201-209.
34. Phillips, RG, Roberts, IJH, Ingham, PW and Whittle, JRS. The *Drosophila* segment polarity gene *patched* is involved in a position-signalling mechanism in imaginal discs. *Development* 1990; 110: 105-114.
35. Simpson, P, El Messal, M, Moscoso Del Prado, J, and Ripoll, P. Stripes of positional homologies across the wing blade of *Drosophila melanogaster. Development* 1988; 103: 391-401.
36. Held, LI, Duarte, CM, and Derakhshanian, K. Extra tarsal joints and abnormal cuticular polarities in various mutants of *Drosophila melanogaster. Roux's Arch. Dev. Biol.* 1986; 195: 145-157.
37. Klingensmith, J and Perrimon, N. The *dishevelled* gene of *Drosophila* functions in multiple modes of positional signalling in imaginal discs. (in preparation).
38. Lawrence, P and Morata, G. The compartment hypothesis. in Insect Development (ed. P. Lawrence). London: Blackwell Scientific Public., 1976.
39. Martinez-Arias, A and Lawrence, P. Parasegments and compartments in the *Drosophila* embryo. *Nature* 1985; 313: 639-642.
40. Hayashi, S and Scott, MP. What determines the specificity of action of *Drosophila* homeodomain proteins? *Cell* 1990; 63: 883-894.
41. Desplan, C, Theis, J and O'Farrell, P. The *Drosophila* developmental gene, *engrailed*, encodes sequence specific DNA binding activity. *Nature* 1985; 318, 630-635.
42. Poole, SJ, Kauvar, LM, Drees, B and Kornberg, T. The *engrailed* locus of *Drosophila*: structural analysis of an embryonic transcript. *Cell* 1985; 40: 37-43.
43. Eaton, S and Kornberg, T. Repression of *ci-D* in posterior compartments of *Drosophila* by *engrailed. Genes and Dev.* 1990; 4: 1068-1077.
44. Hidalgo, A and Ingham, P. Cell patterning in the *Drosophila* segment: spatial regulation of the segment polarity gene *patched. Development* 1990; 110: 291-301.
45. DiNardo, S, Kuner, JM, Theis, J, and O'Farrell, P. Development of embryonic pattern in *D. melanogaster* as revealed by accumulation of the nuclear *engrailed* protein. *Cell* 1985; 43: 59-69.

46. Fjose, A, McGinnis, WJ and Gehring, WJ. Isolation of a homeobox containing gene from the *engrailed* region of *Drosophila* and the spatial distribution of its transcripts. *Nature* 1985; 313: 284-289.

47. Kornberg, T, Siden, I, O'Farrell, P, and Simon, M. The *engrailed* locus of *Drosophila: in situ* localization of transcripts reveals compartment-specific expression. *Cell* 1985; 40: 45-56.

48. Karr, TL, Weir, MJ, Ali, Z and Kornberg, T. Patterns of *engrailed* protein in early *Drosophila* embryos. *Development* 1989; 105: 605-612.

49. Brower, D. *engrailed* gene expression in *Drosophila* imaginal discs. *EMBO J.* 1986; 5: 2649-2656.

50. Hama, C, Ali, Z and Kornberg, T. Region-specific recombination and expression are directed by portions of the *Drosophila engrailed* promoter. *Genes and Dev.* 1990; 4: 1079-1093.

51. Baumgartner, S, Bopp, D, Burri, M and Noll, M. Structure of two genes at the *gooseberry* locus related to the *paired* gene and their spatial expression during *Drosophila* embryogenesis. *Genes and Dev.* 1987; 1: 1247-1267.

52. Cote, S, Preiss, A, Haller, J, Schuh, R, Kienlin, A, Seifert, E and Jackle, H. The *gooseberry-zipper* region of *Drosophila*: five genes encode different spatially restricted transcripts in the embryo. *EMBO J.* 1987; 6: 2793-2801.

53. Orenic, TV, Slusarski, DC, Kroll, KL and Holmgren, RA. Cloning and characterization of the segment polarity gene *cubitus-interruptus Dominant* of *Drosophila. Genes and Dev.* 1990; 4: 1053-1067.

54. Siegfried, E, Perkins, LA, Capaci, TM and Perrimon, N. The *Drosophila* segment polarity gene *zeste-white 3* encodes putative serine/threonine protein kinases. *Nature* 1990; 345: 825-829.

55. Bourouis, M, Moore,P, Puel,L, Grau,Y, Heitzler,P and Simpson, P. An early embryonic product of the gene *shaggy* encodes a serine/threonine protein kinase related to *CDC28/cdc2+* subfamily. *EMBO J.* 1990; 9: 2877-2884.

56. Preat, T, Therond,P, Lamour-Isnard,C, Limbourg-Bouchon,B, Tricoire,H, Erk,I, Mariol,MC and Busson,D. A putative serine/threonine protein kinase encoded by the segment polarity *fused* gene of *Drosophila. Nature* 1990; 347: 87-89.

57. Siegfried, E, Ambrosio, LA and Perrimon, N. Serine/threonine protein kinases in *Drosophila. TIG* 1990; 6:357-362

58. Riggleman, R, Schedl, P and Wieschaus, E. Spatial expression of the *Drosophila* segment polarity gene *armadillo* is post-transcriptionally regulated by *wingless. Cell* 1990; 63: 549-560.

59. Riggleman, R, Wieschaus, E and Schedl, P. Molecular analysis of the *armadillo* locus: Uniformly distributed transcripts and a protein with novel internal repeats are associated with a *Drosophila* segment polarity gene. *Genes and Dev.* 1989; 3: 96-113.

60. Peifer, M and Wieschaus, E. The segment polarity gene *armadillo* encodes a functionally modular protein that is the *Drosophila* homolog of human plakoglobin. *Cell* 1990; 63: 1167-1178.

61. Nakano, Y, Guerrero, I, Hidalgo A, Taylor, A,Whittle, JRS and Ingham, PW. A protein with several possible membrane spanning domains encoded by the *Drosophila* segment polarity gene *patched. Nature* 1989; 341: 508-513.

62. Vinson, CR, Conover, S and Adler, P. A *Drosophila* tissue polarity locus encodes a protein containing seven potential transmembrane domains. *Nature* 1989; 338: 263-264.

63. Gubb, D and Garcia-Bellido, A. A genetic analysis of the determination of cuticular polarity during development in *Drosophila melanogaster*. *JEEM* 1982;

68: 37-57.

64. Vinson, CR and Adler, P. Directional non-cell autonomy and the transmission of polarity information by the *frizzled* gene of *Drosophila. Nature* 1987; 329: 549-551.

65. Baker, NE. Molecular cloning of sequences from *wingless*, a segment polarity gene in *Drosophila*: The spatial distribution of a transcript in embryos. *EMBO J.* 1987; 6: 1765-1773.

66. Cabrera, CV, Alonso, MC, Johnston, P, Phillips, RG and Lawrence, P. Phenocopies induced with antisense RNA identify the *wingless* gene. *Cell* 1987; 50: 658-663.

67. Rijsewijk, F, Schuermann, M, Wagenaar, E, Parren, P, Weigel, D, and Nusse, R. The *Drosophila* homolog of the mouse mammary oncogene *int-1* is identical to the segment polarity gene *wingless. Cell* 1987; 50: 649-657.

68. Simpson, P. Lateral inhibition and the development of the sensory bristles of the adult peripheral nervous system of *Drosophila. Development* 1990; 109: 509-521.

69. Ingham, P. Genetic control of segmental patterning in the *Drosophila* embryo. in Genetics of Pattern Formation and Growth Control. New York: Wiley-Liss, 1990.

70. Morata; G, and Lawrence, PA. The development of *wingless*, a homeotic mutation of *Drosophila. Dev. Biol.* 1977; 56: 227-240.

71. Baker, NE. Embryonic and imaginal requirements for *wingless*, a segment polarity gene in *Drosophila. Dev. Biol.* 1988; 125: 96-108.

72. Babu, P and Bhat, SG. Autonomy of the *wingless* mutation in *Drosophila melanogaster.   Mol. Gen. Genet.* 1986; 205: 483-486.

73. van den Heuvel, M, Klingensmith, J, Perrimon, N and Nusse, R. Cell patterning in the *Drosophila* embryo: distribution of *wingless* and *engrailed* proteins in segment polarity mutants. (submitted).

74. Klingensmith, J, Noll, E, and Perrimon, N. The segment polarity phenotype of *Drosophila* involves differential tendencies toward transformation and cell death. *Dev. Biol.* 1989; 134: 130-145.

75. Klingensmith, J, Zachary, K, Noll, E and Perrimon, N. Cellular basis of a segment polarity phenotype in *Drosophila* embryos. (in preparation).

76. van den Heuvel, M and Nusse, R. (personal communication).

77. Tomlinson, A and Ready, D. *Sevenless*: a cell specific homeotic mutation of the *Drosophila* eye. *Science* 1986; 231: 400-402.

78. Hafen, E, Basler, K, Edstroem, JE and Rubin, G. *Sevenless*, a cell-specific homeotic gene of *Drosophila*, encodes a putative transmembrane receptor with a tyrosine kinase domain. *Science* 1987; 236: 55-63.

79. Reinke, R and Zipursky, SL. Cell-cell interaction in the *Drosophila* retina: the *bride of sevenless* gene is required in photoreceptor cell R8 for R7 cell development. *Cell* 1988; 55: 321-330.

80. Hart, AC, Kramer, H, Van Vactor, DL, Paidhungat, M and Zipursky, SL. Induction of cell fate in the *Drosophila* retina: the *bride of sevenless* protein is predicted to contain a large extracellular domain and seven transmembrane segments. *Genes and Dev.* 1990; 4: 1835-1847.

81. Perrimon, N, Noll, E, McCall, K and Brand, A. Generating lineage specific markers to study *Drosophila* development. *Dev. Genet.* (in press).

82. Klingensmith, J, van den Heuvel, M, Zachary, K, Nusse, R and Perrimon, N. *porcupine* and *dishevelled* function in the signalling pathway of *wingless*, the *Drosophila Wnt-1* homolog. (submitted).

83. Cohen, SM. Specification of limb development in the *Drosophila* embryo

by positional cues from segmentation genes. *Nature* 1990; 343: 173-177

84. Immergluck, K, Lawrence, P and Bienz, M. Induction across germ layers in *Drosophila* mediated by a genetic cascade. *Cell* 1990; 62: 261-268.

85. Gergen, P and Wieschaus, EH. Localized requirements for gene activity in segmentation of *Drosophila* embryos: Analysis of *armadillo, fused, giant* and *unpaired* mutations in mosaic embryos. *Roux's Arch. Dev. Biol.* 1986; 195: 49-62.

86. Wieschaus, E and Riggleman, R. Autonomous requirements for the segment polarity gene *armadillo* during *Drosophila* embryogenesis. *Cell* 1987; 49: 177-184.

87. Wilkinson, DG, Bailes, JA and McMahon, A P. Expression of the proto-oncogene *int-1* is restricted to specific neural cells in the developing mouse embryo. *Cell* 1987; 50: 79-88.

88. Shakleford, GM and Varmus, HE. Expression of the proto-oncogene *int-1* is restricted to post-meiotic male germ cells and the neural tube of mid-gestational embryos. *Cell* 1987; 50: 89-95.

89. Papkoff, J and Shryver, B. Secreted *int-1* protein is associated with the cell surface. *Molec. Cell. Biol.* 1990; 10: 2723-2730.

90. Bradley, RS and Brown, AMC. The proto-oncogene *int-1* encodes a secreted protein associated with the extracellular matrix. *EMBO J.* 1990; 9: 1569-1575.

91. Gavin, BJ, McMahon, JA and McMahon, AP. Expression of multiple novel *Wnt-1/int-1*-related genes during fetal and adult mouse development. *Genes and Dev.* 1990; 4:2319-2332.

92. McMahon, AP and Moon, RT. *int-1* - a proto-oncogene involved in cell signalling. *Development* 1989; 107: 161-167.

93. Deutsch, U, Dressler, GR and Gruss, P. *Pax 1*, a member of a paired box homologous murine gene family, is expressed in segmented structures during development. *Cell* 1988; 53: 617-625.

94. Davis, CA and Joyner, AL. Expression patterns of the homeo-box containing genes *En-1* and *En-2* and the proto-oncogene *int-1* diverge during mouse development. *Genes and Dev.* 1988; 2: 1736-1744.

95. Davis, CA, Noble-Topham, SE, Rossant, J and Joyner, AL. Expression of the homeo-box containing gene *En-2* delineates a specific region of the developing mouse brain. *Genes and Dev.* 1988; 2: 361-371.

Advances in Regulation of Cell Growth, Volume 2;
Cell Activation: Genetic Approaches, **edited by**
James J. Mond, John C. Cambier, and Arthur Weiss.
Raven Press, Ltd., New York.

13

Signal Transduction Through Foreign Growth Factor Receptors

Jacalyn H. Pierce

Laboratory of Cellular and Molecular Biology, National Cancer Institute,
Bethesda, Maryland 20892

The interaction of growth factors with specific membrane receptors triggers a series of intracellular events that are of critical importance in the regulation of normal cell proliferation. An array of cellular responses occur following the binding of growth factors to their specific membrane receptors. Mechanisms of intracellular signal transduction have recently been under considerable investigation since there has been accumulating evidence that imbalances in such second messenger systems may be involved in aberrant cell growth. Major events thought to be involved in signal transduction include activation of protein kinase C and Ca^{+2} influx through phospholipid metabolism, induction of the adenylate cyclase/cyclic AMP pathway, tyrosine and serine/threonine phosphorylation of cellular substrates by activated kinases, immediate cytoplasmic alkalization through Na^+/H^+ exchange, and induction of a family of immediate early response genes (for review see 1-3).

Although the cascade of events which ultimately lead to mitogenesis or differentiation in specific cell types is still not well understood, the introduction of foreign growth factor receptors into naive cells has provided an important tool for dissecting the mechanisms through which growth factors couple with intracellular signalling pathways. There are several advantages which are derived from utilizing artificial introduction of growth factor receptors into naive recipient cells. First, the level of receptor expression can be regulated by the choice of specific promoters in the expression vectors. In particular, overexpression of receptors can easily be achieved and allows for enhanced sensitivity when measuring specific cellular responses to receptor activation. Second, comparisons between different receptor-mediated networks can be made in an identical cell background. Third, the effects of mutations on cloned receptor cDNAs can be evaluated in host cells without the interference of normal endogenous receptor expression. Finally, the ability of an introduced receptor to reconstitute a signalling pathway in a particular cell type can provide insights into the complexity of receptor-mediated activation. For example, one can evaluate whether a single receptor gene product is sufficient for signalling or if multi-unit complex is required.

The molecular cloning of growth factor receptors has been a field of extraordinary progress. Through the recent cDNA cloning of receptors for hematopoietic cytokines, connective tissue growth factors and other ligands, it has become clear that these cell surface glycoproteins can be grouped into a number of families and subfamilies based on structural and/or sequence homologies. Major receptor families which have emerged include the receptors with intrinsic tyrosine kinase activity, the hemotopoietin receptor superfamily, G protein-coupled hormonal receptors which contain multiple transmembrane domains and a recently defined group which includes the nerve growth factor and tumor necrosis factor receptors. This chapter will concentrate on the tyrosine kinase-containing receptors since the majority of studies involving introduction of foreign receptors into recipient cells have been performed utilizing genes from this receptor family.

Tyrosine Kinase-Containing Receptors

Growth factor receptors that possess tyrosine kinase (TK) domains have many similar characteristics (for review, see 1-3). They all possess large glycosylated extracellular ligand binding domains, a hydrophobic transmembrane region, and a ligand regulated intracellular TK domain. The TK domains of these receptors are the most highly conserved regions. In addition to other conserved portions of the TK domain, a consensus sequence of Gly X Gly XX Gly X (15-20) Lys is present which comprises part of the ATP binding site. It appears that many TK-containing receptors undergo receptor oligomerization after ligand binding (1). Receptor clustering may induce conformational change that are translated across the membrane and are involved in TK activation. Ligand binding triggers intracellular receptor TK catalytic activity, which results in receptor autophosphorylation and tyrosine phosphorylation of many cellular substrates. Several cellular substrates have been identified, including GTPase-activating protein, phospholipase C, phosphatidylinositol-3 (PI-3) kinase, the serine/threonine kinase *raf*-1, and members of the *src* family.

Tyrosine kinase-containing receptors can be divided into four subfamilies based on sequence and structural homologies. One subfamily consists of the epidermal growth factor receptor (EGFR), *erb*B-2, and *erb*B-3. These receptors are monomeric and have two cysteine-rich repeat regions in their extracellular ligand-binding domain. A second subclass includes the insulin receptor and the insulin-like growth factor receptors, I and II. These receptors are comprised of disulfide linked tetrameric $\alpha_2\beta_2$ subunits. The β subunit contains a small extracellular region, a transmembrane domain and an intracellular TK catalytic unit. The α subunits are extracellular proteinsthat are disulfide linked to the small extracellular portions of the β subunits. They contain the binding domain and have one cysteine-rich cluster. The third subfamily includes the α and β platelet-derived growth factor receptors (PDGFR), the colony stimulating factor-1 receptor (CSF-1R) and the c-*kit* gene. These receptors are monomeric and contain five immunoglobulin-like repeats in their extracellular binding domains. Unlike the first two subclasses, their intracellular TK domain is interrupted by hydrophillic sequences designated as kinase inserts. The final subgroup includes the fibroblast growth factor receptors (FGFR), *bek* and *flg*. They are similar in structure to the PDGFR subfamily except they have three immunoglobulin-like repeats instead of five in their extracellular domain. A schematic representation depicting the structural characteristics of the TK-containing receptor families is shown in Fig. 1.

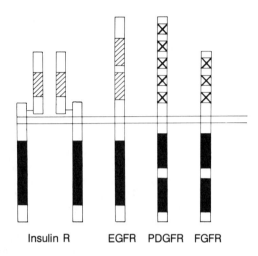

Insulin R   EGFR   PDGFR FGFR

FIG. 1. Structural Features of the Four Subclasses of Tyrosine Kinase-Containing Receptors. The hatched bars represent cysteine-rich clusters; the solid bars represent tyrosine kinase domains, and the cross-hatched bars represent immunoglobulin-like repeats.

The Insulin Receptor

Insulin is a unique growth hormone that is required for optimal proliferation of all cell types. The other structurally related factors, insulin-like growth factor (IGF) I and II, are able to replace the functions of insulin for cell growth. Virtually all cell types possess either insulin or IGF receptors. The cloning of the human insulin receptor cDNA provided the first opportunity to determine the role of a TK domain in growth factor-mediated signal transmission (4).

In two separate studies, the ATP-binding site (lysine 1018) within the β-subunit TK domain was replaced with other amino acids resulting in a kinase defective β-subunit (5-8). When transfected into either Chinese hamster ovary (CHO) or RAT-1 cells, the mutant receptors failed to mediate multiple insulin-induced cellular responses. Since CHO and RAT-1 cells express low levels of endogenous wild-type receptors, comparisons were made after stable transfection with either wild type or the ATP-binding mutants. In both studies, expression of the wild type receptor enhanced the stimulatory effects of insulin on deoxyglucose uptake, glycogen synthesis, S6 kinase activity, substrate tyrosine phosphorylation and thymidine incorporation. By contrast, the mutant receptors had no such enhancing effects despite their capacity to bind insulin. In addition, the mutant receptors failed to undergo tyrosine autophosphorylation. Although the mutant receptors were properly processed and able to bind insulin, they also failed to undergo receptor down modulation and endocytosis. Thus, the ability of insulin to transmit intracellular signals is dependent on activation of an intact tyrosine kinase domain.

Mutation of insulin receptor β-subunit at tyrosine residues 1146, 1150 and 1151 inhibited phophostransferase activity and signal transduction, suggesting that receptor autophosphorylation may play a critical role in receptor activation (9). In a separate study, mutation of Tyr-960 to phenylalanine had no effect on insulin-mediated autophosphorylation or phosphotransferase activity (10). However, this mutant was biologically inactive and did not tyrosine phosphorylate a previously identified cellular substrate, pp185. Taken together, the above results suggest that autophosphorylation of the insulin receptor may be required but is not sufficient for signal transduction, while tyrosine phosphorylation of certain cellular substrates, such as pp185, may be critical for biological activity.

The EGF Receptor

EGF is a potent mitogenic polypeptide that binds to a specific receptor expressed on the surface of various epithelial, epidermal, and fibroblastic cell types. At least two other ligands for the EGFR have been identified, transforming growth factor α (TGFα) and amphiregulin (11,12). The receptor for EGF is encoded by the proto-oncogene c-erbB (13). It has been shown that c-erbB was converted to an oncogene upon amino-terminal truncation together with other modifications in two independently isolated acute transforming retroviruses (14,15). The v-erbB protein possesses constitutive ligand-independent tyrosine kinase activity. Several studies have demonstrated that mutations in the EGFR cDNA which inactivate the TK catolytic domain cannot transduce a mitogenic response in transfected cells. Replacement of the lysine residue of the ATP binding site completely abolished the kinase activity and signalling capacity of the receptor (16,7). Additional mutations within the tyrosine kinase domain at Lys 721 and Thr 654 also destroyed receptor signalling activities (17,18). Although these mutants did not display any tyrosine kinase activity, the defective receptors were expressed on the cell surface and could bind EGF. These results confirmed those observed with the insulin receptor that kinase activity is indispensable for all intracellular responses to receptor activation.

The epidermal growth factor receptor (EGFR) gene is frequently amplified and/or overexpressed in human malignancies (19-22). To investigate the biological effects of its overexpression, our laboratory constructed a eukaryotic vector containing human c-erbB cDNA under the control of retroviral long terminal repeats (LTR) (23). Introduction of this construct led to reconstitution of functional EGFR in NR6 fibroblasts, which are normally devoid of this receptor. EGFR expression in NR6 cells resulted in no significant alterations in growth properties. However, EGF addition led to the formation of densely growing transformed foci in liquid culture and colonies in semi-solid medium. We also obtained NIH/3T3 transfectants which expressed the EGFR at levels 500- to 1000-fold over control NIH/3T3 cells. The NIH-EGFR line also demonstrated a marked increase in DNA synthesis and growth in soft agar in response to EGF (Table 1). Thus, EGFR overexpression appeared to amplify normal EGF signal transduction. Finally, high levels of the 170 kilodalton (kD) EGFR protein, which conferred a transformed phenotype to NIH/3T3 cells in the presence of ligand, were demonstrated in the representative human tumor cell lines that contained amplified copies of the EGFR gene, MDAMB468 and A431 (Fig. 2). Similar results were obtained in two independent studies (24,25).

Table 1. Growth Properties of EGF Receptor-Transfected NIH/3T3 Cells

Transfected DNA	Transforming activity (ffu/pM)[a]		Growth in soft agar (%)[b]	
	-EGF	+EGF	-EGF	+EGF
LTR/EGFR	$<10^0$	2×10^2	0.4	19.7
LTR/v-erbB	3×10^2	3×10^2	12.7	15.4
LTR/erbB-2	4×10^4	4×10^4	71.7	73.2
pSV$_2$/gpt	$<10^0$	$<10^0$	<0.01	<0.01

[a]EGF at 20 ng/ml, 18 days after the transfection, was added to the indicated plates and focus formation (ffu) was scored 8 days later on duplicate plates. The specific activity is adjusted to ffu per picomole (pM) of cloned DNA added, based on the relative molecular weights of the respective plasmids.

[b]Visible colonies were scored at 14 days. The results represent the mean values of three independent experiments (23).

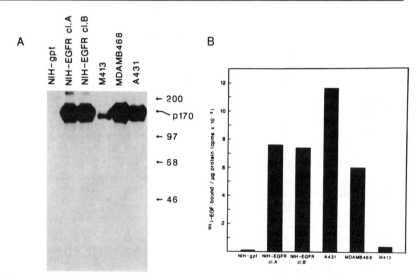

FIG. 2. Comparison of the Levels of EGF Receptor in LTR/EGFR-Transfected Clonal Populations of NIH/3T3 Cells and Human Tumor Cell Lines. (A) Immunoprecipitation and Electrophoretic Analysis. (B) ^{125}I-EGF Saturation Binding. Results are expressed as cpm of ^{125}I-EGF specifically bound after subtraction of nonspecific binding (23).

The gene for TGFα is not transforming or only weakly transforming when expressed in responsive cells displaying moderate number of EGFR, such as NIH/3T3 or RAT-1 (26,27). If transduction of the mitogenic signal must reach a critical threshold to induce the transformed state, this might occur only when there is abundance of both ligand and receptors. In this regard, our laboratory demonstrated that TGFα acted as a potent oncogene when transfected into NIH/3T3 cells overexpressing EGFR (28) (Table 2). Analysis of human tumor cell lines revealed a strong correlation between expression of TGFα and overexpression of EGFR (Fig. 3). Moreover, high levels of EGF-independent tyrosine phosphorylation of the EGFR were detected both in NIH-EGFR expressing TGFα and in high EGFR and TGFα coexpressing human tumor cell lines. Thus, two events which created the EGFR/TGFα autocrine loop resulting in transformation *in vitro* may play a role in the development of some human malignancies.

Table 2. Transformed Phenotype of TGFα Virus-Infected NIH/3T3 Cell Lines Expressing Different Levels of EGFR

Cell line	EGF addition (ng/ml)	Growth in soft agar[1], (%)	Tumorigenicity[2] (cell required for 50% tumor incidence)
NIH/3T3	0	<0.01	>10^6
	20	<0.01	
NIH-TGFα	0	0.1	>10^6
	20	0.2	
NIH-EGFR	0	1	>10^6
	1	1	
	10	13	
	20	50	
	50	66	
NIH-EGFR-TGFα	0	52	10^4
	20	50	

[1]Cells were plated at 10-fold serial dilutions in 0.33% soft agar medium containing 10% calf serum. Visible colonies comprising >100 cells were scored at 14 days.

[2]NFR nude mice were inoculated subcutaneously with each cell line. At least 10 mice were tested for each concentration, at cell concentrations ranging from 10^3 to 10^6 cells/mouse. Tumor formation was monitored for four months (28).

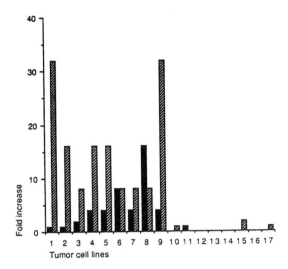

FIG. 3. Coexpression of High Levels of TGFα and EGFR mRNAs in Human Tumor Lines. Quantitation of EGFR (▨) or TGFα (■) mRNA in cell lines (1) A431; (2) MDAMB4; (3) BT20; (4) A2047; (5) A388; (6) A1698; (7) A2781; (8) A704; (9) A253; (10) M413; (11) MDA MB125; (12) DV4475; (13) HBL100; (14) BT474; (15) A172; (16) ZR7530; (17) A1207 (28).

The transforming capacity of recombinant human EGFR constructs was also analyzed in primary chicken fibroblasts and erythroblasts (29). In chicken embryo fibroblasts, overexpression of the normal human EGFR led to a ligand-dependent transformed phenotype. Amino- and/or carboxyl-terminal truncations of EGFR resulted in weak ligand-independent fibroblast and erythroblast transformation. Interestingly, overexpression of the normal EGFR led to a ligand-dependent induction of erythroblast self-renewal. These results suggest that activation of the EGFR can result in phosphorylation of critical cellular substrates in erythroblast target cells, even though these cells do not normally utilize this pathway.

The murine 32D cell line is strictly dependent on interleukin 3 (IL-3) for growth, possesses a normal diploid karyotype, and is nontumorigenic in nude mice (30). Although these cells maintain an immature myeloid phenotype when propagated in IL-3, they can be programmed to differentiate to mature neutrophils when exposed to granulocyte-CSF (G-CSF) (31). Since IL-3-dependent 32D cells are devoid of EGFR, our laboratory investigated the effects of introducing either c-*erb*B or v-*erb*B expression vectors into 32D cells by electroporation (32). Expression of the EGFR in 32D cells conferred the ability to utilize EGF for transduction of a mitogenic signal (Table 3).

Table 3. Mitogenic Response and Cloning Efficiency of 32D Transfectants

Cell line	Treatment	SI of [^{3}H]TdR incorp.[1]	Cloning efficiency(%)[2]
32D	IL-3 + serum	234.5	35.5
	EGF + serum	1.0	<0.1
	Serum	1.0	<0.1
32D-v-*erb*B	IL-3 + serum	225.5	28.0
	EGF + serum	231.0	30.5
	Serum	228.0	28.5
32D-EGFR	IL-3 + serum	215.5	27.5
	EGF + serum	185.5	25.0
	Serum	1.0	<0.1

[1]The cell proliferation assay was performed on cells cultured in Iscove's modified Dulbecco's medium supplemented with 15% fetal calf serum (FCS). IL-3 (50 u/ml) or EGF (100 ng/ml) was added when specified.

[2]Results are expressed as a stimulation index (SI): cpm [^{3}H]thymidine (TdR) incorporation with treatment/cpm [^{3}H]TdR incorporation of 32D with serum. Data are the mean of duplicate samples.

[3]The cloning efficiency was established by plating cells at various concentrations in 5 ml of growth medium and 0.48% sea plaque agarose. IL-3 (50 U/ml) or EGF (100 ng/ml) was included when specified. Visible colonies were scored at 12 days after plating (32).

Exposure to EGF led to a rapid stimulation of phosphoinositol (PI) metabolism, while IL-3 had no detectable effect on PI turnover either in control or EGFR-transfected 32D cells (Fig. 4). When the transfected cells were propagated in EGF, they exhibited a more mature myeloid phenotype than was observed under conditions of IL-3-directed growth. Interestingly, in an independent study, when the human EGFR was introduced into the murine IL-3-dependent pre-mast cell line, IC2, they also differentiated into more mature mast cells in the presence of EGF, as demonstrated by increased granularity and histamine content (33).

Although the 32D transfected cells exhibited high levels of functional EGFRs, they remained nontumorigenic. In contrast, transfection of v-*erb*B not only abrogated the IL-3 growth factor requirement of 32D cells but caused them to become tumorigenic in nude mice. These results revealed that naive hematopoietic cells expressed all of the intracellular components of the EGF-signalling pathway necessary to evoke a mitogenic response and sustain continuous proliferation. Expression of the oncogenic counterpart of the EGF receptor, v-*erb*B, led to IL-3 independence, possibly by constitutively deregulating the EGF network found to be present in these cells.

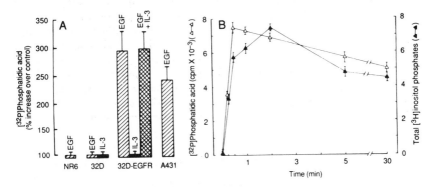

FIG. 4. (A) Effects of EGF on Phosphatidic Acid Formation in Different Cell Lines. Cells were incubated with EGF (100 ng/ml) or IL-3 (500 U/ml) where specified. Results are expressed as percent increases in comparison with unstimulated values. (B) Time Course of EGF-Induced Phosphatidic Acid and Inositol Phosphate Formation in 32D-EGFR Cells. The formation of phosphatic acid induced by EGF is expressed as radioactivity (cpm) after subtraction of basal values. The accumulation of ^{3}H inositol phosphate induced by EGF is expressed as the radioactivity in "total inositol phosphates" fraction per 10^5 cpm in total inositol phospholipids after subtraction of basal values (32).

In a separate study, the human EGFR was introduced into either an IL-3-dependent line, FDCP1, or primary mouse bone marrow cells cultivated in IL-3 (34). Like the 32D cells, EGF was able to maintain proliferation of the FDCP1 cell line expressing EGFR. In the primary myeloid cells expressing EGFR, EGF acted in a synergistic manner with IL-3 to stimulate DNA synthesis and cell proliferation. However, EGF alone was not sufficient to replace the requirement for IL-3 in the primary cultures. It is possible that the levels of EGFR present on the bone marrow cells were not high enough to transduce an independent mitogenic signal. Alternatively, more stringent control over cell proliferation may be present in primary cultures as compared to established lines.

The *ErbB-2* Receptor

*Erb*B-2 was first identified by its amplification in human mammary (35) and salivary (36) carcinomas using probes of the closely related tyrosine kinase domain of the epidermal growth factor receptor. Subsequently, human *erb*B-2 was shown to be amplified in many adenocarcinomas and overexpressed in approximately 30% of primary breast carcinomas (37-39). A putative ligand for gp185^{erb}B-2 has recently been identified, but has not yet been molecularly cloned (40). Several studies have been undertaken to investigate the effects of *erb*B-2 expression in model systems. In our laboratory, *erb*B-2 eukaryotic expression vectors were engineered under the control of either retroviral long terminal repeats or the SV40 early promoter in an attempt to express the *erb*B-2 cDNA at different levels in NIH/3T3 cells (41). The LTR/*erb*B-2 expression vector induced transformed foci at high efficiency, while the SV40/*erb*B-2 construct failed to induce any detectable morphological alteration of NIH/3T3 cells (Table 4). Immunological analysis revealed that SV40/*erb*B-2 transfectants expressed gp185^{erbB-2} at 10-fold higher levels than control NIH/3T3 cells. A further 10-fold increase in gp185^{erbB-2} expression was detected in LTR/*erb*B-2 transfectants. These results demonstrated that the high levels of *erb*B-2 expression under LTR influence correlated with its ability to exert transforming activity.

Table 4. Transfection Analysis of *erb*B-2 Expression Vectors

DNA clone	Transformed foci/plate with DNA added (mg)[1]				Specific transforming activity[2] (ffu/pM)
	1	0.1	0.01	0.001	
LTR-1/*erb*B-2	TMC	TMC	>100	7,3	4.1×10^4
LTR-2/*erb*B-2	TMC	TMC	30,21	4,1	2.0×10^4
SV40/*erb*B-2	0,0	0,0	0,0	0,0	$<10^0$
LTR/*ras*	TMC	TMC	45,50	6,3	3.6×10^4
pSV2/*gpt*	0,0	0,0	NT	NT	$<10^0$

[1]Focus formation on NIH/3T3 cells was scored at 14 to 21 days on duplicate plates. DNA added is shown as micrograms per plate.
[2]Focus-forming units were adjusted to ffu per pM of cloned DNA added, based on the relative molecular weights of the respective plasmids.
NT - not tested.
TMC - too many to count (41).

The level of overexpression of gp185^{erbB-2} in human mammary tumor cell lines possessing amplified *erb*B-2 genes was compared with that of NIH/3T3 cells transformed by the *erb*B-2 coding sequence. An anti-*erb*B-2 peptide serum detected *erb*B-2-specific 185 kD proteins in extracts of the MDA-MB361 and SK-BR-3 mammary tumor cell lines, as well as in LTR/*erb*B-2 NIH/3T3 transformants. The relative levels of gp185^{erbB-2} product were similar in each of the cell lines and markedly elevated over that expressed by MCF-7 cells, where the 185 kD *erb*B-2 protein was not detectable under these assay conditions. Thus, human mammary tumor cells which overexpressed the *erb*B-2 gene demonstrated levels of the *erb*B-2 gene product capable of inducing malignant transformation in a model system. These findings established a mechanistic basis for normal *erb*B-2 gene amplification as a causal driving force in the clonal evolution of a tumor cell rather than being an incidental consequence of tumorigenesis.

While the transforming potential of the overexpressed normal human gp185^{erbB-2} has been independently confirmed (42), contradictory results have been reported for its normal rat homolog, *neu*. A number of studies have found no detectable transformation of murine fibroblasts resulting from overexpression of *neu* (43,44). However, there was no attempt to estimate the levels of overexpression of gp185^{neu} obtained in the various model systems, as compared to the levels of gp185^{erbB-2} needed to induce transformation *in vitro*. To address this question, we systematically compared the effects of different levels of expression of *erb*B-2 and *neu* in NIH/3T3 cells (45). Our results showed that the *neu* gene also acted as a potent oncogene when expressed in NIH/3T3 cells at levels similar to those required for *erb*B-2-induced transformation.

A number of genetic alterations have been shown to enhance transforming potential to *erb*B-2/*neu*. In fact, the transforming *neu* gene initially isolated from rat neuroblastomas was shown to contain an activating point mutation within its transmembrane domain (Val659 to Glu) (43). Subsequently, several point mutations were genetically engineered for this same amino acid (44). Only certain mutations were able to induce potent transformation. Similar mutations engineered in the human normal *erb*B-2 cDNA also activated the oncogenic potential of *erb*B-2 to a similar extent (46) (Table 5). Another alteration capable of activating *erb*B-2/*neu* was amino-terminal truncation. Expression vectors encoding a protein, gp88-96$^{\Delta NerbB-2}$, in which the extracellular ligand-binding domain was deleted, showed increased transforming efficiency at both low and high levels of expression (Table 5). Similar results have been obtained in the rat system (47). Activation of *erb*B-2/*neu* by amino-terminal truncation resembles activation of EGFR by a similar alteration in v-*erb*, suggesting that the extracellular domains of both these proteins exert a negative regulatory influence that can be abolished by deletion or modulated in response to ligand binding.

Based on its predicted primary structure, the *erb*B-2/*neu* product was thought to be endowed with intrinsic TK activity. Indeed, gp185^{erbB-2} and gp185^{neu} undergo *in vitro* autophosphorylation on tyrosine residues and phosphorylate exogenous substrates on tyrosine residues. Unlike other TK-containing receptors, the addition of exogenous ligand was not required to induce tyrosine kinase activity of the *erb*B-2 protein. Whether all cell types synthesize the *erb*B-2 ligand remainds to be determined. The relationship between the intrinsic kinase activity of gp185^{erbB-2} and its transforming ability has been addressed by analysis of tyrosine kinase activities of mutant proteins encoded by genetically altered *erb*B-2 gene constructs in comparison to the normal gp185^{erbB-2} (46) (Table 6). Immunoblot analysis of steady state levels of tyrosine phosphorylation revealed increased phosphotyrosine content per unit protein for such mutants as opposed to the wild type *erb*B-2 molecule. By *in vitro* kinase assay, we found that each of the mutant proteins also displayed both increased autokinase and kinase activity toward exogenous substrates with respect to the wild type protein. Thus, the increased transforming potential of mutant *erb*B-2 proteins correlates with increased *in vivo* and *in vitro* kinase activity.

Table 5. Comparison of Transforming Activities of Different *erb*B-2 Mutants

*erb*B-2 cDNA clone	Transforming efficiency[a] with:	
	SV40 vector	LTR vector
*erb*B-2	$<5 \times 10^0$	2×10^3
*erb*B-2ΔN	1.5×10^2	4×10^4
*erb*B-2 Glu	2×10^2	3×10^4
*erb*B-2 Asp	2×10^2	3.5×10^4

[a]Expressed as ffu per pM.

Table 6. Comparison of Tyrosine Kinase activities of *erb*B-2 Mutant Proteins and Normal *erb*B-2 gp185

Mutant protein	Fold Increase over normal *erb*B-2 gp185 activity[a]			
	Autokinase		Polyglutamic acid-tyrosine kinase	
	2 min	10 min	5 min	15 min
*erb*B-2ΔN gp86-98	7.2	5.3	3.2	3.5
*erb*B-2 Glu gp185	5.6	4.1	5.5	2.6
*erb*B-2 Asp gp185	4.1	3.7	3.1	2.5

[a]For each reaction, equal amounts of *erb*B-2 protein were immunoprecipitated from cell lysates transfected with different LTR vectors. Reactions were run on 10% sodium dodecyl sulfate-polyacrylamide gel electrophoresis, and specific radioactivity was counted after excision of the specific bands from the gel (46).

Clues to the specificity of the *erb*B-2-activated signalling pathway have come from comparisons of its transforming potential to that of the EGFR (48). In NIH/3T3 cells, overexpression of as many as 1-2 x 10^6 surface EGF receptors per cell led to transformation only in the presence of EGF. Yet, at similar levels of overexpression, the *erb*B-2 gene exhibited around 100-fold higher transforming activity, which could be further augmented if the *erb*B-2 kinase was maximally upregulated by structural alterations as described above (47) (Fig. 5). Therefore, despite its close structural similarities to the EGFR, the *erb*B-2/*neu* kinase exerts a much stronger mitogenic effect in NIH/3T3 cells. As previously described, the EGFR can efficiently couple with mitogenic signalling pathways when it is transfected into IL-3-dependent 32D hematopoietic cells (32). When expression vectors for *erb*B-2 or its truncated counterpart, ΔN*erb*B-2, were introduced into 32D cells, neither was capable of inducing autonomous proliferation (48) (Fig. 5). This was despite overexpression and constitutive TK activity of their products at levels associated with potent transformation of fibroblast target cells. Thus, EGFR and *erb*B-2 couple with distinct mitogenic signalling pathways. The region responsible for the specificity of intracellular signal transduction was localized to a 270-amino acid stretch encompassing their respective TK domains (48) (Fig. 6). Thus, tissue- or cell-specific regulation of growth factor receptor signalling can occur at a point after the initial interaction of growth factor with receptor. Such specificity in signal transduction may account for the selection of certain oncogenes in some malignancies.

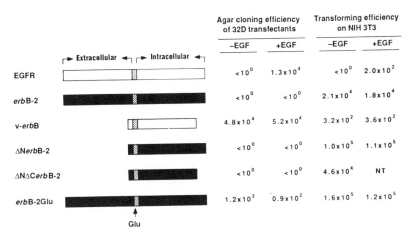

	Agar cloning efficiency of 32D transfectants		Transforming efficiency on NIH 3T3	
	–EGF	+EGF	–EGF	+EGF
EGFR	$<10^0$	1.3×10^4	$<10^0$	2.0×10^2
*erb*B-2	$<10^0$	$<10^0$	2.1×10^4	1.8×10^4
v-*erb*B	4.8×10^4	5.2×10^4	3.2×10^2	3.6×10^2
ΔN*erb*B-2	$<10^0$	$<10^0$	1.0×10^5	1.1×10^5
ΔNΔC*erb*B-2	$<10^0$	$<10^0$	4.6×10^4	NT
*erb*B-2Glu	1.2×10^2	0.9×10^2	1.6×10^5	1.2×10^5

FIG. 5. Comparison of Biological Activity of EGFR and *erb*B-2 Expressed in 32D and NIH/3T3 Cells. The values of agar cloning efficiency are presented as a fraction of the number of colonies that developed in medium with serum compared to medium with serum and IL-3 (colonies/IL-3 colonies x 10^5). The transformation efficiency on NIH/3T3 cells was calculated in ffu/pM DNA. White bars are EGFR sequences; black bars are *erb*B-2 sequences; dotted bars are v-*erb*B sequences; and Glu, glutamic mutation in the transmembrane hatched region of *erb*B-2 Glu (48).

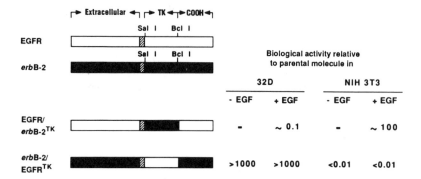

FIG. 6. Biological Activity of Tyrosine Kinase Chimeric Molecules between EGFR and *erb*B-2. Values are presented as biological activity of the chimeric molecules relative to the parental molecules. White bars are EGFR sequences, black bars are *erb*B-2 sequences, and hatched bars are the transmembrane region (48).

The CSF-1 Receptor

Expression of CSF-1R is restricted to hematopoietic cells of the myeloid lineage and placental trophoblasts (for review, see 49). It has been shown that the CSF-1R is encoded by the c-*fms* proto-oncogene (51). Since NIH/3T3 fibroblasts are devoid of CSF-1R and do not respond to CSF-1, an expression vector containing the human c-*fms* cDNA was introduced into these cells (51). The transfected NIH/3T3 cells were able to form colonies in semi-solid medium containing recombinant human CSF-1. In addition, secondary transfection of these cells with a expression vector encoding human CSF-1 led to a transformed phenotype and tumorigenicity via an autocrine mechanism. These results provided evidence that NIH/3T3 fibroblasts possess the intracellular components required for eliciting a mitogenic response to the hematopoietic cytokine, CSF-1. The c-*fms*-transfected NIH/3T3 cells were not contact inhibited and appeared morphologically transformed when grown under serum-free conditions in the presence of recombinant CSF-1. By contrast, these same cells grown in chemically defined medium containing PDGF and insulin remained contact inhibited and morphologically normal under these conditions. These data suggested that activation of the CSF-1-mediated signalling pathway may lead to deregulated growth in this system. Expression of human CSF-1R in Chinese hamster lung fibroblasts also resulted in a CSF-1-mediated mitogenic signal (52). Moreover, the transfected cells exhibited an increased steady state level in DNA synthesis in the absence of ligand that was inhibited by a monoclonal antibody whch blocks CSF-1 signalling. These results also implied that overexpression of normal human CSF-1R may cause a ligand-independent proliferative effect.

Further studies utilizing the NIH/3T3 transfection system revealed that introduction of a point mutation in the extracellular domain (Leu-301 to Ser) of c-*fms* induced a transformed phenotype to the transfected cells (53). A second mutation in the carboxyl-terminal (Tyr-969 to Phe) upregulated this transforming potential. Only certain amino acid substitutions at position 301 activated the transforming potential of CSF-1R, suggesting that activating mutations may induce a conformational change that mimics CSF-1 binding. Treatment of these cells with monoclonal antibody directed against the extracellular domain of the CSF-1R reversed the transformed phenotype, indicating that the activating mutations may enhance receptor dimerization in the absence of ligand. As previously mentioned, the CSF-1R contain an insert in the TK domain. To determine the role of the kinase insert (ki), 58 of 64 amino acids were deleted in murine c-*fms* receptor cDNA by oligonucleotide-directed mutagenesis (54). Transfection of the mutant receptor into NIH/3T3 cells revealed that the ki domain was dispensable for enzymatic and transforming activities in fibroblasts. In another report, a 67 amino acid ki deletion in human c-*fms* resulted in a partially diminished response confirming that this region is not absolutely essential for CSF-1-induced signal transduction (55). Whether this region plays some confirmational role which enhances CSF-1 signalling capacity remains to be determined.

Two studies utilizing IL-3-dependent murine cell lines as recipients for the CSF-1R have provided insights into the differentiation potential of CSF-1 for hematopoietic cells (56,57). The fact that 32D is a clonal myeloid precursor cell line raised the question as to whether signals transduced through the CSF-1R might interact with substrates capable of promoting a sustained differentiation as well as a proliferation signal. Thus, our laboratory inserted expression vectors containing either the normal or point-mutated human c-*fms* genes into 32D cells to provide a model system for determining the effects of CSF-1 on biological signalling pathways in a homogeneous clonal population of myeloid progenitor cells (56). CSF-1 was shown to induce partial monocytic differentiation of the 32D-c-*fms* cells (Fig. 7). Monocytic differentiation was reversible upon removal of CSF-1, implying that CSF-1 was required for maintenance of the monocytic phenotype but was not sufficient to induce an irrevocable commitment to differentiation. Human CSF-1 but not IL-3 was also shown to be a potent chemoattractant for 32D-c-*fms* cells, suggesting that CSF-1 may serve to recruit monocytes from the circulation to tissue sites of inflammation or injury. Although c-*fms* did not release 32D cells from factor dependence, point-mutated c-*fms*[S301,F969] was able to abrogate their IL-3 requirement and induce tumorigenicity. Factor-independent 32D-c-*fms*[S301,F969] cells also displayed a mature monocyte phenotype, implying that differentiation did not interfere with progression of these cells to the malignant state. In a separate study, the murine c-*fms* gene was introduced into IL-3-dependent FDCP1 cells by retroviral infection (57). Stimulation of FDCP1-c-*fms* cells by CSF-1 led to the formation of colonies with altered morphology and expression of mature myeloid surface marker proteins. The morphological and phenotypic alterations were reversible, supporting the concept that CSF-1 does not irreversibly commit myeloid progenitor cells to terminally differentiate. All of these findings demonstrate that a single growth factor receptor can specifically couple with multiple intracellular signalling pathways and play a critical role in modulating cell proliferation, differentiation and migration.

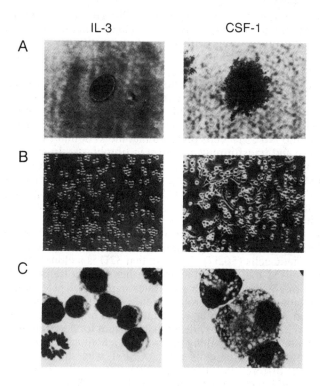

FIG. 7. Comparison of Morphological Features of 32D-c-*fms* Cells in Medium with IL-3 or CSF-1. (A) Semisolid medium (X 24); (B) Liquid medium (X 60); (C) Wright-Giemsa-stained preparation of cells (X 600) (56).

The α and β PDGF Receptors

Two PDGF receptors, designated the α and βPDGFR, have been identified and molecularly cloned (58-60). PDGF is known to induce a variety of functional responses, including DNA synthesis, chemotaxis, membrane ruffling, and PI breakdown in connective tissue cells which generally coordinately express both PDGFR (for review, see 61,62). We have investigated specific functions mediated by the products of two independent PDGFR-encoding genes (63). By use of a strategy involving introduction of expression vectors for α and βPDGFR cDNAs into the naive 32D hematopoietic cell line, we showed that each receptor could independently couple with mitogenic signal transduction pathways inherently present in these cells (Fig. 8). Moreover, both receptors could induce a readily detectable chemotactic response (Table 7). Finally, activation of either receptor rapidly stimulated inositol phospholipid metabolism and mobilization of intracellular Ca^{2+} (Table 8). All of these findings establish that the major biological and biochemical responses seen in cells normally triggered by PDGF can be reconstituted in the IL-3-dependent 32D cell line by expression of either α or βPDGFRs. The ability of these receptors to independently induce these responses argues that each receptor is inherently capable of doing so in the absence of interaction with the other receptor.

Table 7. Comparison of Chemotaxis in 32D-αR and -βR Cells in Response to Different PDGF Isoforms

32D transfectant	PDGF isoform	Chemotaxis (fold increase in cell migration)[a]	
		10 ng/ml	300 ng/ml
αR	AA	2.0	8.0
βR		ND[b]	ND
αR	AB	2.7	8.7
βR		ND	5.1
αR	BB	3.7	8.0
βR		3.2	7.8

[a]Cell migration was assayed by means of a modified Boyden chamber technique using 0.5 micron nucleopore filters. Results are expressed as fold increase in cell migration compared with that seen using medium without PDGF and are mean values of triplicate samples.
[b]ND, no difference compared with levels in medium without PDGF (63).

PDGF consists of AA, AB and BB isoforms, which arise as dimeric products of two independent PDGF-encoding genes. Our present results indicate that a major level of regulation of the spectrum of PDGF functional responses resides in the relative affinities of the three PDGF isoforms for either receptor. PDGF-BB, which exhibited similar high affinity for binding α and β PDGFRs, induced each of the functional responses analyzed with comparable efficiency in 32D-αR and -βR cells. Similarly, human PDGF-AB and PDGF-AA, which bound the α receptor with much better affinity than the β receptor, induced functional responses preferentially in 32D-αR cells. Thus, the availability of specific PDGF isoforms and the relative expression of each receptor gene product appear to be major determinants of the PDGF response.

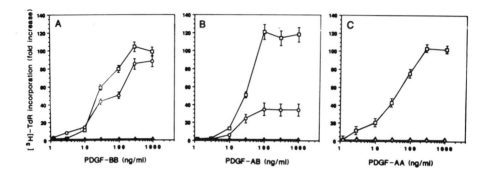

FIG. 8. Mitogenic Response of 32D Transfectants to Different PDGF Isoforms. DNA synthesis by 32D-αR (□), 32D-βR (○) and 32D (△) cells in response to PDGF-BB (A), PDGF-AB (B) or PDGF-AA (C). Results are expressed as fold increase in ^{3}H TdR incorporation over that without ligand (63).

Table 8. Comparison of Inositol Phospholipid Turnover and Calcium Influx in 32D-αR and -βR Cells in Response to Different PDGF Isoforms.

32D transfectant	PDGF isoform	Inositol phospholipid turnover (fold increase)[a]		Calcium mobilization, $(\Delta[Ca^{2+}]i)$[b]
		250 ng/ml	1000 ng/ml	250 ng/ml
αR	AA	NT	4.5	90 nM
βR		ND	ND	0
αR	AB	5.5	5.5	180 nM
βR		ND	4.5	0
αR	BB	6.0	6.0	200 nM
βR		5.0	5.0	190 nM

[a]Results are expressed as fold increase in total inositol phosphates over unstimulated values and represent mean values of triplicate samples.
[b]The fluorescent indicator fura-2 was used to determine cytosolic Ca^{2+} concentration $[Ca^{2+}]i$ in 32D transfectants and was calculated on the 340 nM/380 nM ratio (63).

A characteristic feature of the α and βPDGFR is the presence of an insert sequence in the TK domain. Activation of the βPDGFR induces autophosphorylation at two major tyrosine residues, Tyr-751 and Tyr-857, and is also accompanied by complex formation with and tyrosine phosphorylation of a variety of cellular substrates (64). Known substrates which are found in the PDGFR complex include *raf*-1, phopholipase C, PI-3 kinase, GTPase-activating protein and three *src* family kinases (65-69). In one study, the Tyr-751 residue, which lies in the ki region, was mutated to Phe or Gly (64). When the mutant receptors were expressed in dog kidney epithelial cells, they did not associate with PI-3 kinase or other tyrosine phosphorylated substrates in the complex. These results suggest that autophosphorylation in this region regulates the binding of activated PDGFR to specific cellular proteins. Thus, autophosphorylation in the ki region may play a critical role in signal transduction. In a separate study, an 82-amino acid deletion in the 104-amino acid ki region of βPDGFR resulted in a receptor that did not transduce a mitogenic signal in CHO cells (70). However, PDGF was still capable of inducing receptor autophosphorylation and PI hydrolysis. In anothere study, a receptor mutant which lacked the ki of 97 amino acids from the ki region was expressed in CHO cells (71). This mutant receptor bound PDGF-BB with high affinity but did not transduce a mitogenic signal. However, this mutant displayed decreased autophosphorylation after ligand stimulation and decreased ability to phosphorylate exogenous substrates. Taken together, these results suggested that the ki region of the βPDGFR may be required for transduction of many known signals that are induced by PDGF.

In order to define the role of the ki sequences in the human αPDGFR, our laboratory generated deletion mutants, designated αRDki-1 and αRDki-2, which lacked 80 (710-789) and 95 (695-789) amino acids of the 104-amino acid ki region, respectively (72). Their functional characteristics were compared to those of the wild type αPDGFR following introduction into the naive hematopoietic cell line, 32D. Biochemical responses, including PDGF-stimulated αPDGFR tyrosine phosphorylation, PI turnover, and receptor-associated PI-3 kinase activity, were differentially impaired by the deletions. Despite lack of any detectable receptor-associated PI-3 kinase activity, 32D cells expressing αRDki-1 showed only partially impaired chemotactic and mitogenic responses and were capable of sustained proliferation *in vitro* and *in vivo* under conditions of autocrine stimulation by PDGF-BB. The 32D transfectants expressing the larger ki deletion (αRDki-2) showed markedly decreased or abolished biochemical and biological responses. However, insertion of the highly unrelated smaller c-*fms* (685-750) ki domain into αRDki-2 restored each of these activities to wild type αPDGFR levels. Since PI metabolism is not induced by activation of the CSF-1R, the ability of the c-*fms* ki domain to reconstitute PI turnover in the αRDki-2 transfectant provided evidence that the ki domain of the αPDGFR does not directly couple with this pathway. Taken together, all of these findings imply that their ki domains have evolved to play very similar roles in the known signalling functions of PDGF and CSF-1 receptors.

The FGF Receptors

The FGF growth factor family consists of at least seven related heparin-binding proteins which include acidic FGF, basic FGF, *int*-2, *hst*/KFGF,

KGF, FGF-5, and FGF-6 (for review, see 73). Previous studies had indicated that FGF ligands stimulate TK activity in intact cells. The recent cDNA cloning of two FGF receptors, *flg* and *bek*, confirmed that they were TK-containing receptors. The receptor for basic FGF was purified from chicken embryos by affinity chromatography (74). Amino acid sequences derived from tryptic peptide fragments facilitated the design of oligonucleotide probes used to screen a chick embryo cDNA library. Analysis of this cDNA revealed that it was the avian counterpart of a partial human cDNA for a *fms*-related gene (*flg*) (75). It was then determined that acidic FGF was able to bind to and activate the intrinsic TK activity of *flg*. A second FGF receptor, *bek*, was isolated by screening a mouse liver cDNA expression library with antiphosphotyrosine antibodies (76).

Recently, complete cDNA clones were isolated for both human *bek* and *flg* (77). The amino acid sequences of these genes were very homologous. However, their pattern of mRNA expression in various cell types differed. Expression vectors containing the cDNA of *bek* or *flg* were transfected into NIH/3T3 cells in order to determine the specificity of ligand binding to these putative FGF receptor gene products. Interestingly, acidic and basic FGF specifically bound with high affinity to both *flg* and *bek* transfectants. These results provided the first demonstration that two related ligands can bind to two related receptors. These transfectants should be instrumental in determining whether the *bek* and *flg* gene products also represent the receptors for all seven known FGF peptides.

Conclusions

The importance of growth factor receptor activation and its role in signal transduction have become increasingly evident over the past decade. The mechanisms by which ligands activate receptors and how triggered receptors transduce signals to the nucleus are still not completely understood. However, considerable progress in this field has resulted from studies involving the expression of foreign receptor cDNAs in recipient cells by transfection methods. Although a relatively small number of studies have been described in this chapter, one can envision how instrumental such studies will be in the future for determining the fundimental mechanisms involving growth factor-mediated signal transmission and the implications these results may have on controlling abnormal cell growth.

This chapter has concentrated on the insights gained in certain studies concerning the introduction of TK-containing receptors in recipient cells which are normally devoid of such receptors. One general concept which has been elucidated from these studies is that the intracellular components required to couple with activated receptor tyrosine kinases are frequently available in cell types that do not normally express such receptors. Thus, major determinants controlling cell proliferation through TK receptors may be at the level of growth factor receptor expression and ligand availability. Of course, it is likely that cell-specific components still play a deterministic role in controlling certain aspects of proliferation, differentiation and functional activition which result from TK receptor-mediated signalling.

Another basic concept which has been confirmed by reconstitution studies is that the retention of an active tyrosine kinase catalytic domain is an absolute requirement for signal transduction through these receptors. Interestingly, several studies have demonstrated that ligand stimulation of hematopoietin receptors which do not possess TK domains still leads to the phosphorylation of cellular substrates on tyrosine. These results suggest that cytokine receptors may couple with and activate intracellular tyrosine kinases which may be essential for signal transmission. Although critical links in the cascade which result in cell division after tyrosine phosphorylation of cellular substrates have not yet been clearly defined, it is possible that when candidate genes are identified, reconstitution of a receptor- mediated pathway in a naive cell type will provide definitive evidence that such genes are crucial for a particular growth factor regulatory network.

It is becoming increasingly evident that both autocrine and nonautocrine aberrations in growth factor receptor-mediated pathways lead to subversion of normal growth control. Different mutations or truncations of receptor genes have been identified which lead to constitutive TK activity suggesting that certain regions of receptor molecules exert a negative regulatory effect on receptor activation. Whether such activating mutations within receptor genes might contribute to human malignancies remains to be determined. As suggested by certain studies described in this chapter, overexpression of growth factor receptors may confer increased sensitivity of malignant cells to ligand *in vivo* or may even lead to ligand-independent receptor activation. Finally, genetic alterations which deregulate intracellular genes are part of the receptor-mediated signalling cascade may short-circuit growth factor regulatory networks and lead to malignancy. Thus, the identification of new receptors or receptor-mediated signalling genes through reconstitution studies such as those described here may potentially help to define the mechanisms which results in neoplasia.

References

1. Ullrich A. Schlessinger K. *Cell* 1990; 61:203-212.
2. Yarden Y, Ullrich A. *Annu. Rev. Biochem.* 1988; 57:443-478.
3. Williams LT. *Science* 1989; 243:1564-1510.
4. Ullrich A, Bell J, Chen E, et al. *Nature* 1985; 313:756-761.
5. Ebina Y, Ellis L, Jamagin K, et al. *Cell* 1985; 46:747-758.
6. Chou CK, Dull TJ, Russell DS, et al. *J. Biol. Chem.* 1987; 262:1842-1847.
7. McClain DA, Maegawa H, Lee J, Dull TJ, Ullrich A, Olefsky JM. *J. Biol. Chem.* 1987; 262:14663-14671.
8. Russell DS, Gherzi R, Johnson EL, Chou C-K, Rosen OM. *J. Biol. Chem.* 1987; 262:11833-11840.
9. White MF, Shoelson SE, Kewtmann H, Kahn CR. *J. Biol. Chem.* 1988; 263:2969-2980.
10. White MF, Livingston JN, Backer JM, et al. *Cell,* 1988; 54:641-649.
11. Carpenter G, Stoscheck C, Preston Y, DeLauro J. *Proc. Natl. Acad. Sci. USA* 1983; 80:5627-5630.
12. Shoyab M, Plowman GD, McDonald VL, Bradley JG, Todaro GJ. *Science* 1989; 243:1074-1076.
13. Downward J, Yarden Y, Mayer E, et al. *Proc. Natl. Acad. Sci. USA* 1977; 74:565-569.
14. Debuire B, Henry C, Bernissa, et al. *Science* 1984; 224:1456-1459.
15. Yamamoto T, Nishida T, Miyajima N, Kawai S, Ooi T, Yoyoshima K. *Cell* 1983; 35:71-78.

16. Chen WS, Lazar CS, Poenie M, Tsien PY, Gill GM, Rosenfeld MC. *Nature* 1987; 328:820-823.
17. Honegger A, Dull TJ, Bellot F, et al. *EMBO J.* 1988; 7:3045-3052.
18. Glenney JR, Chen WS, Lazar CS, et al. *Cell* 1988; 52:675-684.
19. Xu YH, Richert N, Ito S, Merlino GT, Pastan IH. *Proc. Natl. Acad. Sci. USA* 1984; 81:7308-7312.
20. Libermann TA, Nusbaum HR, Razon N, et al. *Nature* 1985; 313:144-147.
21. King CR, Kraus MH, Williams LT, Merlino GT, Pastan IH, Aaronson SA. *Nucleic Acids Res.* 1988; 13:8447-8486.
22. Kraus MH, Popescu NC, Amsbaugh SC, King CR. *EMBO J.* 1987; 6:605-610.
23. Di Fiore PP, Pierce JH, Fleming TP, et al. *Cell* 1987; 51:1063-1070.
24. Velu TJ, Beguinot L, Vass WC, et al. *Science* 1987; 237:1408-1410.
25. Riedel H, Massoglia S, Schlessinger J, Ullrich A. *Proc. Natl. Acad. Sci. USA* 1988; 85:1477-1481.
26. Finzi E, Fleming TP, Segatto O, et al. *Proc. Natl. Acad. Sci. USA* 1987; 84:3733-3737.
27. DeLaro JE, Todaro GJ. *Proc. Natl. Acad. Sci. USA* 1978; 75:4001-4005.
28. Di Marco E, Pierce JH, Fleming TP, et al. *Oncogene* 1989; 4:831-838.
29. Khazase K, Dull TJ, Graf T, et al. *EMBO J.* 1988; 7:3061-3071.
30. Greenberger JS, Sakakeeny MA, Humphries RK, Eaves CJ, Eckner RJ. *Proc. Natl. Acad. Sci. USA* 1983; 80:2931-2935.
31. Valtreri M, Tweardy DJ, Caracciolo D, et al. *J. Immunol.* 1987; 138:3829-3835.
32. Pierce JH, Ruggiero M, Fleming TP, et al. *Science* 1988; 239:628-631.
33. Wang LM, Collins M, Arai KI, Miyajima A. *EMBO J.* 1989; 8:3677-3684.
34. Von Ruden T, Wagner EF. *EMBO J.* 1988; 7:2749-2756.
35. King CR, Kraus MH, Aaronson SA. *Science* 1985; 229:974-976.
36. Semba K, Kamata K, Toyoshima K, Yamamoto T. *Proc. Natl. Acad. Sci. USA* 1985; 82:6497-6501.
37. Kraus MH, Popescu, NC, Amsbaugh SC, King CR. *EMBO J.* 1987; 6:605-610.
38. Slamon DC, Clark GM, Wong SG, Levin WJ, Ullrich A, McGuire WL. *Science* 1987; 235:177-182.
39. Van de Vijver M, Van de Bersselaar R, Deville P, Cornelisse C, Peterse J, Nusse R. *Mol. Cell. Biol.* 1987; 7:2019-2023.
40. Lupu R, Colomer R, Zugmaier G, et al. *Science* 1990; 249:1552-1554.
41. Di Fiore PP, Pierce JH, Kraus MH, Segatto O, King CR, Aaronson SA. *Science* 1987; 237:178-182.
42. Hudziak RM, Schlessinger J, Ullrich A. *Proc. Natl. Acad. Sci. USA* 1987; 84:7159-7163.
43. Bargmann CI, Hung MC, Weinberg RA. *Cell* 1986; 45:649-657.
44. Bargmann CI, Weinberg RA. *EMBO J.* 1988; 7:2043-2052.
45. Di Marco E, Pierce JH, Knicely CL, Di Fiore PP. *Mol. Cell. Biol.* 1990; 6:3247-3252.
46. Segatto O, King CR, Pierce JH, Di Fiore PP, Aaronson SA. *Mol. Cell. Biol.* 1988; 8:5510-5514.

47. Bargmann C, Weinberg RA. *Proc. Natl. Acad. Sci. USA* 1988; 85:5396-5398.
48. Di Fiore PP, Segatto O, Taylor WG, Aaronson SA, Pierce JH. *Science* 1990; 248:79-83.
49. Sherr CJ. *Blood* 1990; 75:1-12.
50. Sherr CJ, Rettenmier CW, Sacca R, Roussel MF, Look AT, Stanley ER. *Cell* 1988; 41:665-676.
51. Roussel MF, Dull TJ, Rettenmier CW, Ralph P, Ullrich A, Sherr CJ. *Nature* 1987; 325:549-552.
52. Hartmann T, Seuben K, Roussel MF, Sherr CJ, Pouyssegur J. *Growth Factors*, in press.
53. Roussel MF, Downing JR, Rettenmier CW, Sherr CJ. *Cell* 1988; 55:979-988.
54. Taylor G, Reedijk M, Rothwell V, Rohrschneider L, Pawson T. *EMBO J.* 1989; 8:2029-2037.
55. Shurtleff SA, Downing JR, Rock CO, Hawkins SA, Roussel MF, Sherr CJ. *EMBO J.* 1990; 9:2415-2421.
56. Pierce JH, Di Marco E, Cox G, et al. *Proc. Natl. Acad. Sci. USA* 1990;
57. Rohrschneider L, Metcalf D. *Mol. Cell. Biol.* 1989; 9:5081-5092.
58. Matsui T, Heidaran M, Mike T, et al. *Science* 1989; 243:800-804.
59. Yarden Y, Escobedo JA, Kuang WT, et al. *Nature* (London) 1986; 323:226-232.
60. Claesson-Welsh L, Eriksson A, Moren A, et al. *Mol. Cell. Biol.* 1988; 8:3476-3486.
61. Heldin CH, Westermark B. *Trends Genet.* 1989; 5:108-111.
62. Ross R, Raines EW, Bowen-Pope DF. *Cell* 1986; 46:155-169.
63. Matsui T, Pierce JH, Fleming TP, et al. *Proc. Natl. Acad. Sci. USA* 1989; 86:8314-8318.
64. Kazlauskas A, Cooper J. *Cell* 1989; 58:1121-1133.
65. Morrison DK, Kaplan DR, Escobedo JA, Rapp UR, Roberts TM, Williams LT. *Cell* 1989; 58:649-657.
66. Kumjian DA, Wahl MI, Rhee SG, Daniel TO. *Proc. Natl. Acad. Sci. USA* 1989; 86:8232-8236.
67. Meisenhelder J, Suh PG, Rhee SG, Hunter T. *Cell* 1989; 57:1109-1122.
68. Molloy C, Bottaro D, Fleming TP, Marshall M. Gibbs J, Aaronson SA. *Nature* 1989; 342:711-714.
69. Kazlauskas A, Ellis C, Pawson T, Cooper J. *Science* 1990; 247:1578-1581.
70. Escobedo J, Williams LT. *Nature* 1988; 335:85-87.
71. Severinsson L, Ek B, Mellström K, Claesson-Welsh L, Heldin CH. *Mol. Cell. Biol.* 1990; 10:801-809.
72. Heidaran M, Pierce JH, Lombardi D, et al. *Mol. Cell. Biol.* 1990, in press.
73. Burgess WH, Maciag T. *Annu. Rev. Biochem.* 1989; 58:575-606.
74. Lee PL, Johnson D, Cousens LS, Fried VA, Williams LT. *Science* 1989; 245:57-59.
75. Ruta M, Houk R, Ricca G, et al. *Oncogene* 1988; 3:9-15.
76. Kornbluth S, Paulson KE, Hanafusa H. *Mol. Cell. Biol.* 1988; 8:5541-5544.

77.   Dionne CA, Crumley G, Bello TF, et al. *EMBO J*. 1990; 9:2685-2692.

Advances in Regulation of Cell Growth, Volume 2;
Cell Activation: Genetic Approaches, edited by
James J. Mond, John C. Cambier, and Arthur Weiss.
Raven Press, Ltd., New York © 1991.

14

Somatic Cell Genetic Approach to Understanding T Cell Antigen Receptor Signal Transduction

Gary A. Koretzky, M.D., Ph.D.,
Mark A. Goldsmith, M.D., Ph.D., Martha Graber, M.B.,
Joel Picus, M.D., and Arthur Weiss, M.D., Ph.D.

Departments of Medicine, Microbiology and Immunology, and Cancer Research
Institute, Howard Hughes Medical Institute and University of California,
San Francisco, CA 94143

T lymphocytes play several crucial roles in the immune system. In addition to being the primary effectors in cell-mediated immunity, T cells are critical regulators of a variety of immune responses. Thus, it is essential that activation of T cells be a carefully regulated event. The primary structure responsible for the initiation of T cell stimulation is the T cell antigen receptor (TCR) which contains both antigen recognition and signal transduction components. In this review we will present a combination of genetic approaches, utilizing both random mutagenesis and the targeting of specific genes, to address questions regarding control of signal transduction via the TCR.

Structure of the TCR

The TCR is a complex structure consisting of at least 7 transmembrane proteins (1-4). Two of these chains, termed Ti, comprise a disulfide-linked heterodimer. These proteins exhibit extraordinary structural diversity, deriving their repertoire through a combinatorial association of rearranging gene segments (1). All of the information necessary for antigen and major histocompatibility complex antigen recognition resides in the Ti dimer (5,6). A collection of at least 5 proteins (the CD3 complex) is associated noncovalently with Ti (4). CD3 consists of the γ, δ, and ϵ chains associated with either a ζ-ζ homodimer (7), a ζ-η heterodimer (8), or a heterodimer of ζ with the Fc_ϵ receptor γ chain (9). The predominant form of the CD3 complex on at least one murine hybridoma appears to contain the ζ-ζ homodimer (10).

The first study reported to use somatic cell genetics to address the mechanism of signal transduction via the TCR demonstrated that there is an obligatory association of Ti and CD3 for their expression on the cell surface (11). Selection for mutant cells lacking the surface Ti chains resulted in cells that also lacked surface CD3. Reconstitution of Ti expression by gene transfer simultaneously reconstituted expression of surface CD3 (12). More recent studies have suggested that all chains, with the exception of η and Fc receptor γ, must be present in order for the TCR to be efficiently expressed on the cell surface (13,14).

While it is clear that the Ti chains are the antigen recognition component of the receptor complex, the role of CD3 remains speculative. Several lines of evidence suggest, however, that CD3 is responsible for signal transduction. First, the Ti chains contain very little information in their cytoplasmic domains (each has only 3-5 intracellular amino acids). In contrast, each of the CD3 chains, with ~40 to ~155 cytoplasmic amino acids, have considerable potential to interact with other cytosolic elements (2-4). Second, monoclonal antibodies (mAbs) directed against the CD3 complex are able to activate T cells in a fashion which mimics antigen stimulation (reviewed in (4)). Third, signal transduction mutants which have lost responsiveness to mAbs directed against Ti still respond to direct stimulation of CD3 (reference (15) and see below). Finally, recent reports have suggested that the loss of η expression and mutagenesis of the ζ chain of the CD3 complex result in alterations of signalling characteristics of the receptor complex (reference (16) and see below).

TCR stimulation results in the activation of at least two second messenger pathways.

Several biochemical changes are apparent in T cells within seconds of stimulation of the TCR by antigen or mAb. The first signalling pathway found to be linked to the TCR was the phosphatidylinositol (PI) pathway (17). Activation of the TCR results in the stimulation of an as yet unidentified phospholipase C (PLC) isoenzyme specific for the minor membrane phospholipid, phosphatidylinositol 4,5-bisphosphate (PIP_2). PIP_2 hydrolysis results in the formation of inositol 1,4,5-trisphosphate (IP_3) and diacylglycerol (DG) (18,19). IP_3 interacts with specific receptors on the endoplasmic reticulum resulting in the release of intracellular stores of calcium (20) while DG functions as a physiological activator of protein kinase C (PKC) (21) (Figure 1). The importance of the PI signal transduction pathway in T cell activation was underscored by the observation that the use of pharmacologic agents which mimic the effects of IP_3 and DG lead to distal events associated with T cell activation (such as the production of lymphokines). Thus, calcium ionophores which increase cytoplasmic free calcium ($[Ca^{+2}]_i$) and phorbol esters, such as phorbol 12-myristate 13-acetate (PMA), which directly activate PKC, can substitute for TCR stimulation to induce interleukin 2 (IL-2) production by T cells (4).

More recently it has been shown that stimulation of the TCR also results in the activation a protein tyrosine kinase (PTK) (22) (Figure 1). Thus, soon after ligation of the TCR with antigen or mAb a series of proteins newly phosphorylated on tyrosine residues appear (23,24). However, unlike other receptors which transduce a PTK signal, none of the TCR molecules themselves possess intrinsic kinase activity. Interestingly, a component of the

TCR, the ζ chain, is one of the new phosphotyrosine containing proteins seen after stimulation (25). The significance of this phosphorylation event remains to be determined.

The identity of the PTK and its relevant substrates have not been determined. Candidates for the PTK include two *src* related kinases, *fyn* and *lck*. *Fyn* has, in fact, been coprecipitated with the TCR (26); however the stoichiometry of this association is low and there is no evidence that the kinase activity of *fyn* is altered by TCR activation. The other candidate PTK, *lck*, is physically associated with two other T cell surface antigens, CD4 and CD8, which may play a role in antigen recognition and modulate signal transduction (27,28). There are no conclusive data, however, to indicate that *lck* is the PTK activated by TCR stimulation.

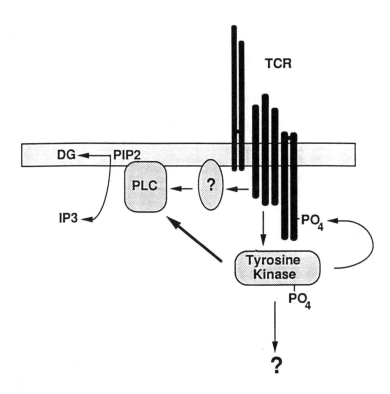

Fig. 1. Stimulation of the TCR activates at least two second messenger pathways. Ligation of the TCR by antigen or mAb results in the activation of a PLC specific for PIP$_2$ yielding IP$_3$ and DG. TCR stimulation also results in the activation of an, as yet, unidentified PTK. Candidates for the kinase are *lck* and *fyn*, two members of the *src* family.

The relationship between the PI and PTK second messenger pathways and their relative importance in the initiation of distal biologic events remains unclear. As mentioned above, pharmacologic agents which mimic activation of the PI pathway can induce production of lymphokines such as IL-2. A preliminary report has also shown that transfection of constitutively active *src* into a T cell hybridoma results in constitutive IL-2 production; however the effect on the PI pathway by transfection of this gene was not described (29).

Recent evidence suggests that the PTK and PI second messenger pathways may be causally linked in T cells. These studies have made use of pharmacologic agents to show that inhibition of PTK activity interferes with TCR-mediated PI signalling (30,31). That this is not due to nonspecific effects of the drugs was shown in a T cell transfected with the human muscarinic receptor type 1 (HM1). This receptor, which is not endogenously present in T cells, couples to the PI pathway in its native environment of neuronal and smooth muscle cells via a guanine-nucleotide binding protein (32). There is no evidence to suggest that HM1-mediated PIP_2 hydrolysis is dependent upon PTK activity (33). The PTK inhibitor, herbimycin A, was found to inhibit TCR-, but not HM1-mediated PI signal transduction in the HM1 transfected T cells (manuscript submitted). Another piece of evidence suggesting that the two pathways may be linked is a kinetic study showing that PTK activity may precede the formation of PI-derived second messengers (24). Finally, studies of several growth factor receptors which transduce both PI and PTK signals have demonstrated that one substrate of the PTK is a PLC isoenzyme (PLC γ_1) (34,35). Although the physiological relevance of this phosphorylation event remains unclear, these findings suggest that PTK activity may be essential to activate the PLC responsible for generation of PI-derived second messengers.

SELECTION FOR SIGNAL TRANSDUCTION DEFECTIVE CELLS AFTER RANDOM MUTAGENESIS.

Several years ago our laboratory undertook a project to define the components necessary for TCR-mediated signal transduction by randomly mutagenizing the human T cell leukemic line, Jurkat, and then selecting for signal transduction defective cells (36,37). For some of our mutants our selection scheme involved mutagenesis followed by incubation with a TCR-dependent lectin or anti-TCR mAb for two weeks. We had previously noted that prolonged stimulation of Jurkat with such agents resulted in significant cell death, thought to be due to prolonged activation of the cells' signal transduction machinery (36). The population which survived the treatment with activating stimuli was thus presumed to be enriched for cells defective in signal transduction. Because these cells showed variable expression of the TCR (one class of TCR signal transduction mutant is a low or TCR-negative phenotype), selection was continued for cells with high level of TCR expression that failed to increase $[Ca^{+2}]_i$ in response to TCR stimulation. Enrichment for such cells was performed utilizing the calcium sensitive dye, Indo-1. When excited by ultraviolet light, this dye has maximum emission at 486 nm when unbound to calcium in the cytosol and at 404 nm when bound to calcium. A qualitative response to activating stimuli can be determined for individual cells loaded with Indo-1 in a fluorescence activated cell sorter (FACS) by examining changes in the fluorescence ratio. In addition to determining whether or not a given cell responds to activating stimuli, this technique allows for the isolation of cells of a given phenotype by sterile sorting. Thus, it was possible to rapidly and specifically select for TCR-expressing cells which did not increase $[Ca^{+2}]_i$ in response to activating mAb (36).

Using slightly varied techniques for stimulating the cells in order to select for non-responders, we have isolated three variants of the parental line, Jurkat. These clones are designated J.CaM1, J.CaM2, and J.CaM3 (Jurkat calcium mutants 1-3). Although they express similar high levels of surface TCR when compared to the wild type, the mutants each are defective in their ability to respond to anti-TCR mAb (Table 1). As expected from our selection strategy, the clones fail to mobilize $[Ca^{+2}]_i$ when stimulated with the anti-TCR mAb used in the selection process. There were, however, differences in the signalling phenotypes of each of the three clones obtained. For example, while J.CaM1 and 3 did not respond at all to the anti-Ti mAb, C305, these cells did exhibit a positive calcium response to some, but not all, anti-CD3 mAbs and to combinations of non-activating mAb. On the other hand, J.CaM2 did not respond with a calcium increase to any mAb, either alone, or in combination. Furthermore, when the ability of the TCR to couple to the PTK pathway was evaluated, we found that no TCR stimuli activated this second messenger pathway in J.CaM1 or 3. However, there clearly was the rapid appearance of new phosphotyrosine containing proteins in J.CaM2 after stimulation with anti-TCR.

Table I summarizes some of the TCR-related signal transduction characteristics of the three clones. In every situation where a defect in calcium mobilization was seen, we observed a defect in the generation of soluble inositol phosphates (IP). In order to evaluate the status of the PI pathway in these cells further, we transfected the HM1 into each of the clones (38). In each case, stimulation of this heterologous receptor resulted in IP generation, whereas the TCR remained uncoupled from this second messenger pathway. Thus the defect in each clone was intrinsic to the TCR signal transduction machinery, and not due to a global dysfunction in the PI second messenger pathway.

The J.CaM clones have provided insight into the role of second messenger production in other TCR-mediated events. Studies from many laboratories have demonstrated that receptor occupancy by mAb results in rapid internalization of the TCR (36). It was speculated that this is a result of receptor-mediated PKC activation with subsequent serine phosphorylation of components of the CD3 complex. Additional support for this model came from the observation that direct activation of PKC by PMA also results in TCR internalization (39). However, incubation of each J.CaM clone with anti-TCR mAb results in receptor internalization similar to that seen in wild-type Jurkat (Table I). Thus, receptor occupancy (and possibly crosslinking) without the generation of known second messengers appears to be a sufficient signal for internalization.

We evaluated signalling characteristics of the mutant clones further by assessing distal events associated with T cell activation. Bypassing the TCR in each clone with PMA and ionomycin resulted in IL-2 production, however no IL-2 was produced after stimulation of the TCR in any of the cells. This was true even in J.CaM2, although stimulation of its TCR resulted in at least partial activation of a PTK (manuscript submitted). While we can not be sure that the

TABLE I

SIGNAL TRANSDUCTION CHARACTERISTICS OF J.CAM1-3

	JURKAT	J.CaM1	J.CaM2	J.CaM3
$[Ca^{+2}]_i$ increases to:				
Anti-Ti Vβ (C305)	++++	-	-	-
anti-CD3 (OKT3)	+++	-	-	-
anti-CD3 (235)	++++	+++	-	+++
IP increase to C305[1]	++++	-	-	-
PTK acativation[2]	++++	-	++[3]	-
TCR-ζ phosphorylation	++++	-	++++	-
TCR internalization[4]	++++	++++	++++	++++

1. Soluble inositol phosphates measured after 30 minute stimulation.
2. Analysis of whole cell lysates after TCR stimulation by immunoblotting with anti-phosphotyrosine mAb.
3. Only some of the substrates seen in Jurkat lysates are phosphorylated in J.CaM2 after TCR stimulation.
4. Internalization seen after incubation with PMA or anti-TCR mAb.

PTK activity in J.CaM2 is completely normal, these data suggest that PTK stimulation may not be sufficient for IL-2 production and that PI-derived second messengers play a critical role in the production of this lymphokine. Further support for the role of the PI second messenger pathway in IL-2 production comes from a study demonstrating production of IL-2 after carbachol stimulation of the HM1 transfected cells. In this situation significant amounts of PI-derived second messengers are produced, however no PTK activity is observed (33).

The next step in our analysis was to attempt to define the sites of the mutations in the three clones and to investigate how the mutant molecules normally contribute to TCR-mediated signal transduction. Initial biochemical characterization of all of the TCR chains revealed no gross abnormalities in any of the mutant cells. Because we could not exclude point mutations in the TCR chains from our biochemical analysis and because it was impossible to predict which other genes were likely to have been mutated in the generation of the clones, we developed a technique to rapidly assign the signal transduction mutants into complementation groups by use of a transient heterokaryon assay (Figure 2) (37,40). This technique again makes use of the FACS to analyze Indo-1 loaded cells. One partner clone is loaded with the calcium sensitive dye while the second mutant partner is stained with a fluorescein-conjugated (FITC) mAb directed against a surface antigen not involved in TCR-mediated signal transduction. The two clones are then fused with polyethylene glycol. Successfully fused heterokaryons are isolated electronically by gating on FITC positive and Indo-1 loaded cells in the FACS. A TCR stimulus is then applied and the response of the heterokaryons is monitored. If the mutations in the two clones lie in separate complementation groups, the heterokaryons will exhibit a calcium response. Conversely, if the mutations lie within the same gene, the heterokaryons will remain non-responsive.

We have used this heterokaryon assay to determine whether the mutations in the J.CaM clones complement each other and if any cells with defined mutations (i.e. those lacking expression of Ti α or β) are capable of reconstituting the J.CaM clones (40). These studies have demonstrated that none of the J.CaM mutations lie in the Ti proteins and that each of the J.CaM clones have mutations affecting distinct structural components. We conclude that at least three non-Ti proteins are essential for the TCR to couple with the PI second messenger pathway.

Although our approach of random mutagenesis followed by selection for functional mutants is unbiased in the targeting of specific components in the signal transduction pathway, it suffers from the difficulty of defining the sites of the mutations. Understanding the J.CaM mutants will probably require the development of a genetic reconstitution system. However, as more is learned of the intracellular components of the signal transduction pathway and as other defined TCR signal transduction mutants become available (see below), it will be possible to use biochemical analyses and the heterokaryon assay with the J.CaM clones as partners to characterize further their defects.

CD45 is essential for TCR-mediated PTK activity.

Because the TCR in wild-type Jurkat cells is such a potent stimulator of a PTK (33), we were able to easily assess the requirement for CD45 expression in this activation event. We compared the ability of anti-TCR mAb to induce the appearance of new phosphotyrosine containing proteins in whole cell lysates of Jurkat and J45.01 by immunoblotting with a mAb specific for phosphotyrosine containing proteins (figure 5.) As can be seen from these Western blots, although several new phosphotyrosine containing proteins are evident in Jurkat after TCR stimulation, none are seen in J45.01 To eliminate the possibility that PTK activity occurs, but is delayed in J45.01, an experiment looking at TCR stimulation for periods of 5 seconds to 15 minutes was performed. This study failed to demonstrate PTK activity in J45.01 at any time point tested (57). The defect in TCR-associated PTK activity is not due to a global dysfunction of all PTK's in J45.01 because tyrosine phosphorylation of a 42-44 kd protein (likely to be microtubule associated protein 2 kinase (58,59)) occurs in this cell after direct PKC activation with PMA (57). Thus, it appears that in addition to being required for coupling the TCR to the PI second messenger pathway, CD45 is necessary for TCR-mediated PTK activity to occur. These findings are consistent with recent data suggesting that PTK activation is required for subsequent PI second messenger generation in T cells (30,31).

CD45 is essential for CD2-mediated T cell activation.

Several laboratories have shown that in addition to the TCR complex, other surface molecules are able to transduce activating signals in T lymphocytes. One such receptor is CD2, which was originally identified as the sheep erythrocyte receptor (60) and later shown to specifically bind to the lymphocyte associated function antigen 3 (LFA-3) (60-62). Stimulation of T cells with combinations of mAb directed against CD2 results in increases in PI-derived second messengers (63,64) and in the appearance of new phosphotyrosine containing proteins (65). Interestingly, studies using TCR loss variants and the J.CaM mutants demonstrate that CD2 ligation results in T cell activation only in cells which express a functional TCR (63). Since Jurkat and J45.01 have nearly equivalent levels of surface CD2 and TCR, we were able to address the role of CD45 in signal transduction via CD2.

In experiments testing the ability of CD2 ligation to stimulate second messenger production in Jurkat and J45.01 we found that increases in PI-derived second messengers and the appearance of new phosphotyrosine containing proteins only occurred in Jurkat when activating combinations of anti-CD2 mAb were added to cultures (figure 4). Thus, it appears that CD45 expression is required in order for both the TCR and CD2 to couple to their signal transduction machinery.

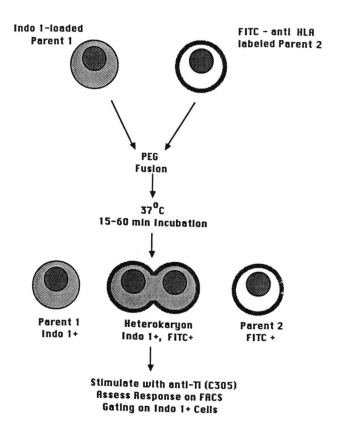

Fig. 2. Formation of transient heterokaryons to assign signal transduction mutants into complementation groups. One mutant clone is loaded with the calcium sensitive dye, Indo-1, the other is surface stained with a fluorescein-conjugated mAb. The clones are subjected to polyethylene glycol and successfully fused heterokaryons are isolated in a fluorescence activated cell sorter. Calcium responses of the heterokaryons to activating stimuli can then be assessed.

ISOLATION OF SIGNAL TRANSDUCTION MUTANT BY TARGETING SPECIFIC GENES.

In addition to random mutagenesis and selection for signalling defective cells, a complementary approach of targeting specific genes by mutagenesis to address the role of their products in TCR-mediated signal transduction has been used by several laboratories. The most obvious targets for this approach are the genes encoding the CD3 molecules. Several studies have now described CD3-ζ deficient variants of a murine T cell hybridoma, 2B4 (14,16). These cells express only low levels of TCR on their surface and do not respond normally to antigen. In contrast, however, the ζ-deficient cells respond when the CD3 complex is directly stimulated with cross-linked mAb. Reconstitution of the ζ-deficient cells with wild type and mutated ζ cDNA's has led to the hypothesis that the role of the ζ chain is to couple the Ti components to the other CD3 elements, thus transferring information from the antigen recognition domains to signal transduction domains of the complex.

Another variant of this T cell hybridoma which lacks expression of CD3-η, an alternatively spliced transcript of the ζ gene (41), has been described. TCR expressed by this cell contains only the form with a ζ-ζ homodimer. Stimulation of the TCR in this cell results in apparent normal PTK activity, but much reduced PI second messenger generation (10). Studies of cells with variable amounts of η led to the speculation that the TCR complexes containing a ζ-η heterodimer were capable of coupling to the PI pathway, while those with the ζ-ζ homodimer activated the PTK. Now that the cDNA encoding the η chain has been cloned (41), definitive experiments involving reconstitution of the η-deficient cells to assess the role of the ζ-η heterodimer in coupling the TCR to the PI pathway can be performed.

Another recent experiment designed to address the function of the CD3 molecules demonstrated that when murine cells transfected with a cDNA encoding a truncated human CD3 ε are stimulated with mAb directed specifically against human CD3 ε, IL-2 is produced (42). Thus, it appears that the cytoplasmic domain of CD3 ε may not be essential for TCR signal transduction. However, this conclusion must be reinterpreted in light of recent studies suggesting that two CD3 ε chains are expressed per TCR complex (43).

In addition to the TCR complex itself, other molecules are likely to be important in the regulation of TCR signal transduction. Recently our laboratory has undertaken a project to address the role of another T cell surface molecule, CD45, in this regulation.

CD45.

CD45, also known as LCA (leukocyte common antigen) or T200, is a family of closely related membrane glycoproteins found on all hematopoietic cells except for red blood cells and platelets (44). Several different isoforms have been identified which differ in both primary sequence and degree of glycosylation (44,45). The heterogeneity found in their different isoforms rests entirely in the extracellular domains and results from alternative splicing of multiple exons while the transmembrane and cytoplasmic regions are identical in all forms within a given species. Furthermore, while there exist significant differences among species in the extracellular domains of CD45 (only 34% protein sequence identity comparing mouse, rat, and human) the cytoplasmic domains are highly conserved (85% amino acid homology) (45). Recently it has been shown that the highly conserved cytoplasmic domain of CD45 bears striking homology to a protein tyrosine phosphatase (PTPase 1b) isolated from human placenta (46). The ~700 amino acid intracellular residues include two tandem repeats of ~300 amino acids which share approximately 35% homology with the placental phosphatase. Furthermore, immunoprecipitates of CD45 from T lymphocytes have intrinsic tyrosine phosphatase activity in *in vitro* assays (47,48).

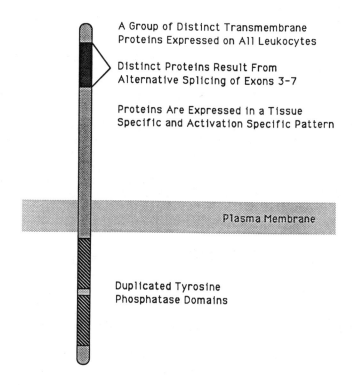

Fig. 3. CD45, also known as T200 or leukocyte common antigen is expressed on all hematopoietic cells except for red blood cells and platelets. The several isoforms, arising from alternative splicing of a single gene, are expressed in a tissue and activation specific manner. All isoforms of CD45 express the same cytoplasmic domain which has intrinsic tyrosine phosphatase activity.

Several lines of evidence suggest that CD45 may be involved in regulation of signalling by the TCR, however, the precise role of CD45 in lymphocyte activation remains unclear. Several laboratories have presented conflicting data regarding the effects of mAbs against CD45 on cellular activation (49-52). One recent study described a mutant murine T cell line which lacked surface expression of CD45 (53). This cell line failed to respond by proliferation to stimulation by antigen but did proliferate in response to exogenous IL-2. Other recent information has linked CD45 to the state of tyrosine phosphorylation (and hence degree of activation) of the *src*-like kinase *lck* (54,55) It is interesting to speculate that CD45, by controlling the state of phosphorylation of the TCR-stimulated PTK, may play a major role in the ability of this PTK to become activated. Control of the PTK may, in turn, be critical in the regulation of the PI second messenger pathway. Thus, CD45 is an attractive molecule to target in genetic studies of the regulation of TCR signal transduction.

Isolation and characterization of CD45-deficient variants of human T cell lines.

To address the role of CD45 in the regulation of TCR signal transduction it was necessary to obtain cells deficient in expression of this surface antigen. We first screened a number of human T cell lines carried in our laboratory for CD45 by indirect immunofluorescence. While most of the cells tested stained brightly with a mAb directed against a determinant common to all isoforms of CD45, one subline of HPB-ALL was heterogeneous and largely negative for CD45 expression. This was not merely an epitope loss since a panel of mAb, directed against various sites on the CD45 molecule, all failed to stain the cells. We subcloned the HPB-ALL cells by limiting dilution and obtained clonal populations of CD45-negative (HBP.45.0) and CD45-positive (HPB.45.1) cells (56). Northern analysis revealed that no CD45 mRNA was present in the CD45-negative clone (data not shown).

Analysis of the CD45-negative cells derived from HPB-ALL was limited by the fact that the TCR in the CD45-positive clone couples poorly with the PTK second messenger pathway. This is probably due to a sensitivity problem, as TCR expression is low in HPB.45.1 (56). Our studies were additionally limited by the fact that wild type HPB-ALL does not produce IL-2 (a valuable marker for distal events in T cell activation) and several T cell surface antigens which can deliver activation signals (see below) are absent in this cell line. We therefore mutagenized a clone of the leukemic line, Jurkat, and used several negative selection techniques to obtain CD45-deficient variants. One clone obtained, J45.01, expressed less than 8% as much CD45 as did the parent Jurkat clone. It should be noted that at no time during our selection protocol did we assess the signalling characteristics of the cells.

Northern analysis revealed nearly equivalent amounts of CD45 mRNA in Jurkat and J45.01 (unpublished data.) We therefore compared the amount of CD45 protein found in the two clones by immunoblotting with mAb 9.4 which is directed against a determinant found on all CD45 isoforms. Whereas substantial amounts of CD45 are detected in whole cell lysates from Jurkat, none is detectable in lysates from comparable numbers of J45.01 (57). However, when immunoprecipitation from large numbers of cells was performed, a small amount of CD45 was found in the mutant cells. We therefore compared the tyrosine phosphatase activity of membranes derived from Jurkat and J45.01 in an *in vitro* assay. Whereas membranes from both clones had similar amounts of alkaline phosphatase activity, membranes from J45.01 exhibited several fold less tyrosine phosphatase activity than did membranes from Jurkat (57). Thus, it appears by immunofluorescence and enzymatic studies that J45.01 is functionally deficient in CD45-associated tyrosine phosphatase activity.

Having obtained variants of two human T cell lines which were deficient in CD45 expression allowed us to investigate the role of this tyrosine phosphatase in TCR-mediated signal transduction. Initial studies focused on the integrity of the PI second messenger pathway in the CD45-deficient and -positive cells. Stimulation of the TCR in the CD45-positive Jurkat and HPB-ALL clones resulted in large and sustained increases in $[Ca^{2+}]_i$ whereas the same stimuli resulted in no change from a similar baseline $[Ca^{2+}]_i$ in both HPB.45.0 and J45.01 (56,57). As expected, in every case increases in $[Ca^{2+}]_i$ correlated with the generation of soluble IP (figure 4).

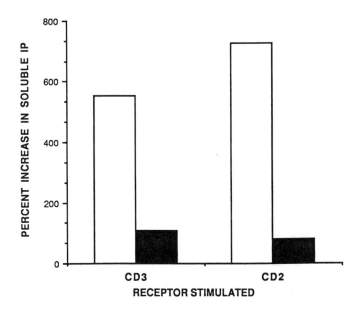

Fig. 4. CD45 is essential for TCR- and CD2-mediated PLC activation. Jurkat (open bars) and J45.01 (stippled bars) were loaded with ^{3}H-myoinositol then stimulated with either anti-CD3 mAb or a combination of anti-CD2 mAbs. Production of soluble inositol phosphates was measured at 30 minutes.

Although our selection scheme for J45.01 required only that the cells be CD45-deficient, it was formally possible that the signalling mutation was independent of the defect resulting in CD45 loss. We therefore made use of the transient heterokaryon assay described above to assign J45.01 into complementation groups with our other mutants (Table II). Each of the J.CaM mutants described above are CD45-positive. When fused to J45.01 each is able to reconstitute signal transduction. Conversely, our other non-signalling mutant, HBP.45.0, which is CD45-negative is not capable of reconstituting signalling via the TCR expressed on J45.01. That this is not due to the inability of cells from HPB-ALL origin to form functional heterokaryons with Jurkat-derived cells was demonstrated by the successful fusion of HPB.45.1 to J45.01 resulting in heterokaryons with J45.01-derived TCR capable of inducing $[Ca+2]_i$ increases.

TABLE II

CD45 EXPRESSION IS REQUIRED TO COMPLEMENT THE MUTATION IN J45.01[1]

Fusion Partner	CD45 expression	% Response[2]
Jurkat	+	65
J.Cam1	+	66
J.Cam2	+	69
HPB.45.0	-	8
HPB.45.1	+	72

1. J45.01 cells were loaded with Indo-1, then fused with a FITC labelled partner as described.
2. Heterokaryons were stimulated for one minute with the C305 mAb and analyzed for Indo-1 ratio. Percent of heterokaryons increasing $[Ca^{+2}]_i$ is indicated.

Reconstitution of CD45 expression and restoration of signalling competence is the best demonstration that the signalling defect in the CD45-negative cells is due to the lack of CD45 expression. In preliminary experiments we noted that treatment of HPB-ALL with PMA increased surface expression of CD45 (data not shown). The CD45-negative clones were therefore incubated with PMA for 48 hours. Although no increase in CD45 expression was noted in J45.01, PMA treatment resulted in the expression of a small, but reproducible amount of surface CD45 in HPB.45.0 (56). The PMA-treated cells were then stimulated with anti-TCR mAb with a resultant large increase in PI-derived second messengers and $[Ca^{2+}]_i$.

Additionally, we have been successful in transfecting a cDNA encoding murine CD45 into HPB.45.0 (56). The transfectants express small amounts of surface murine CD45, but remain negative for human CD45 expression. These cells, when stimulated with anti-TCR mAb, also show a large increase in $[Ca^{2+}]_i$. Thus, it appears that CD45 expression is required in order for the TCR to couple to the PI second messenger pathway.

To evaluate the global status of the PI second messenger pathway in the CD45-negative cells, HPB.45.0 was also transfected with cDNA encoding the HM1 (56). This receptor, when transfected into human T cells, couples to the PI pathway independently of the TCR (38). Stimulation with carbachol in an HM1-transfected HPB.45.0 line resulted in a large increase in PI second messengers (56). The TCR in this cell remained uncoupled from the PI signal transduction pathway. Thus, CD45 is not generically required to generate PI-derived second messengers, rather CD45 expression is necessary to specifically couple the TCR to the PI pathway.

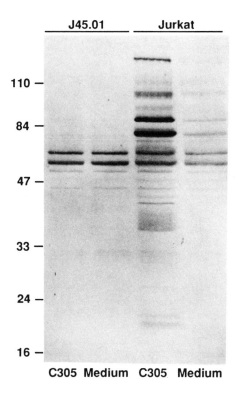

Fig. 5. CD45 expression is essential for TCR-mediated PTK activity. Jurkat and J45.01 were left unstimulated or incubated with anti-TCR mAb (C305) for two minutes. Whole cell lysates were subjected to SDS polyacrylamide gel electrophoresis then transferred to nitrocellulose and immunoblotted with a mAb specific for phosphotyrosine containing proteins.

The role of CD45 in distal events associated with T cell activation.

The isolation of a CD45-deficient variant of Jurkat enabled us to examine the role of the phosphatase in distal events in T cell activation. Two markers of distal activation studied in our laboratory are the production of the lymphokine, IL-2, and the expression of CD69, a surface antigen which appears on T cells after PKC stimulation (66). Table III summarizes the results of several experiments where the expression of CD69 and the production of IL-2 were assessed after CD2 and TCR stimulation. CD45 is essential in order for both

TABLE III

CD45 EXPRESSION IS ESSENTIAL FOR CD2- AND TCR-MEDIATED IL2 PRODUCTION AND CD69 INDUCTION

STIMULUS	CD69 EXPRESSION[1]		IL2 PRODUCTION[2]	
	CLONE		CLONE	
	Jurkat	J45.01	Jurkat	J45.01
MEDIUM	18	5	<2	<2
PMA	638	496	<2	<2
ANTI-CD3	264	6	6	<2
ANTI-CD2	57	6	11	<2
IONOMYCIN	ND	ND	14	10
ANTI-CD28 + IONOMYCIN	ND	ND	43	46

1. CD69 expression in arbitrary units calculated after analysis by indirect immunofluorescence after treating cells for 24 hours with the indicated reagents.
2. IL2 production measured by a bioassay after 24 hours of stimulation. Anti-CD3, anti-CD2, ionomycin, and anti-CD28 + ionomycin stimulations were all done in the presence of PMA.

of these activation events to occur after either CD2 or TCR stimulation. It is important to note that J45.01 is capable of producing IL-2 when stimulated with the combination of PMA and ionomycin and that it expresses high levels of CD69 when PKC is activated by PMA.

Table III demonstrates further that not all signal transduction pathways require CD45 in human T lymphocytes. Several laboratories have shown that another T cell surface antigen, CD28, is able to transduce activating signals (as manifested by augmenting IL-2 production over levels seen with PMA and ionomycin alone) (67,68). There is no evidence to indicate that CD28 functions via either the PI or PTK second messenger pathways (31,68). Table III illustrates that CD45 expression is not essential in order for CD28 to transduce its activating signal since IL-2 production is augmented similarly in Jurkat and J45.01 when the anti-CD28 mAb, 9.3, is added to cultures.

CD45: Implications.

The mechanism by which CD45 regulates TCR-mediated stimulation of the PI and PTK second messenger pathways remains speculative. Because recent data have indicated that the generation of PI-derived products may require prior activation of the PTK, it is plausible that CD45 plays a role only in the regulation of PTK activity. The most likely candidates for the TCR-related PTK are members of the *src* family of PTK's. Several experiments have shown that the activity of some members of this family can be modified by the state of phosphorylation of critical tyrosine residues (69,70). Thus, it is possible that CD45 exerts its regulatory function by controlling the degree of phosphorylation of the TCR-associated PTK. Only when critical tyrosines are dephosphorylated by the CD45 phosphatase can receptor occupancy be translated into an activation signal. This model is currently being tested in our laboratory.

SUMMARY.

Genetic approaches have proven to be invaluable in attempts to understand the mechanisms of signal transduction via cell surface receptors. Our laboratory has taken two concurrent approaches in an attempt to identify the essential components regulating TCR-mediated signalling. First we have generated a family of signal transduction deficient clones by random mutagenesis of a well characterized T cell line and are now in the process of attempting to identify the genes which have given rise to the altered phenotypes. Concurrently, we have identified genes of potential interest, such as CD45, to be targeted directly. Having demonstrated the importance of this molecule in TCR signal transduction, we are currently in the process of establishing some of the structure/function rules which govern the interaction of CD45 and the TCR regulated signal transduction pathways.

ACKNOWLEDGEMENTS.

This work was supported, in part, by grants from the Arthritis Foundation and the Rosalind Russel Arthritis Center (G.A.K.), the American Cancer Society (J.P.), and the National Institutes of Health (A.W. and M.G.)

References.

1. Allison JP, Lanier LL. Structure, function, and serology of the T-cell antigen receptor complex. Ann. Rev. Immunol. 1987; 5:503-540.

2. Ashwell JD, Klausner RD. Genetic and mutational analysis of the T cell antigen receptor. Ann. Rev. Immunol. 1990; 8:139-168.

3. Clevers H, Alarcon B, Willeman T, Terhorst C. The T cell receptor/CD3 complex: A dynamic protein ensemble. Ann. Rev. Immunol. 1988; 6:629-662.

4. Weiss A, Imboden JB. Cell surface molecules and early events involved in human T lymphocyte activation. Adv. Immunol. 1987; 41:1-38.

5. Dembic Z, Haas W, Weiss S, et al. Transfer of specificity by murine α and β T-cell receptor genes. Nature 1986; 320:232-238.

6. Saito T, Weiss A, Miller J, Norcross MA, Germain RN. Specific antigen-Ia activation of transfected human T cells expressing murine Ti $\alpha\beta$-human T3 receptor complexes. Nature 1987; 325:125-130.

7. Samelson LE, Harford JB, Klausner RD. Identification of the components of the murine T cell antigen receptor complex. Cell 1985; 43:223-231.

8. Baniyash M, Garcia-Morales P, Bonifacino JS, Samelson LE, Klausner RD. Disulfide linkage of the ζ and η chains of the T cell receptor. J. Biol. Chem. 1988; 263:9874-9878.

9. Orloff DG, Ra C, Frank SJ, Klausner RD, Kinet J-P. Family of disulphide-linked dimers containing ζ and η chains of the T-cell receptor and the γ chain of Fc receptors. Nature 1990; 347:189-191.

10. Mercep M, Weissman AM, Frank SJ, Klausner RD, Ashwell JD. Activation-driven programmed cell death and T cell receptor $\zeta\eta$ expression. Science 1989; 246:1162-1165.

11. Weiss A, Stobo JD. Requirement for the coexpression of T3 and the T cell antigen receptor on a malignant human T cell line. J. Exp. Med. 1984; 160:1284-1299.

12. Ohashi P, Mak T, Van den Elsen P, et al. Reconstitution of an active surface T3/T-cell antigen receptor by DNA transfer. Nature 1985; 316:606-609.

13. Bonifacino JS, Suzuki CK, Klausner RD. A peptide sequence confers retention and rapid degradation in the endoplasmic reticulum. Science 1990; 247:79-82.

14. Sussman JJ, Bonifacino JS, Lippincott-Schwartz J, et al. Failure to synthesize the T cell CD3-ζ chain: Structure and function of a partial T cell receptor complex. Cell 1988; 52:85-95.

15. Goldsmith MA, Weiss A. Generation and analysis of a T lymphocyte somatic mutant for studying molecular aspects of signal transduction by the antigen receptor. Ann. N.Y. Acad. Sci. 1988; 91-103.

16. Frank SJ, Samelson LE, Klausner RD. The structure and signalling functions of the invariant T cell receptor components. Sem. Immunol. 1990; 2:89-97.

17. Imboden JB, Stobo JD. Transmembrane signalling by the T cell antigen receptor: Perturbation of the T3-antigen receptor complex generates inositol phosphates and releases calcium ions from intracellular stores. J. Exp. Med. 1985; 161:446-456.

18. Berridge MJ, Irvine RF. Inositol trisphosphate, a novel second messenger in cellular signal transduction. Nature 1984; 312:315-321.

19. Berridge MJ. Inositol trisphosphate and diacylglycerol: Two interacting second messengers. Ann. Rev. Biochem 1987; 56:159-193.

20. Ferris CD, Huganir RL, Supattapone S, Snyder SH. Purified inositol 1,4,5-trisphosphate receptor mediates calcium flux in reconstituted lipid vesicles. Nature 1989; 342:87-89.

21. Nishizuka Y. Studies and perspectives of protein kinase C. Science 1986; 233:305-312.

22. Samelson LE, Patel MD, Weissman AM, Harford JB, Klausner RD. Antigen activation of murine T cells induces tyrosine phosphorylation of a polypeptide associated with the T cell antigen receptor. Cell 1986; 46:1083-1090.

23. Hsi ED, Siegel JN, Minami Y, Luong ET, Klausner RD, Samelson LE. T cell activation induces rapid tyrosine phosphorylation of a limited number of cellular substrates. J. Biol. Chem. 1989; 264:10836-10842.

24. June CH, Fletcher MC, Ledbetter JA, Samelson LE. Increases in tyrosine phosphorylation are detectable before phospholipase C activation after T cell receptor stimulation. J. Immunol. 1990; 144:1591-1599.

25. Baniyash M, Garcia-Morales P, Luong E, Samelson LE, Klausner RD. The T cell antigen receptor ζ chain is tyrosine phosphorylated upon activation. J. Biol. Chem. 1988; 263:18225-18230.

26. Samelson LE, Phillips AF, Luong ET, Klausner RD. Association of the fyn protein-tyrosine kinase with the T-cell antigen receptor. Proc. Natl. Acad. Sci. USA 1990; 87:4358-4362.

27. Veillette A, Bookman MA, Horak EM, Bolen JB. The CD4 and CD8 T cell surface antigens are associated with the internal membrane tyrosine-protein kinase p56lck. Cell 1988; 55:301-308.

28. Veillette A, Bookman MA, Horak EM, Samelson LE, Bolen JB. Signal transduction through the CD4 receptor involves the activation of the internal membrane tyrosine-protein kinase p56lck. Nature 1989; 338:257-259.

29. O'Shea JJ, Ashwell JD, Mercep M, Cross SL, Samelson LE, Klausner RD. Expression of the oncogene, pp60v-src, in T cells results in constitutive interleukin-2 production. Clin. Res. 1989; 37:558A.

30. Mustelin T, Coggeshall KM, Isakov N, Altman A. T cell antigen receptor-mediated activation of phospholipase C requires tyrosine phosphorylation. Science 1990; 247:1584-1587.

31. June CH, Fletcher MC, Ledbetter JA, et al. Inhibition of tyrosine phosphorylation prevents T cell receptor-mediated signal transduction. Proc. Natl. Acad. Sci. USA 1990; 87:7722-7726.

32. Nathanson NM. Molecular properties of the muscarinic acetylcholine receptor. Ann. Rev. Neurosci. 1987; 10:195-236.

33. Desai D, Newton ME, Kadlecek T, Weiss A. Stimulation of the phosphatidylinositol pathway can induce T cell activation. Nature 1990; 348:66-68.

34. Meisenhelder J, Suh P-G, Rhee SG, Hunter T. Phospholipase C-γ is a substrate for the PDGF and EGF receptor protein-tyrosine kinases *in vivo* and *in vitro*. Cell 1989; 57:1109-1122.

35. Wahl MI, Daniel TO, Carpenter G. Antiphosphotyrosine recovery of phospholipase C activity after EGF treatment of A-431 cells. Science 1988; 241:968-970.

36. Goldsmith MA, Weiss A. Isolation and characterization of a T-lymphocyte somatic mutant with altered signal transduction by the antigen receptor. Proc. Natl. Acad. Sci. USA 1987; 84:6879-6883.

37. Goldsmith MA, Dazin PE, Weiss A. At least two non-antigen-binding molecules are required for signal transduction by the T cell antigen receptor. Proc. Natl. Acad. Sci. USA 1988; 85:8613-8617.

38. Goldsmith MA, Desai DM, Schultz T, Weiss A. Function of a heterologous muscarinic receptor in T cell antigen receptor signal transduction mutants. J. Biol. Chem. 1989; 264:17190-17197.

39. Shackelford DA, Trowbridge IS. Identification of lymphocyte integral membrane protein as substrates for protein kinase C. J. Biol. Chem. 1986; 261:8334-8341.

40. Goldsmith MA, Weiss A. New clues about T-cell antigen receptor complex function. Immunology Today 1988; 9:220-222.

41. Jin Y-J, Clayton LK, Howard FD, et al. Molecular cloning of the CD3η subunit identifies a CD3ζ-related product in thymus-derived cells. Proc. Natl. Acad. Sci. USA 1990; 87:3319-3323.

42. Transy C, Moingeon P, Stebbins C, Reinherz EL. Deletion of the cytoplasmic region of the CD3ε subunit does not prevent assembly of a functional T-cell receptor. Proc. Natl. Acad. Sci. USA 1989; 86:7108-7112.

43. Blumberg RS Ley, S., Sancho, J., Lonberg, N., Lacy, E., McDermott, F., Schad, V., Greenstein, J.L.,Terhorst, C. Structure of the T-cell antigen receptor: Evidence for two CD3 ε subunits in the T-cell receptor-CD3 complex. Proc. Natl. Acad. Sci. USA 1990; 87:7220-7224.

44. Thomas M. The leukocyte common antigen family. In: Paul WE, Fathman CG, Metzger H ed. Ann. Rev. Immunology. Palo Alto, Ca.: Annual Reviews Inc., 1989: 339-370.

45. Johnson NA, Meyer CM, Pingel JT, Thomas ML. Sequence conservation in potential regulatory regions of the mouse and human leukocyte common antigen gene. J. Biol. Chem. 1989; 264:6220-6229.

46. Charbonneau H, Tonk NK, Walsh KA, Fischer EH. The leukocyte common antigen (CD45): A putative receptor-linked protein tyrosine phosphatase. Proc. Natl. Acad. Sci. USA 1988; 85:7182-7186.

47. Kiener PA, Mittler RS. CD45-protein tyrosine phosphatase cross-linking inhibits T cell receptor CD3-mediated activation in human T cells. J. Immunol. 1989; 143:23-38.

48. Tonks NK, Charbonneau H, Diltz CD, Fischer EH, Walsh KA. Demonstration that the leukocyte common antigen CD45 is a protein tyrosine phosphatase. Biochem. 1988; 27:8695-8701.

49. Martorell J, Vilella R, Borche L, Rojo I, Vives J. A second signal for T cell mitogenesis provided by monoclonal antibodies CD45 (T200). Eur. J. Immunol. 1987; 17:1447-1451.

50. Ledbetter JA, Tonks NK, Fischer EH, Clark EA. CD45 regulates signal transduction and lymphocyte activation by specific association with receptor molecules on T or B cells. Proc. Natl. Acad. Sci. USA 1988; 85:8628-8632.

51. Mittler RS, Greenfield RS, Schacter BZ, Richard NF, Hoffmann MK. Antibodies to the common leukocyte antigen (T200) inhibit an early phase in the activation of resting human B cells. J. Immunol. 1987; 138:3159-3166.

52. Takeuchi T, Rudd CE, Schlossman SF, Morimoto C. Induction of suppression following autologous mixed lymphocyte reaction; role of a novel 2H4 antigen. Eur. J. Immunol. 1987; 1797-103.

53. Pingel JT, Thomas ML. Evidence that the leukocyte-common antigen is required for antigen-induced T lymphocyte proliferation. Cell 1989; 58:1055-1065.

54. Mustelin T, Coggeshall KM, Altman A. Rapid activation of the T-cell tyrosine protein kinase pp56lck by the CD45 phosphotyrosine phosphatase. Proc. Natl. Acad. Sci. USA 1989; 86:6302-6306.

55. Ostergaard HL, Shackelford DA, Hurley TR, et al. Expression of CD45 alters phosphorylation of the lck-encoded tyrosine protein kinase in murine lymphoma T-cell lines. Proc. Natl. Acad. Sci. USA 1989; 86:8959-8963.

56. Koretzky GA, Picus J, Thomas ML, Weiss A. Tyrosine phosphatase CD45 is essential for coupling T cell antigen receptor to the phosphatidyl inositol pathway. Nature 1990; 346:66-68.

57. Koretzky GA Picus, J, Schultz, T, Weiss, A. The tyrosine phosphatase CD45 is required for T cell antigen receptor and CD2 mediated activation of a protein tyrosine kinase and IL2 production. Proc. Natl. Acad. Sci. USA, in press, 1990.

58. Nel AE, Hanekom C, Rheeder A, et al. Stimulation of map-2 kinase activity in T lymphocytes by anti-CD3 or anti-Ti monoclonal antibody is partially dependent on protein kinase C. J. Immunol. 1990; 144:2683-2689.

59. Anderson NG, Maller JL, Tonks NK, Sturgil TW. Requirement for integration of signals from two distinct phosphorylation pathways for activation of MAP kinase. Nature 1990; 343:651-653.

60. Springer TA, Dustin ML, Kishimoto TK, Marlin SD. The lymphocyte function-associated LFA-1, CD2, and LFA-3 molecules: Cell adhesion receptors of the immune system. Ann. Rev. Immunol. 1987; 5:223-252.

61. Dustin ML, Sanders ME, Shaw S, Springer TA. Purified lymphocyte function-associated antigen 3 binds to CD2 and mediates T lymphocyte adhesion. J. Exp. Med. 1987; 165:677-692.

62. Dustin ML, Olive D, Springer TA. Correlation of CD2 binding and functional properties of multimeric and monomeric lymphocyte function-associated antigen 3. J. Exp. Med. 1989; 169:503-517.

63. Bockenstedt LK, Goldsmith MA, Dustin M, Olive D, Springer TA, Weiss A. The CD2 ligand LFA-3 activates T cells but depends on the expression and function of the antigen receptor. J. Immunol. 1988; 141:1904-1911.

64. Meuer SC, Hussey RE, Fabbi M, et al. An alternative pathway of T-cell activation: A functional role of the 50 kd T11 sheep erythrocyte receptor protein. Cell 1984; 36897-906.

65. Monostori E Desai, D., Brown, M.H., Cantrell, D. A., & Crumpton, M.J. Activation of human T lymphocytes via the CD2 antigen results in tyrosine phosphorylation of T cell antigen receptor ζ-chains. J. Immunol. 1990; 144:1010-1014.

66. Testi R, Phillips JH, Lanier LL. Leu 23 induction as an early marker of functional CD3/T cell antigen receptor triggering: Requirement for receptor crosslinking, prolonged elevation of intracellular [Ca2+] and stimulation of protein kinase C. J. Immunol. 1989; 142:1854-1860.

67. Thompson CG, Lindstein T, Ledbetter JA, et al. CD28 activation pathway regulates the production of multiple T-cell-derived lymphokines/cytokines. Proc. Natl. Acad. Sci. USA 1989; 86:1333-1337.

68. Weiss A, Manger B, Imboden J. Synergy between the T3/antigen receptor complex and Tp44 in the activation of human T cells. J. Immunol. 1986; 137:819-825.

69. Amrein KE, Sefton B. Mutation of a site of tyrosine phosphorylation in the lymphocyte-specific tyrosine protein kinase, p56lck, reveals its oncogenic potential in fibroblasts. Proc. Natl. Acad. Sci. USA 1988; 85:4247-4251.

70. Cooper JA, Gould KA, Cartwright CA, Hunter T. Tyr 527 is phosphorylated in pp60c-src: implication for regulation. Science 1986; 231:1431-1434.

Subject Index

A

A23187, 207
Adenosine, *Dictyostelium* differentiation, 24, 29
Adenovirus late promoter, 218
Adenylate cyclase, *Dictyostelium* aggregation, 11, 13, 15, 17
Adenylyl cyclase, 167, 168
Adrenergic receptors, 161–163
   α-1, 163, 164
   α-2, 163, 165, 172–173
   β-2, 163–170, 172–175
   membrane topology, 161, 162
   binding, catechol ring, 173–174
   cell-free expression, 170–171
   cloning, 163–164
   expression technology, 166–170
      altered mammalian cells, 167–168
      Escherichia coli, 170, 174
      insect cells, 168–169
      Xenopus laevis oocytes, 168
      yeast, 169–170
   G protein binding, 163, 168–172, 175
   mutagenesis, 166, 171–175
      chimeric receptors, 172–173
      deletion, 171–172
      saturation mutagenesis, 174–175
      single amino acid substitution, 173–174
   site-directed antibodies, 166
   structural analysis, 164–166
      amino acid sequence, 164, 165
   subtypes, 163
Aggregation phase, signal transduction, *Dictyostelium discoideum*, 9, 10, 11–18, 22, 23, 30
   molecular genetics, 14–17
Amphiregulin, 278
Antigen presenting cell(s), 141, 150
   T cell activation, 184–185, 190
Anti-μ antibodies, *egr-1* gene, B cells,
204–206, 209–211, 216, 219
AP-1, 105
   and *egr-1* gene expression, 203, 216
AP-2, 53, 54, 64, 69
armadillo gene, *Drosophila,* 256–260, 262, 264–266, 268, 269
Astrocytes, TPA-induced sequences, induction by
   mitogens, 94–96
   neurotransmitters, 98–100
ATF2 (CRE-BP1), 64, 66, 71, 73, 75
ATP-binding sites, receptors, 277, 278
Attenuator domain, G protein α subunits, 134, 136
Autophosphorylation
   erb B-2 receptor, 285
   insulin receptor, 278
5-Azacytidine, 213–214, 218

B

Baculovirus, 168–169
BAL-17 cells, 208, 212–216
B7/BB1 natural ligand, CD28 accessory molecule, 145–146, 150, 154
B cells
   activation signalling development, 221
   sIgM signalling pathways, 201–202, 204–205, 207, 216, 217, 220–221
   immature, WEHI-231 cell as model for, 217, 219
   lymphoma cells, *egr-1* gene, 207, 208
   -restricted cell surface activation antigen B7/BB-1, 141, 146, 150, 154
   see also egr-1 gene, B cells
Benzodiazepines, astrocyte TIS induction, rat, 100
Bombesin, 94
bride-of-sevenless gene, *Drosophila,* 263
bZIP (leucine zipper) superfamily of DNA-binding phosphoproteins,

64–65, 66, 69, 71, 74–75

C

Calcium
  channels, 119, 122
  Dictyostelium aggregation, 13
  intracellular, 150, 151
    egr-1 gene expression, 204, 207
    PDGF receptors, 291, 292
    T cell activation, 183
    T cell receptors, 300, 302, 303, 311, 312
Calcium–calmodulin kinases, CRE binding, 61, 62
Carbachol, astrocyte TIS induction, rat, 99–100, 101, 103, 104
CA-rich elements, prespore cell, *Dictyostelium*, 27–29, 30
Casein kinase II, CRE binding, 65, 67, 68, 72, 73
Catechol ring, adrenergic receptor binding, 173–174
CAT reporter gene, *egr-1* gene, B cells, 205, 211, 214, 215, 216
CCAAT box, CRE binding, 69–70
CD2, T cell receptors, 306–307, 313–315
CD3 complex. *See under* T cell receptor
CD4, 142, 148, 149, 189–195
  mature T cells, 189–193
  MHC, 183–184
  and p56^{lck} complex, 186, 187–189, 190, 193, 232–234, 237, 238, 241, 242–244
  and T cell receptors, 183, 184–185, 301
  thymus ontogeny, 182, 193–195
  tyrosine kinase, 185–187
CD8, 142
  mature T cells, 189–193
  MHC, 183–184
  and p56^{lck}, 232–234, 237, 238, 241–244
  and T cell receptors, 183–185, 301
  thymus ontogeny, 182, 193–195
  tyrosine kinase, 185–187
CD28 accessory molecule
  B7/BB1 natural ligand, 145–146, 150, 154
  distribution and phenotype, 141–143
  inositol phosphates, 153, 154

lymphokine production, 149, 151, 153
molecular and genetic characterization, 143–145
  human cf. mouse, 144
  PMA/ionomycin, 142, 147–149, 151–153
  protein kinase C, 147, 150
  signal transduction pathway, 152–154
  stimulation, functional effects, 146
    effector function, 148–152
    proliferation, 147–148
  T cell activation, 141, 148–154
  T cells, peripheral, 142, 147–148
  thymus, 142, 147
CD45 (T200), 190–192
  p56^{lck}, 234–235
  T cell receptor, 308–310
CD69 expression, T cell receptor, 313–315
C/EBP, 64, 65, 69
Cell–cell communication, segment polarity genes, *Drosophila melanogaster*, 257, 260, 261, 266
Cell cycle
  Dictyostelium differentiation, 24
  egr-1 gene, B cells, 204–205
  mitogens and, 84, 100
Cell density and *Dictyostelium* aggregation, 17
Cell-free expression, adrenergic receptors, 170–171
Cell-specific restriction of expression, TPA-induced sequences (TIS), 106–110, 111
Chimeric adrenergic receptors, 172–173
CHO cells, 136–137
Cholera toxin-catalyzed ADP ribosylation, G protein α subunits, 132, 134–135
c-H-ras, 125
Chromosome 2, human, *creb*, 69
cis-acting factors
  Dictyostelium differentiation, 19, 27, 31
  DNA methylation, *egr-1* gene, B cells, 202, 203, 211–215, 217–219, 221
Colony-stimulating factor(s) (CSF)
  CSF-1, 288, 289, 290
  c-*fes* proto-oncogene protein tyrosine kinase (p93c-^{fes}), 51–53